Polymer-Carbonaceous Filler Based Composites for Wastewater Treatment

Polymer-Carbonaceous Filler Based Composites for Wastewater Treatment serves as the first book to offer a concise treatment of the use of these materials in the treatment of wastewater. It provides a systematic and comprehensive account of recent developments and encompasses novel methods for the synthesis of carbonaceous derivatives-based fillers for polymer composites, their characterization techniques, and applications for the remediation of water contamination. This book seeks to:

- Introduce novel concepts in wastewater treatment with poly-carbonaceous composites
- Describe modern fabrication methods and characterization techniques
- Present information on processing, safety, and disposal
- Discuss current research, future trends, and applications

Filling the void for a one-stop reference book for researchers, this work includes contributions from leaders in the industry, academia, government, and private research institutions across the globe. Academics, researchers, scientists, engineers, and students in the fields of materials and polymer engineering and wastewater treatment will benefit from this application-oriented book.

Polymer-Carbonaceous Filler Based Composites for Wastewater Treatment

Edited by
Jyotishkumar Parameswaranpillai, Poushali Das,
Sayan Ganguly, Murthy Chavali, Nishar Hameed

CRC Press
Taylor & Francis Group
Boca Raton London New York

CRC Press is an imprint of the
Taylor & Francis Group, an **informa** business

First edition published 2024
by CRC Press
2385 Executive Center Drive, Suite 320, Boca Raton, FL 33431

and by CRC Press
4 Park Square, Milton Park, Abingdon, Oxon, OX14 4RN

CRC Press is an imprint of Taylor & Francis Group, LLC

ISBN: 978-1-032-35090-5 (hbk)
ISBN: 978-1-032-35699-0 (pbk)
ISBN: 978-1-003-32809-4 (ebk)

DOI: 10.1201/9781003328094

Typeset in Times
by codeMantra

Contents

Editors

Jyotishkumar Parameswaranpillai is an Associate Professor at Alliance University, Bangalore, India. He is a prolific editor and researcher who has published more than 35 edited books, 140 high-quality international research articles, 65 book chapters, and 1 patent. He has received numerous awards and recognitions including prestigious KMUTNB best researcher award 2019 and INSPIRE Faculty Award 2011.

Poushali Das is a postdoctoral researcher at School of Biomedical Engineering, Faculty of Engineering, McMaster University, Canada. Previously, she worked as senior post-doctoral researcher at Bar-Ilan University, Israel for 2 years. She has completed her Ph.D. degree in 2019 from the Indian Institute of Technology Kharagpur, India. She has more than 50 research publications in reputed international journals. She has been serving as a topic editorial board member and reviewer of reputed journals and an international consultant. She has presented papers in many international conferences. She has received prestigious awards including DST INSPIRE Scholar Award, Government of India, Horizon Europe Marie Skłodowska-Curie Postdoctoral Fellowship Award by European Commission, and Seal of Excellence from European Commission.

Sayan Ganguly is a postdoctoral researcher in the Department of Chemistry, University of Waterloo, Waterloo, Canada. He worked as senior postdoctoral researcher at Bar-Ilan University, Israel for more than 2 years. He obtained his Ph.D. from the Indian Institute of Technology, Kharagpur. His primary research interests include superabsorbent hydrogels, composite hydrogels, polymer-graphene nanocomposites, MXene-polymer systems, polymer composites for EMI shielding, and conducting polymer composites. He has published more than 80 papers and chapters in international journals and books. He is currently editing books with Elsevier and CRC Press.

Murthy Chavali is Professor and Dean, at Research & Development, Dr. Vishwanath Karad MITWPU, Maharashtra, India. A former fellow of NSD-OEAD, NSF, JSPS, VBL, and NSC, he also acts as a Visiting Professor/Researcher/Scientist/Fellow to 13 universities/research institutes abroad. He has over 29 years of research experience, 26 years of teaching experience, and 12 years of industrial experience. During this time, he established nine research laboratories (including four Centres of Excellence). Prof. Murthy is a recipient of several awards and travel grants. He has published extensively in international journals and completed over 38 sponsored research projects both at national and international levels. He made a team contribution towards 12 patents as well with help from international collaborators.

Nishar Hameed is an Associate Professor at Swinburne University of Technology, Australia. He has published nearly 100 high-impact journal papers, 6 book chapters, 3 edited books, and 2 patents.

Contributors

Mohd Ali
Department of Physics
Institute of Science, Banaras Hindu
   University
Uttar Pradesh, India

M. Anupama Ammulu
Department of Civil Engineering
Prasad V Potluri Siddhartha Institute of
   Technology
Andhra Pradesh, India

Anil K. Bahe
Department of Chemistry
Dr. Harisingh Gour Central University
Madhya Pradesh, India

Elango Balaji T.
Department of Chemistry
Utkal University
Odisha, India

Mokae F. Bambo
Advanced Materials Division
DSI/Mintek-Nanotechnology Innovation
   Centre
Johannesburg, South Africa

Akhilesh N. Bendre
Centre for Research in Functional
   Materials (CRFM)
JAIN University
Karnataka, India

Madhusmita Bhuyan
School of Physics
Sambalpur University
Odisha, India

Asit Kumar Das
Department of Chemistry
Krishnath College
Murshidabad, India

Himadri Tanaya Das
Centre of Excellence for Advance
   Materials and Applications
Utkal University
Odisha, India

Megha Das
Department of Education
Dr. Harisingh Gour Central University
Madhya Pradesh, India

Nigamananda Das
Centre of Excellence for Advance
   Materials and Applications
Utkal University
Odisha, India
and
Department of Chemistry
Utkal University
Odisha, India

Payaswini Das
Department of Geology
Utkal University
Odisha, India

Ratnesh Das
Department of Chemistry
Dr. Harisingh Gour Central University
Madhya Pradesh, India

Tushar Kanti Das
Institute of Physics – Center for Science
   and Education
Silesian University of Technology
Katowice, Poland

Tukaram D. Dongale
School of Nanoscience and
  Biotechnology
Shivaji University
Kolhapur, India

Swapnamoy Dutt
Centre of Excellence for Advance
  Materials and Applications
Utkal University
Odisha, India
and
Department of Chemistry
Utkal University
Odisha, India

Santosh Ganguly
Bharat Pharmaceutical Technology
Tripura, India

Sayan Ganguly
Department of Chemistry
Bar-Ilan Institute for Nanotechnology
  and Advanced Materials, Bar-Ilan
  University
Ramat-Gan, Israel

Avishek Ghatak
Department of Chemistry
Dr. A. P. J. Abdul Kalam Government
  College (West Bengal State
  University)
Kolkata, India

Anil H. Gore
Tarsadia Institute of Chemical Science,
  UkaTarsadia University
Gujarat, India

Debanjan Guin
Department of Chemistry
Institute of Science, Banaras Hindu
  University
Uttar Pradesh, India

Smita Jadhav
Engineering Science and Allied
  Engineering
Bharati Vidyapeeth's College of
  Engineering for Women
Pune, India

Dipika Jaspal
Department of Applied Science
Symbiosis Institute of Technology
  (SIT), Symbiosis International
  (Deemed University)
Maharashtra, India

Ashish Kaushal
Department of Materials science and
  Engineering
National Institute of Technology
Hamirpur, India

Vemuri Praveen Kumar
Department of Biotechnology
Koneru Lakshmaiah University
Andhra Pradesh, India

Mahaveer D. Kurkuri
Centre for Research in Functional
  Materials (CRFM)
JAIN University
Karnataka, India

Chandra Shekhar Kushwaha
Bhaskaracharya College of Applied
  Sciences
University of Delhi
Delhi, India

Arti Malviya
Department of Engineering Chemistry
Lakshmi Narain College of Technology
Madhya Pradesh, India

Kgabo P. Matabola
Advanced Materials Division
DSI/Mintek-Nanotechnology Innovation
    Centre
Johannesburg, South Africa

Arunesh K. Mishra
Department of Chemistry
Dr. Harisingh Gour Central University
Madhya Pradesh, India

Pratibha Mishra
Department of Chemistry
Dr. Harisingh Gour Central University
Madhya Pradesh, India

Teboho C. Mokhena
Advanced Materials Division
DSI/Mintek-Nanotechnology Innovation
    Centre
Johannesburg, South Africa

Ishita Mukherjee
Department of Chemistry
The University of Burdwan
West Bengal, India

M. S. R. Niranjan Kumar
Department of Mechanical Engineering
Prasad V Potluri Siddhartha Institute of
    Technology
Andhra Pradesh, India

Shakuntala Ojha
Department of Mechanical Engineering
Kakatiya Institute of Technology &
    Science (KITS)
Telangana, India

Aparajita Pal
Centre of Rubber Technology
Indian Institute of Technology
Kharagpur, India

Chandrashekhar S. Patil
Centre for Research in Functional
    Materials (CRFM)
JAIN University
Karnataka, India

Amit Pramanik
Department of Chemistry
A.B.N Seal College
Cooch Behar, India

G. Raghavendra
Department of Mechanical Engineering
National Institute of Technology (NIT)
Warangal, India

Atish Roy
Department of Chemistry
Dr. Harisingh Gour Central University
Madhya Pradesh, India

Dibakar Sahoo
School of Physics
Sambalpur University
Odisha, India

Priyanka Sahu
School of Physics
Sambalpur University
Odisha, India

Priyatosh Sarkar
Bharat Pharmaceutical Technology
Tripura, India

Keshav Sharma
Department of Chemistry
Institute of Science, Banaras Hindu
    University
Uttar Pradesh, India

Rahul Sharma
Department of Chemistry
CSIR-National Physical Laboratory
Delhi, India

Saroj Kr. Shukla
Bhaskaracharya College of Applied
    Sciences
University of Delhi
Delhi, India

Ashutosh Kumar Singh
Department of Biotechnology and
    Medical Engineering
National Institute of Technology
Odisha, India

Vishal Singh
Department of Materials science and
    Engineering
National Institute of Technology
Hamirpur, India

M. Somaiah Chowdary
Department of Mechanical Engineering
National Institute of Technology (NIT)
Warangal, India
and
Department of Mechanical Engineering
Prasad V Potluri Siddhartha Institute of
    Technology
Andhra Pradesh, India

Manoj K. Tiwari
Bhaskaracharya College of Applied
    Sciences
University of Delhi
Delhi, India

Chandra Shekhar Pati Tripathi
Department of Physics
Institute of Science, Banaras Hindu
    University
Uttar Pradesh, India

1 Introduction to Polymer-Carbonaceous Filler-Based Composites

Sayan Ganguly
Bar-Ilan University

CONTENTS

1.1   INTRODUCTION

Composites are often formed by combining many elements, each of which has its own unique set of features, with the intention of producing a single material that possesses all of those properties. Matrixes and reinforcements are the two components of composites. There is a potential for a wide variety of permutations involving the matrix and the reinforcing components. The incorporation of the characteristics of two different types of materials into a single product is the primary benefit that comes from using composites in a variety of contexts [1]. In the present day and age, composite materials are becoming increasingly used in many different industries, including the building, aviation, automotive, biomedical, and industrial sectors, amongst others. Researchers are increasingly turning their attention to the potential use of the aforementioned materials in the treatment of wastewater. The particular strength, processability, and design flexibility of composites are among the features given by these materials [2,3]. This book covers research on several essential and relevant categories of composites, and includes those categories.

The treatment of wastewater and the provision of clean water may be accomplished through a variety of different procedures. For instance, membrane separation, adsorption, chemical precipitation, electrochemical treatment, ion-exchange, chlorination, and ozonation, as well as a number of other processes, are among the methods that are typically utilised in order to remove harmful substances to ensure a clean water supply [4]. Because of its excellent energy efficiency and cheap cost in comparison to the other methods, membrane separation has been the most popular choice in the recent years. The separation achieved by using membranes is accomplished by moving molecules of interest through pores in the membranes in a manner

DOI: 10.1201/9781003328094-1

1

that is selectively determined by the sizes of those molecules, and then observing the interaction of those molecules with the pores in the absence of phase change and additional chemicals. In general, the membrane systems are straightforward, and they require a minimal amount of auxiliary apparatus [5]. The separation of ions in electrochemical processes and the filtration of particulates from liquid suspensions are two applications that make extensive use of membrane technology because of its low cost and high efficiency. Other applications include the production of different kinds of water, the dialysis of blood and urine, and the filtration of particulates. There are three distinct processes that bring about the separation that happens across membranes: (i) size exclusion induced by the pores across the membrane, which allows passage of compounds smaller than the pore size; (ii) pore flow caused by the interaction between pore surface and passing molecules, which induces selective transportation of molecules with similar size to the pores; (iii) solution diffusion induced by the diffusion of molecules into the membrane, resulting in migration of the molecules across the membrane, which occurs exclusively in polymeric membranes; and (iv) solution diffusion induced by the diffusion of molecules The flux rate, selectivity, mechanical/chemical/thermal stabilities under working circumstances, fouling qualities, and service durability are the primary properties that define membrane performance. Other basic properties include the membrane's service durability. Because membranes play a barrier role in the separation process, the total membrane performance is heavily influenced by the surface features of the membranes themselves. These surface parameters include pore size, pore shape, surface roughness, and physicochemical properties. The fouling behaviour of the membrane is heavily impacted by the roughness and hydrophilicity of the membranes, while the mechanical, chemical, and thermal stabilities of the membranes are the factors that decide how long the membranes will remain functional.

1.2 MATERIALS IN DEMAND

A number of states are rapidly running out of available water. Only 10% of the world's water supply is suitable for usage in homes and other residential settings. According to a study titled "Progress on Sanitation and Drinking Water" (2013) published by the World Health Organization and UNICEF, there are around 768 million people who do not have access to clean water. According to the report titled "Climate Change: Impacts, Adaptation and Vulnerability (2014)" written by the Intergovernmental Panel on Climate Change (IPCC), a warning has been issued stating that 80% of the world is currently facing a significant risk of having insufficient access to water [6]. An increase in population and the beginning of an era of industrialisation, which has led to widespread contamination of the world's water bodies, are two of the primary factors that have contributed to the current state of water scarcity. In such a precarious situation, it is very necessary to make responsible use of water in addition to transforming wastewater into a form that may be put to productive use [7]. As water bodies become increasingly polluted, there is a growing need for materials that are highly effective, cost effective, and adaptable to a variety of various types of contaminants.

Polymers are organic materials that possess a variety of great qualities, including high mechanical strength, amazing flexibility, chemical stability, and high surface area. Polymers may be broken down into smaller units called monomers. Because of these characteristics, polymers can serve as hosts for a wide variety of organic and inorganic compounds. As a result, we were able to synthesise many composites with the qualities we were looking for. As a result of this, polymer composites have garnered a lot of interest in the field of water treatment and desalination. By mixing, crosslinking, and functionalising the surface, polymer–polymer composites gave the opportunity to fine-tune the adsorptive capabilities. The simplicity of preparation and applications, high chemical stability even under harsh operating conditions, the ability to remove a wide variety of pollutants, and good recyclability combined with high adsorption capacities are the primary benefits offered by this category of composites [8]. Other advantages include: the ability to use the composites to remove pollutants, and use for the removal of pollutants. On the other hand, the most significant detriment is still the relatively high cost of manufacture. For the goal of removing colours from wastewater, a great deal of polymer–polymer composites have been synthetically produced and made available. For instance, Elkady et al. utilised a procedure known as solution polymerisation in order to manufacture a copolymer consisting of styrene and acrylonitrile. Electrospinning was the method that was utilised in the process of transforming the produced copolymer into a nanofiber. The surface of this nanofiber was then functionalised by the chemical addition of carboxylic acid groups in order to enhance the dyes' capacity to be absorbed by the nanofiber [9]. The scanning electron microscopy (SEM) was used to confirm the changes that occurred in the nanofiber following the functionalisation process. The results showed that the carboxylated nanofibers had a more uniform shape when compared to the nanofibers that had not been functionalised. The process of adsorption of dyes from wastewater was impacted as a result of these morphological modifications. The low specific surface area and the absence of surface porosity of these two polymers cannot, without a shadow of a doubt, be the driving forces behind the achievement of such adsorption. Therefore, it is possible to deduce that the charged groups that were formed on the surface of the copolymer, in addition to the morphological changes that took place inside the nanofiber structure, were significant contributors to the achievement of such an impressively high adsorption value. Immobilising ternary carboxylic acid and acrylamide units on straw particles resulted in the synthesis of a low-cost, multifunctional, and straw-based adsorbent. In this way, the adsorbent was manufactured. According to the findings of this research, the incorporation of several carboxyl and amino groups into the composite structure has resulted in an increased capacity for the adsorption and removal of dyes. Specifically, the adsorption capacities for MB and MO reached 120.84 and 3053.48 mg/g, which were three and fifty-four times greater than those of unmodified straw, respectively [10]. Due to their remarkable physicochemical characteristics and cavities, cyclodextrin-based composites are a new generation of adsorbents for colour removal from wastewater [11]. These enhanced structural characteristics of cyclodextrin composites facilitated industrial use [12]. Zhou et al. examined current achievements in these new composites [13]. In order to make an environmentally safe composite for the removal of various dyes, for instance, cyclodextrin polymer was crosslinked with polydopamine before the process began. It was determined that the novel structural properties and the functional groups of the polymer composite were responsible for the remarkable adsorption effectiveness of these dyes [14]. In the same vein, cellulose derived from agricultural waste was grafted with the monomers of 2-acrylamido-2-methylpropane sulfonic acid and acrylic acid in the presence of a crosslinker in order to produce extremely effective polymer composites for the removal of colour. The SEM image of the obtained polymer composite (Figure 1.1) showed smooth and long fibre-like threads for untreated cellulose. On the other hand, in the case of the crosslinked graft copolymers, flaky and highly thick threads were obtained, indicating surface functionalisation of the cellulose. Figure 1.1 is the SEM image of the obtained polymer composite. The adsorption behaviour of this cellulose composite was studied in order to determine the influence that various adsorption parameters had on the removal of cationic (malachite green and crystal violet) and anionic (Congo red; CR) dye from an aqueous solution. As was previously said, the cationic dyes were eliminated at a pH of 7.0 in a period of 90 minutes, but the anionic dye was adsorbed at a pH of 2.2 over the course of 8 hours. The Langmuir isotherm and the pseudo-second-order model accounted for the adsorption data rather satisfactorily. According to the findings of this research, cellulose-based copolymers have the capability of functioning as promising adsorbents for the absorption of anionic as well as cationic dyes from industrial wastewater [15]. For the purpose of selectively removing and separating various colours, cellulose, which is a naturally occurring polymer, was further functionalised with a hyperbranched polyethylenimine. Specifically, a Schiff-base formation was utilised in order to establish a covalent bond between polyethylenimine and the backbone of cellulose molecules. In order to create a polymer composite, the NH_2 groups of polyethylenimine and the CHO groups of oxidised cellulose were connected to one another and joined. The adsorption capabilities of the newly produced

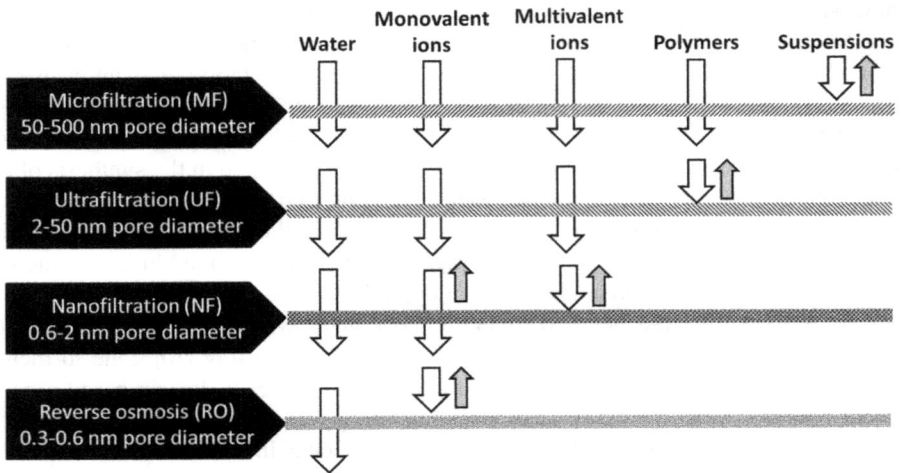

FIGURE 1.1   The pore size ranges of different membranes and their capacities to repel specific pollutants.

composite were analysed with regard to the removal of CR and basic yellow dyes from aqueous mediums. The highest adsorption capacity of Congo red dye was 2100 mg/g, while the maximum adsorption capacity of basic yellow dye was 1860 mg/g. The composite demonstrated selective adsorption of a variety of dyes, such as brilliant blue and reactive red; however, the absorptivity of eosin and brilliant yellow dyes was quite low [16]. It is important to note that the hyperbranched structure of the polymer composite acted as a brush, which improved the interparticle diffusion of certain dye molecules into the composite. This is something that should not be overlooked (depending on the dye size). As a result, both the adsorption capacity and selectivity of the material are improved. The primary characteristics that determine membrane properties are the pore size, pore size distribution, morphology, and surface properties of membranes. For instance, the selectivity is mostly determined by the pore size and how it is distributed, but the permeability, fouling, and selectivity are all influenced by the characteristics of the membrane materials. The membrane flow is also affected by the overall thickness of the membranes as well as the form of the pores. Microfiltration (MF), ultrafiltration (UF), nanofiltration (NF), and reverse osmosis (RO) are the four subcategories that can be applied to the membrane separation process that is used for water purification and desalination. Each subcategory is named after the smallest pore size that it can accommodate (Figure 1.1).

The molecular sieving and size exclusion process is the primary phenomena responsible for solute rejection in macroporous MF and mesoporous UF membranes, respectively. These suspended particles, asbestos, and biological components, such as particles, bacteria, protozoa, and red blood cells, are often rejected by MF membranes. MF membranes also reject asbestos. On the other hand, UF membranes contain smaller holes and are, therefore, able to reject smaller particles, microsolutes like sugars and salts, and macromolecules like viruses, proteins, and pyrogens. NF membranes, like MF and UF membranes, can reject low-molecular-weight uncharged solutes through the size exclusion mechanism, whereas NF membranes can reject charged molecules through a combination of the size exclusion mechanism, Donnan exclusion/equilibrium, and dielectric exclusion. NF membranes can also reject low-molecular-weight charged solutes through the Donnan exclusion/equilibrium mechanism (electrostatic interactions). They consist of the vast majority of organic compounds, as well as viruses and salts. In particular, NF membranes have the ability to repel divalent ions, and they are frequently employed in the process of water softening. RO membranes, on the other hand, have holes that are between 0.3 and 0.6 nm in size, making it possible to classify them as non-porous. This is in contrast to MF, UF, and NF membranes. The movement of molecules across RO membranes is controlled by a technique known as solution diffusion, in which solutes are dissolved into the membrane material and then diffuse along with the concentration gradient. The separation is carried out in environments where the solutes have varying degrees of solubility and diffusibility. The desalination of brackish groundwater and saltwater is the most prevalent use for RO membranes.

It is usual practice to characterise composite membranes as polymer membranes that either integrate inorganic nanoparticles, on which inorganic nanoparticles are formed, or the ones that are supported by ceramic substrates, as shown in Figure 1.2. There has been a consistent increase in the number of articles published

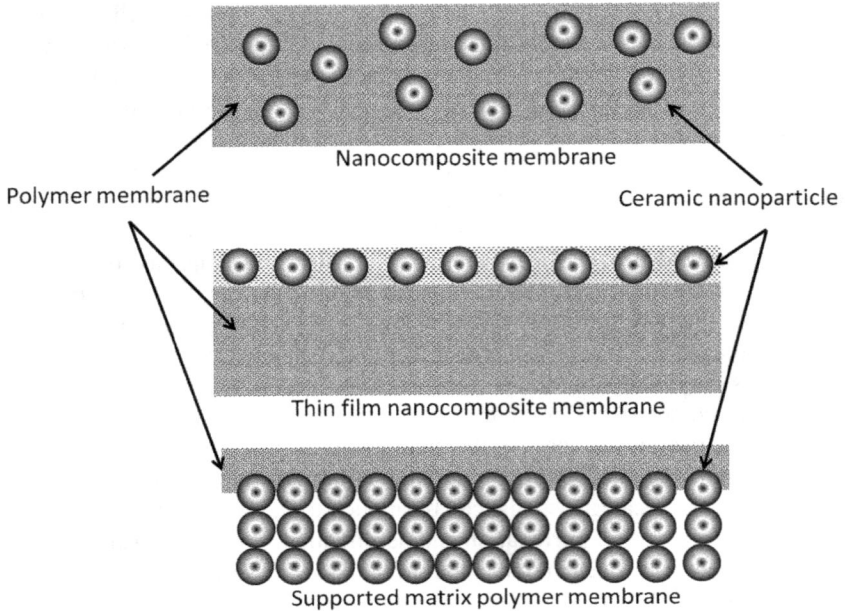

FIGURE 1.2   Membrane structures composed of ceramic and polymer composites.

on ceramic–polymer composite membranes, which have been utilised in a wide variety of applications. This is due to the fact that these membranes have become increasingly popular. In more recent times, new forms of composite membranes have begun to emerge. These new composite membranes include those in which MOFs (Metal-Organic Frameworks) or COFs (Covalent-Organic Frameworks) have been inserted or integrated. In addition, there are also the composite-type membranes that comprise three or more components, such as ceramic–(different) ceramic–polymer, GO–CNT–polymer, carbon–ceramic–polymer, MOF–carbon–polymer, and so on. These membranes are used for a variety of applications.

1.3   SUMMARY AND FUTURE PERSPECTIVE

In this chapter, a variety of cutting-edge composite materials and the momentum they have achieved for removing heavy metals, pesticides, and other hazardous pollutants from wastewater and for purifying household water were discussed. It was discovered that almost all of the adsorption processes are endothermic, leading to an increase in spontaneity. This chapter presents a discussion on the numerous isothermal models that are utilised for the investigation of the adsorption phenomena. An objective analysis found that activated carbon composite is a promising material for the purification of wastewater due to the fact that it is cost effective. It has been proven by a number of researchers that the change of the activated carbon's structure results in an increase in the contaminant-capturing propensity of the activated carbon. Before processing these composites for use in mass purification applications, more research is required to investigate how the mechanical and thermal characteristics of the composites may be strengthened. In order to have a more complete understanding of the performance characteristics of the composites, the modelling design concepts are essential. Carbon nanocomposite, polymer composite, graphene composite, and other similar materials have specific elimination characteristics, the ability to retain pollutants, greater durability, and regeneration capacity than conventional adsorbent materials. This gives them an advantage over traditional adsorbent materials. However, further research has to be done to investigate cost-effective high-performance alternatives, with a particular emphasis on the alternatives' large-scale usefulness and environmentally friendly purification technique. The chapter makes it abundantly evident that there has been a significant increase in the usage of composite materials in contrast to the solitary materials for the purpose of water purification over the course of the previous 10 years. Because of this, it is certain that in the not-too-distant future, the process of water purification will undergo a revolution thanks to the introduction of new, better, and more effective materials. This would contribute to the solution of the worldwide problem of insufficient potable water.

In particular for the manufacturing of such adsorptions on a wide scale, the polymer composites that were being used for water treatment and desalination did not reach their ideal level of performance. The following is a list of future views that might be considered in order to achieve an optimal performance of the adsorbents:

a. The low-cost integration of nanoadsorbent materials with biosorbent carriers.

b. Enhancing the number of functional groups that are capable of increasing the adsorption efficiency, which may be done by investigating novel functional materials and/or inventing hybrid technologies. This can be accomplished by adding to or increasing the number of functional groups.

c. Finding additional agro-waste products that can be used as adsorbents in order to lower manufacturing costs and maintain the market for adsorbent-based products.

d. Researching in great depth the process of adsorption in order to get an understanding of the scientific principles behind its behaviour and, as a result, to enhance the effects of various adsorbents under varying environmental circumstances.

e. Enhancing the selectivity of the adsorbents by carefully choosing the functional groups that are formed on the adsorbents.

f. Decreasing the costs of operations in order to achieve large-scale uses of adsorbents, which may be accomplished by introducing innovative approaches for the removal of contaminants in the environment.

REFERENCES

1. Jaspal D, Malviya A. Composites for wastewater purification: A review. *Chemosphere.* 2020;246:125788.
2. Mahajan GV, Aher VS. Composite material: A review over current development and automotive application. *International Journal of Scientific and Research Publications.* 2012;2:1–5.

3. Huang Y, Li J, Chen X, Wang X. Applications of conjugated polymer based composites in wastewater purification. *Rsc Advances*. 2014;4:62160–78.
4. Yadav D, Dutta J. *Green Approaches to Prepare Polymeric Composites for Wastewater Treatment. Green Composites*: Springer; 2021. p. 531–70.
5. Koros WJ, Zhang C. Materials for next-generation molecularly selective synthetic membranes. *Nature Materials*. 2017;16:289–97.
6. Gao Q, Demissie H, Lu S, Xu Z, Ritigala T, Yingying S, et al. Impact of preformed composite coagulants on alleviating colloids and organics-based ultrafiltration membrane fouling: Role of polymer composition and permeate quality. *Journal of Environmental Chemical Engineering*. 2021;9:105264.
7. Lakkimsetty NR, Feroz S, Karunya S, Motilal L, Saidireddy P, Suman G. Synthesis, characterization and application of polymer composite materials in wastewater treatment. *Materials Today: Proceedings*. 2022.
8. Dutt MA, Hanif MA, Nadeem F, Bhatti HN. A review of advances in engineered composite materials popular for wastewater treatment. *Journal of Environmental Chemical Engineering*. 2020;8:104073.
9. Elkady MF, El-Aassar MR, Hassan HS. Adsorption profile of basic dye onto novel fabricated carboxylated functionalized co-polymer nanofibers. *Polymers*. 2016;8:177.
10. Liu Q, Li Y, Chen H, Lu J, Yu G, Möslang M, et al. Superior adsorption capacity of functionalised straw adsorbent for dyes and heavy-metal ions. *Journal of Hazardous Materials*. 2020;382:121040.
11. Zhou Y, Hu Y, Huang W, Cheng G, Cui C, Lu J. A novel amphoteric β-cyclodextrin-based adsorbent for simultaneous removal of cationic/anionic dyes and bisphenol A. *Chemical Engineering Journal*. 2018;341:47–57.
12. Liu Q, Zhou Y, Lu J, Zhou Y. Novel cyclodextrin-based adsorbents for removing pollutants from wastewater: A critical review. *Chemosphere*. 2020;241:125043.
13. Zhou Y, Lu J, Zhou Y, Liu Y. Recent advances for dyes removal using novel adsorbents: A review. *Environmental Pollution*. 2019;252:352–65.
14. Chen H, Zhou Y, Wang J, Lu J, Zhou Y. Polydopamine modified cyclodextrin polymer as efficient adsorbent for removing cationic dyes and Cu2+. *Journal of Hazardous Materials*. 2020;389:121897.
15. Kumar R, Sharma RK, Singh AP. Removal of organic dyes and metal ions by cross-linked graft copolymers of cellulose obtained from the agricultural residue. *Journal of Environmental Chemical Engineering*. 2018;6:6037–48.
16. Zhu W, Liu L, Liao Q, Chen X, Qian Z, Shen J, et al. Functionalization of cellulose with hyperbranched polyethylenimine for selective dye adsorption and separation. *Cellulose*. 2016;23:3785–97.

2 Different Types of Hazardous Materials and Their Analysis Techniques

Ashutosh Kumar Singh
National Institute of Technology Rourkela

CONTENTS

LIST OF ABBREVIATIONS

AC	Activated carbon
BSFL	Black soldier fly larvae
CNT	Carbon nanotubes
ECs	Emerging contaminants
MFB	Membrane filtrations bioreactors
SRT	Sludge retention time
UV	Ultraviolet
WWT	Wastewater treatment
WWTPs	Wastewater treatment plants

2.1 INTRODUCTION

Prime component for life survival is water. Getting clean and pure water is a big deal to some human acts, such as release or disposal of untreated wastes, chemicals and other effluents. Poor/weak administration policies towards environment adversely affect water quality. Due to implausible engineering activities, heavy metals are created and proved to be prime mover hazardous cause to local terrain. Human life

DOI: 10.1201/9781003328094-2

autonomy is related to clean water availability for drinking—availability of pure water for drinking, agriculture and industrial application. Provision of clean water and sanitation are driving agents for zero hunger, good health and well-being. The global recognition for suitable water parameters, and having competent and reliable WWT led to considerations of waste water effluents as biogas resource and an alternative source of portable water. Increasing research efforts were directed to improve performance treatment of biological, chemical and physical processes that are used in primary, secondary and tertiary treatment of these effluents to meet stringent regulatory requirements on quality of processed wastewater. Diverse family of hazardous materials are toxic, reactive and corrosive by-products of manufactured or discarded containments (either organic or inorganic) in solid, liquid, gas or sludge phases, which have been already proved to be threat to all flora–fauna. Whole global village is facing water pollution, and is in an alarming condition due to industrialisation and urbanisation. Particularly, industrial waste discharge brings about life-threatening pollution issue in environment (water, land and air), and has become a challenge for researchers and scientists. In the 21st century, emerging containments (human excretion such as urban or rural sewage, improper industrial disposal, landfill leachate, heavy metals such as cadmium, lead, phosphates, sulphates or other ions, pharmaceuticals pollutants such as dye materials) have been posed as environmental burning problem due to exponential growth in land, water and air contamination. Heavy metals are noxious substance with high soluble affects in living beings in various forms and creates serious natural threat. Unique solutions to this problem are turning into safe formats and admiring for other safe methods before disposal of their waste treatment. For this treatment not only techniques need to be treated, but also creations of additional problems that the wastewater plant processing itself adds pollution to environment. Emerging contaminants (ECs) enter water system from different sources of degraded remnants. ECs can accrue in the lipid-rich tissues of organisms by the agency of their hydrophobic strength, threaten to the endocrine systems of biotic and abiotic animals, and advances antimicrobial resistance. ECs that have been identified as endocrine-disrupting chemicals, then after highly associated with instigating endometriosis, prostate, testicular and breast cancers, and give rise to severe complications in the reproductive system for humans and animals as to reduce sperm count in humans, cause to the creation of fragile eggs and compromise with aquatic wildlife. The long-haul hazards of ECs are missing further exploration and monitoring, where various government, regional and international agencies are actively winded up in tackling the accoutrements of ECs on human health and the environment practicing EC deportation technologies. WWTPs effluents are being incrementally recognised as predominantly from where ECs are commuted into the environment as wastewater from both industrial and residential areas aggregated in WWTPs. Industries and industrial sources (paper, textile, paint, electroplating industries and plastic formation industries that utilise water in different processing units and discharge solution for pullulated water into sewage, river and lakes. Their proper management is the demand of modern science, because their mishandling impacts over water, climate and land. Regular development in environmental way of life tends to deplete natural resources and increased waste generation per capita. Highly polluted water (higher concentration of metals and heavy metal ions, bacteria, swages, viruses, toxic organisms, garbage and detergents). The fundamental metals or their alloys with molecular weight (63–200) g/mol or densities more than (5 g/m³) are said to be heavy metals. The presence of heavy metals in wastewater is threatening our health to environment, because their low concentration can cause serious problem to human's health and ecosystem. That is why heavy metals are said to be chemical components having low density with high toxicity. Toxic metals are tough to segregate due formation of third groups with available protein in water. With respect to environment safety, wastewater from industries factories or other processing units from industries, factories or other processing units must be treated before being released into environment. Procreation of new industrially generated toxins in the environment continues to cause boundless anxiousness. Pharmaceuticals, organic pollutants, heavy metal ions, microorganisms and others are examples of incessant organic chemicals whose fallouts are anonymous, because they have recently infiltrated into environment and displaying up in wastewater treatment facilities. Human medicine's potential for harmful ecotoxicological consequences in aquatic environment, exclusively in sublethal bulk quantity, has been a source of concern. The continuation of pharmaceutical synthetic materials and their modalities in the aquatic habitat regarded as emergent organic micropollutants is a significant issue globally. The health care system's adulthood to administer and counter illness has resulted in rapid increase in pharmaceutical use. Pollution is described as the release of toxins to surroundings. Human-made wastes such as domiciliary food trashes, industrial and floricultural wastes, fertilisers used by farmers, oil spills, pharmaceuticals and radioactive materials are often culpable for water and soil infection. They have Cu (II), Fe, Ni (II), Cr (VI), Pb (II), Hg (II), Ag (I) and Ar ions as heavy toxic metals in flora–fauna. Major electroplating extracts are Ni, Ag, Cr and Cd. Iron and steels are used in colour and paint industry for concoction of electromagnetic accessories, and vehicle exhausts for metallurgy and as a catalyst. These metal ions constitute human body with more than concentration that might create major health problems. Chromium is dangerous for aquatic life; with development of industries, it is possible to restore or reuse wastewater. According to the Waste Index Report 2021, the 931 million tons of food wastes produced in 2019 accounted for about 17% of all food accessible to customers [1]. Crop residues are described as undesired waste generated by agricultural operations (horticultural plastics and veterinary medicines, wastes from a slaughterhouse, poultry houses and farms, herbicides, pesticides, fertilisers, manure). Soil contaminations inputs include natural nutrients (fertilisers) and pesticides' input, sewage sludge and manure application, air deposition of exhaust fumes and particles from wastewater discharges in surface waterways. The physical characteristics of soil may be affected by oil contaminants in which pore gaps of soil molecules may get blocked, reducing soil freshening and sedimentation of diffuse particles from wastewater diffusion while also increasing bulk density, limiting plant development. Oil spills may suffocate fish, tangle birds and animal's plumage, and block light photosynthetic plants in water. Drugs from humans and animals end up in rivers, lakes and even drinking water. Radioactive wastes are a kind of hazardous wastes that contains radioactive materials such as nuclear medicine, nuclear research, nuclear power generation, rare-earth mining and nuclear weapons reprocessing. Due to huge mechanisation and profit-making developments or melodramatic adjustments

TABLE 2.1
List of Water Contamination Sources [3]

Water Toxic Wastes	Related Sources	Illustrations
Plants minerals	Sewage, manure	Phosphates, nitrates & sulphates
Thermal energy	Power plants, industries	Heat wave
Oxygen demanding	Garbage & agricultural leavings	Agricultural wastes
Infection assistants	Animals excrete	Microbes such as virus, fungus, parasites & bacteria
Radioactive materials	Mining & power plants	Thorium & ceramics
Inorganic & organic salts	Domestics & smokestack sewerage	Acids, salts, metals, plastics & oil

in style of living and growing population density, it is easy to observe significant increase of hazardous solid waste in India. As per National Policy Act of 1969, it is an indispensable part of all industrial and social activities to perform measurements for effective managements of hazardous waste [2]. Urban local executive communities have a lot of pollution-control laws, still hazardous solid waste disposal is ineffective. In this cutting age of modernisation, industrial solid waste may contain high contents of hazardous biochemicals. Industrial multiparous waste contains several combinations of chemical, radioactive and biological hazards. That is why it is more necessary to effectively manage that hazardous waste to implement the best strategies to maximise environmental impacts. Aim is to evaluate various new practices and techniques for hazardous waste reduction, segregation, handling, recycling, reusage and treatment including disposal (Table 2.1).

A detailed study of WWT and ECs research from 1998 to 2021 by using bibliometric comprehensive research. This study depends on the remarking basis of web of science core collection database. A database of 10,605 publications have been retrieved. As per the available data China has the highest number of publications. China and the USA have close operational application. After referring a higher number of papers reveal that these wastewater purification or removal technology such as ozonation or membrane filtration can effectively eliminate medical or pharma-related compounds from water bodies. The effective identification of ECs and also their optimised removal methods are current challenges. Introduction of heavy metals at minute quantity is believed to be a risk for human beings. Thus, effectively and deeply remove undesirable metals from water system is still a burning and challenging task for environmental engineers. Most popular methods of WWTPs and their associated unitary operations and progressions are referring to them. Stabilisation techniques are led to origination of contaminated materials with variety of organic compounds. Since this management system normally expects induction of processed sludges to ground state, it causes soil pollution with unknown organic compounds. Thermal processing of raw sewage sludge excludes it. Majorly, organic complexes are transformed into simple and mineralised form. Most burning issue is contamination of sewage sludge ashes with dissolved heavy metals. Identification of heavy metals in various types of ashes is much simpler than the convectional organics. Water and air quality is degrading because of unfavourable materials. Textile industry is one of the globally richest and the largest source of water pollution. Dye ubiquity in wastewater leads to serious environmental issues that have negative impact on their respective quality, including decreased light penetration effects into water and disrupting photosynthesis. Direct or indirect side-effects of emerging contaminants is on progress, because their minute or unsignificant quantities could be risky for all types of animals. Prime ways for humans' or animals' exposure to toxic chemicals is via intake of contaminated foods and drinks in the form of bioaccumulation and biomagnification at the climax of food chain. Still, there are no available data related to toxicity and impacts of industrial or domestic wastes.

2.2   METHODS

Various remediation procedures for contamination removal in industries such as usage of activated carbon or biomass, ion-exchanging modes, reverse osmosis, oxidative degradation, chemical precipitation, ultrafiltration, electrochemical process, ozonation, membranes' separation and absorption have limited wastewater treatment efficiencies, residues, cost, versatility, time or energy consumption. The most promising wastewater treatment techniques are physiochemical techniques. For this, many researchers are in search of promisingly proficient methods for waste remediation in WWT. One of the most promising methods from economic perspective is adsorption with simple designs but leads wide application range for recycled adsorbents, operational security against hazardous metals and low space requirement. These properties have equipped them to employ for both water and wastewater bioremediation. This is mostly recommended method for production sites to treat wastewater. Due to regular advent in nanoscience, plenty of nanomaterials might be used as adsorbents for waste remediation due to their efficiencies. Nanomaterials such as CNT, fullerene, ferric oxides nanoparticles, chitosan, zinc particles, silica, and other polymeric and graphene families have highly performed adsorption, specific surface area, solution mobility, reactivity, and their small size leads to use in wastewater purification. They have completed successful study of remediation related to large number of organic and inorganic ECs. An ideal adsorbent should possess surprising adsorption capacity, selectivity, recyclability and eco-friendliness. Modern research teams are performing constant studies to stupefy all these conditions for wastewater treatment.

The twenty-first-century researchers have believed that nano-adsorbents would prove to be result oriented for these global issues to search for efficient and effective techniques. Among all, adsorption is the most propitious one because of its versatile ability, broad applicability, and ease of economic feasibility and operational cost. Adsorption is a physiochemical behaviour where attachments of molecules to adsorbent surfaces. Adsorbate is followed by material surface, furthermore physical facial facet is adsorbate. In it, the ECs in solution gel added onto sorbent surfaces through effective locations are present on surfaces. The adsorption process using below-valued adsorbents, such as clay soil, zeolites, silicon-related admixtures and horticultural droppings, have been widely known for elimination of various effluents.

```
                          ┌─────────────────────────┐
                          │   Waste-water Treatment  │
                          └─────────────────────────┘
              ┌───────────────────────────┐          ┌──────────────────────┐
              │  Conventional Techniques   │          │   Hybrid Techniques  │
              └───────────────────────────┘          └──────────────────────┘
```

Physical techniques	Chemical techniques	Biological techniques
(*Mass transfer concept*)	(*Chemical reactions) dependent*)	(*Cellular degradation concept*)
Such as Screening, Sedimentation, Adsorption, Aeration, Skimming and Thermal treatments	Such as Ion-Exchange, UV light, Chlorination, Neutralization & Precipitation	Such as Bioremediation, Biofilters, Biosorption

(*Composed of 2 or more conventional methods*)

Such as Constructed Wetlands,

Waste Stabilization Ponds

and Membrane Filtrations Bioreactors

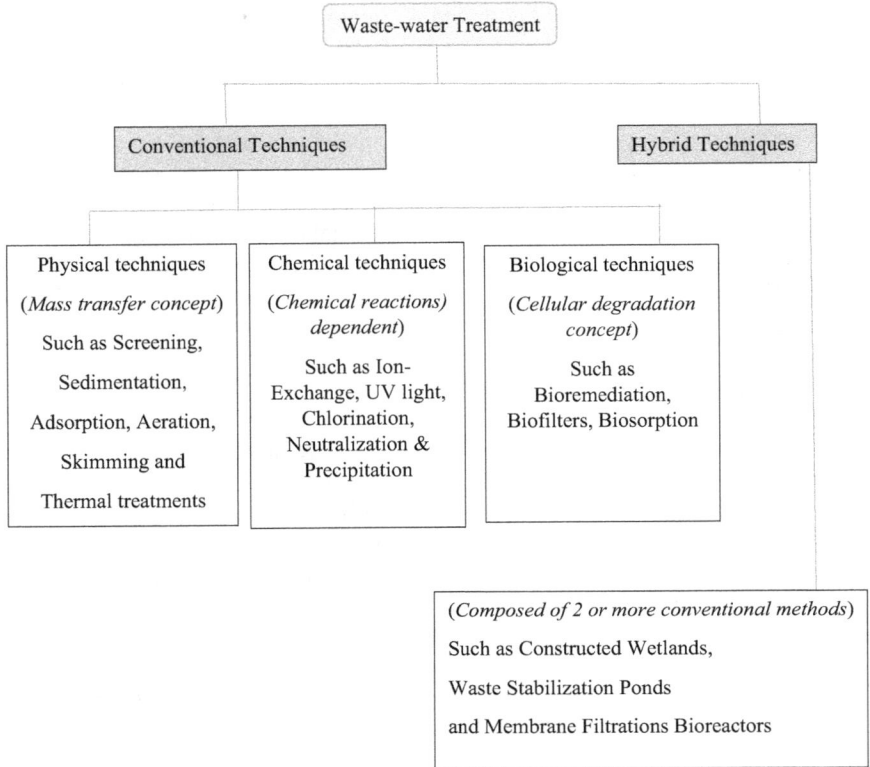

FIGURE 2.1   Structural classification of WWT.

Wastewater conventional and modern treatments are on focus, characterised as below (Figure 2.1):

No biological agents or chemicals are used in physical treatment. Screening, an ancient method for physical segregation of solid wastes in water has been characterised based on size of solid wastes into two parts: fine screening (0.001–6 mm) and coarse screening (>6 mm). When solid waste dimension is more than 6 mm, screening becomes irrelevant; then comminution is used. In comminution, solid wastes are crushed, precipitated and removed from wastewater. Skimming is used to remove liquid or gels like wastes from water such as oil, fat, grease, etc., which should have low weight in comparison to water. Floatation as its alternative has been employed to remove solid or liquid wastes from water by introducing gas in it. Gas entraps contaminates and settles down. It is applied on solid and biological sludges. Aeration is applied on gaseous and volatile organic wastes to remove nutrients from domestic wastes. Aeration and floatation are also used in hybrid techniques to increase its efficiency and effectiveness. Adsorption is applied on solid organic, inorganic and toxic substrates that are based on high specific surface area that needs to be activated. Generally, carbon activation is used for smoothening of adsorbents.

TABLE 2.2

Tabular Comparison among Different Wastewater Treatment Modes [4–7]

Treatment Modes	Achieved Targets	Findings
Physical treatment	Carbon resumption.	Low sewage total processing time.
	Decrease in mass.	Consumption of high power.
	Complete pathogen removal.	Micropollutants emission affects air, water and land.
	Heavy metal counteraction.	Requirement of complex set-ups
	Low nitrogen detention.	and techniques.
Chemical treatment	Heavy metal offsetting.	Easiest mode of sewage treatment.
	Pathogen abolishment.	Heavy investment and operational
	Nitrogen resumption.	cost.
	High nitrogen retention.	Gain in sludge mass.
	Pathogen blockage till pH 12.	
Biological treatment	Carbon resumption.	Long treatment duration.
	Nitrogen resumption.	Requirement of complex
	Decrease in mass.	technology.
	Pathogen and heavy metals counteract due to anaerobic digestion.	Anaerobic digestion demands huge investment.

Biological techniques are more cost effective and can be performed into two or three stages to segregate biodegradable contaminants using biofilters, biofilm reactors and bioremediation. Bioremediation is applied in oxidation ponds, lagoons and biofilters to encourage adoption of phytoremediation processes, plant growth and environmental sustainability. Biofilters are commonly used in case of biofiltration against fungi, worms, yeasts like microorganisms. Biosorption is highly efficient biological process applied to highly graded effluents via physical retention and biodegrading acts. It also enhances efficiency of hybrid system based on biological and chemical principles.

When mechanical, physical or biological techniques are inefficient, we move to chemical techniques. It is applied mainly in industries and some agricultural contaminants. Precipitation is the oldest chemical method where dissolved substances are filtered out. Chemical oxidation is an emerging technique where oxidants are used to turn harmful wastes into harmless ones or negligible harmful contents. These chemical techniques have been broadly divided into two parts: convectional oxidation methods (photolysis, ozonation, Fenton oxidation process) and advanced oxidation process (photocatalysis and photo-Fenton process). Photolysis is the release of hydroxyl free radicals upon breakdown of water molecules via electromagnetic radiations or ultraviolet rays without any usage of catalysts or oxidants. Ozonation has been applied on organics or inorganics over the last decades (Table 2.2).

Among available modes of wastewater treatment strategies, anaerobic digestion is a highly matured method focused on advanced countries such as Germany, Italy, Netherland, United States of America and China with total heavy investments nearby

125 million euros and generates more than two lakhs tons of biogas per year. It turns biodegradable organic matters and sewage sludges into biogas or bio-composts in nearly no-oxygen environment. Decrease in mass of sewage sludge is expressed in terms of volatile solids and total solids reduction. Volatile solid could be reduced nearly 35%–60% or may be ignored if biodegradable organic matter would have been heated up more than 550°C. Total solid is said to have reduced biodegradable and non-biodegradable material up to 30% that is left after being evaporated at 105°C [8].

Hence, to enhance its broad acceptance, vermicomposting and black soldier fly larvae are low-cost waste treatment to control sludge problems, and could eliminate physical or chemical modes of sewage waste treatments. Vermicomposting decomposes the sewage sludge four times that of naturally decomposed disposal. Organic modes treatments using vermin composts on large scale are on priority and cause successful generation of co-substrates. The BSFL incorporates the organic decaying products such as food stuffs and natural composts, and turns nutritious wastes into insect biomass or bio-composts along with high lipid and protein contents. Therefore, the BSFL treatment has also been produced high-protein poultry feed through feeding with organic wastes (Figure 2.2).

In the beginning of anaerobic digestion, large molecules like polysaccharides or complex proteins are simplified into long chains of fatty acids, ethanoic acid or small glucose-like soluble components via hydrolysis using bacteria such as Cellulomonas, Bacteroides, Fibrobacter, Microbispora species, etc. After that, their by-products are formed in small-fatty-acid chains, ammonia, carbon dioxides and hydrogen sulphides using acidogenic bacteria such as Peptoccus, Bacteroides, Eubacterium, Phodopseudomonas, Sarcina species, etc. Further decomposition of remained heavy organic acids are biodegraded into ethanoic acid, hydrogen and carbon dioxide gases using Syntrophobacter, Syntrophus, Pelotomaculum, Syntrophomonas, and Syntrophothermus species. At last, methane is produced using reaction of acetate

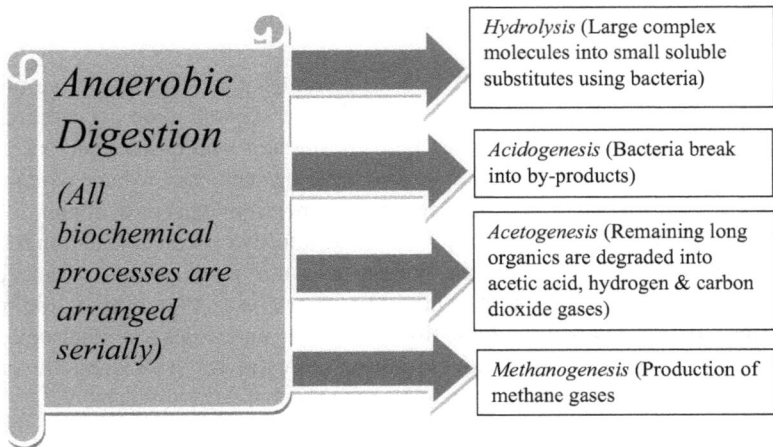

Anaerobic Digestion *(All biochemical processes are arranged serially)*

Hydrolysis (Large complex molecules into small soluble substitutes using bacteria)

Acidogenesis (Bacteria break into by-products)

Acetogenesis (Remaining long organics are degraded into acetic acid, hydrogen & carbon dioxide gases)

Methanogenesis (Production of methane gases

FIGURE 2.2   Stepwise characterisation of anaerobic digestion.

and water or carbon dioxide with hydrogen gas. The associated time related to sewage treatment is said to be sludge retention time. Normal period of SRT varies from 10 to 20 days. Ideal time for SRT is 14 days.

2.2.1 Physical Treatments

Simply physical analysis of micro-contaminants is based on concept of mass reduction. Its associated equipment are flexible to various formats where spawning of solid wastes is below than other ones. This mode of wastewater treatment is subdivided into two parts: primary physical treatment methods and secondary physical treatment methods. Screening, sedimentation and skimming are considered as primary physical treatments. Secondary physical treatments are aeration, thermal treatment, adsorption, and membrane-based technology are successively followed by biological and chemical processes. If screening is not working, then move to comminution (Figures 2.3 and 2.4; Table 2.3).

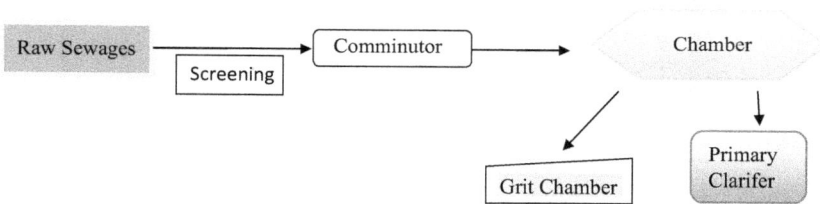

FIGURE 2.3 Primary physical treatment.

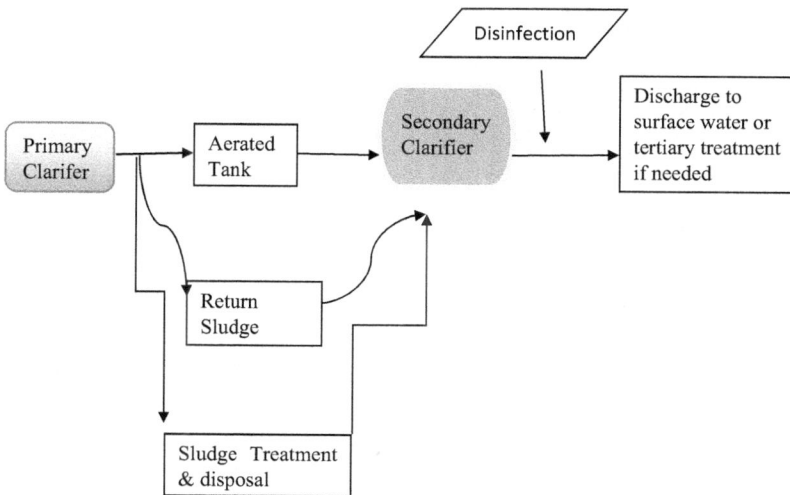

FIGURE 2.4 Secondary physical treatment.

TABLE 2.3
Discussion on Physical Treatment Methods

Physical Treatment Technology	Subdivision	Main Tasks	Findings	Reference
Screening	Coarse screening Fine screening	Eliminates solid wastes size over 6 mm. Eliminates solid wastes size between (0.001 and 6) mm.	Ensures removal of larger and floated wastes for the minimum damage, blockage & interrupted processes.	[9]
Grit chamber	Horizontal flow grit chamber Vortex flow grit Aeration grit	Ensures sedimentation of remained wastes so that they don't clog them with equipment and promotes settlement of floating materials.	Effective for heavier solid materials in comparison to organic solid materials.	[10]
Sedimentation	Horizontal flow Solid contact Inclined surface	Maintains uniform velocity & flow velocity. Solid wastes surge in sludge blanket where liquid go. Surface depth is divided into shallower sections for fast settlement.	Assures better cleaning of wastes, allows sludges to be settled & separates precipitated of heavier materials using gravitational force, which are affected by velocity of water wastes and time of settlement.	[10]
Skimming		Removal of those contaminants that are lighter than water.	Frequent removal oil, grease, fat and oil like contaminants of the skimmer to maintain skimming process efficiency.	[11]
Floating		Uses gas particles to isolate suspended solids, liquid oils & gaseous contaminants. Follows gas trapping hybrid mechanism of treatment.	More efficient and effective than other treatment process.	[12]
Aeration		Based on electrochemical oxidation mechanism. Using air bubble removal of dissolved gaseous, metallic or volatile contaminants.	Low expenses needed than others. Easily coupling with other treatment technology.	[13]
Thermal treatment		Pollutants removal at high temperature. Deteriorate emerging contaminants in sustainable manner.	Need of coupling with other technologies for effective performance.	[14]
Adsorption		Degrades soluble substances using solid materials of high specific surface area.	Ensures contamination removal with accuracy and efficiency. Specific surface area enhances engulfing stamina of adsorption.	[5]
Membrane-based technology		Porous filtration membrane separates pollutants depending upon its types and sizes.	Its driving forces are hydrostatic pressures throughout membrane. Fouling occurs due to poor maintenance.	[15]

2.2.2   Biological Treatments

After physical treatments of industrial or other contaminants, remained toxic contaminants may promote microbial development. So, results of physical treatments demand for other techniques of treatment. Then, we move towards biological treatment after immediate completion of physical treatment. Biotreatment of contaminants using different microbes or different cellular processes in two or three stages have been proved economical and eco-friendly for effective product decomposition with an objective of achieving biodegraded waste treatments. Prime ways for humans' or animals' exposure to toxic chemicals are via intake of contaminated foods and drinks in the form of bioaccumulation and biomagnification at climax of food webs. Still there is no availability of research study related to toxicity and impacts of industrial or domestic wastes. Conventional treatment comprises of all traditional processes such as biofilters, biofilm reactors, nitrification–denitrification, microbial-based treatment, bioremediation and aerobic/anaerobic methods. Out of them, bioremediation is the most widely economically accepted where organisms such as plants and microbes are used to remove contaminants with high degradation and adsorption potential. Usage of plants for bioremediation is said to be phytoremediation. It has been regarded as an effective means to accumulate heavy metals released. It upgrades soil quality, plant surging and global stability. Biofilters are alternate biotreatment set-ups for degradation or separation of heavy or light contaminants. Bio-nitro-dinitro treatment is an oxidation or reduction process that turns ammonia into nitro-related compounds, then further nitrogen-related products. Basically, it removes suddenly salinity or ionic pollutants in wastewater. Anaerobic treatments have high contamination removal efficiency in comparison to aerobic treatments.

Bio-non-conventional treatment study is still on research. Majorly, they are of biosorption, membrane bioreactor (MBR), microbial fuel cells (MFC) and constructed wetland areas. Biosorption is an impactful treatment process that is used in cultured or harvested state to enhance coefficient of strong binding force between pollutants and microbes. MFC is an alternative treatment based on electrochemical hybrid strategy of chemical and biological mechanisms (Table 2.4).

2.2.3   Chemical Treatment

In some cases, physical/mechanical and biological processing techniques have been proved irrelevant; then move towards chemical treatments. Chemical treatments for wastewater treatment stands for modification/alteration/processing of available wastes with the help of chemicals where dissolved pollutants are separated using targeted substances. It is highly appreciable for some agricultural and industrial wastes. Its mechanisms may include ion exchange, disinfection through chlorination, carbon oxidation method, UV radiation, ozone, neutralisation and precipitation. Precipitation mechanism turns prior dissolved substances into dissoluble substances that can be further filtered out. Carbon oxidation method is an emerging approach whereby pollutants are converted into harmless and controllable form using chemical materials along with redox mechanism. Chemical oxidation demand determines

TABLE 2.4
Summary of Bioprocesses for Wastewater Treatments

Biotreatment Processes	Main Tasks	Findings	Reference
Bioremediation	To protect environment via using microbes.	Decreases pollution, dislodges contaminants and improves fouling susceptibility.	[16]
Biofiltration	To filter micro or nano scale contaminants using microbes.	Eco-friendly process. Efficient contamination removal.	[17]
Composting	Organic biodeteration to biomass.	Environment friendly in nature. Lowering energy consumption & sludge production down. Having small reactor volume. Increases contamination removal efficiency.	[18]
Aerobic & anaerobic treatments	Aerobic treatments recede P and N_2 pollutants in O_2 presence. Anaerobic treatment turns those wastes into biogas without oxygen presence.	Degradation of phosphorous- and nitrogen-related pollutants. Anaerobic treatment has high effectiveness for contamination removal.	[19]
Biosorption	Microbes immobilise to capture contaminants via binding mechanism of adsorbates.	Periodic adsorbates cleaning.	[20]
MBR	Contamination degradation via natural confinement and microbial biodegradation.	Highly competent & expensive pollutants segregation, which could easily couple with other advanced techniques with improved performance.	[21]
Microbial fuel cells	Hybrid mechanism of microbiological & chemical concepts of contaminants removal.	Highly monitored microbial development, which might hamper processing of viable resources.	[22]

chemical oxidants-contaminated contents where chemical oxidants remove or reduce weak or smaller by-products. The advanced oxidation process treats high carbonic wastes so as to generate free radicals for oxidation of most of available complexes in matrices of environment. Carbon oxidation processes have been categorised into two broad approaches: conventional and advanced oxidation processes. Mostly convection oxidation approaches comprise of photolysis, ozonation and Fenton oxidation process. These processes mechanisms basically consist of free radicals as driving catalyst to reduce toxicants. As convection oxidation processes are always unable to fulfil their purposes, advanced oxidation processes have come up with simultaneous different oxidation processes. They gratify inflexible organics-contained wastewater with active hydroxyl radicals. Photo-Fenton, photocatalysis and solar-driven processes are some frequently used advanced oxidation approaches (Table 2.5).

TABLE 2.5

Summary of Chemical Approaches for Polluted Water Treatment

Chemical Treatments	Subdivisions	Main Tasks	Findings	Limitations	Reference
Convectional approaches	Photolysis	Electromagnetic radiation breaks or degrades waste compounds.	Generation of free hydroxyl radicals. Colour removal from wastewater.	Enhanced turbid media may restrict UV penetration in polluted water.	[23]
	Ozonation	As strong oxidiser biodegrades wastes without toxic by-products.	Effective deterioration of personal care & pharmaceutical products. Catalytic oxidation method reduces organic toxicity.	Ozone has compromised half-life, utilisation efficiency & oxidative power to mineralise organic complexes.	[24]
	Fenton oxidation process	Ferrous ions interact to hydrogen peroxides and releases free hydroxyl radicals.	Fenton catalysts replace unsaturated Fe complexes and improves organic degradation of compounds. Small pH values. Effectively organic degradation in polluted water. An eco-friendly approach for filtrating, dewatering & conditioning of polluted water.	Some risks related to transporting, handling and storing reagents.	[25]
Advanced oxidation approaches	Photocatalysis	Energy transfer from photon to water molecules using catalysts.	UV photocatalysis has four times intense carbon contents that of colour reduction efficiency. Titania has high photocatalysis effect for pollutants & microorganisms' removal. Reusing of catalysts. Easily operative at normal temperature & pressure.	Difficulty in achieving uniform catalytic effect on large surface area.	[23]
	Photo-Fenton process	(Fenton process + UV radiation = Photo-Fenton process) Free hydroxyl radical removes toxic contents from polluted water against low hydrogen peroxide intake.	Simple technology, economic High antibiotic removal efficiency.	Limited production of chlorinated by-products.	[23]

2.2.4 HYBRID TREATMENT

Integration of two or more contaminated removal methods are caused for energy-efficient, sustainable and stable environment called hybrid treatment. In last 10 years, physical, chemical and biological approaches are prioritised for optimising energy and cost. Hybrid treatments are characterised into physiochemical and electrochemical processes. Physiochemical processes comprise of adsorption and ion-exchange mechanisms, whereas electrochemical processes consist of electrodeposition, electro-floatation and electrocoagulation techniques.

Adsorption is based on concept of mass variation between different sorbents of solid and liquid phases. In it, liquified or gaseous solutes are adsorbed on sorbent's surface such as liquid or solids. This is categorised into three stages: first is pollutant dispersal from solution to adsorbent surface, second is contamination addition to adsorbent surface and third is adsorption expansion adsorbent structure. Suitable methods for WWTPs due to its effective applications as homogeneity, modest, flexibility and high adsorption amplitudes due to large specific surface area. Its main advantage is its high sorption reversibility throughout desorption as well as through natural abstracts, domestic farming and industrial wastes. Due to uneconomic way adsorption materials, they are still only at laboratory level at high scale. Mostly CNT, AC, algal biomass, fungal biomass, bacteria and sawdust are said to be adsorbents. CNT and AC are commonly used eco-friendly bio-adsorbents for toxicants or contaminants removal. In fact, CNT depends on specific surface area, porosity, structure, packing density and purity. CNT removes Cu, Pb, Cr and Ni as major components. Prepared powered or pellet forms of activated carbon from potassium dichromate (from farming) and carboneous materials (from coal and biomass) at 900°C is used for removal of Pb, Ni, Cr and Cu. Pure AC is uneconomic, so its composite is in usage as its alternative, which depends on high toxic removal abilities of ions, ratio of surface area to volume and pore structure.

Ion-exchange method is firstly used in 1995, where heavy metals and their ions are physically absorbed via concept of ion reversibility based upon functional groups and countered-ions formed. It leads to formation of resins, also called magnetic ion exchanger that serves the purpose of elimination of natural organics through electrolysis. Reversible actions between two solids or two liquid phases are used to remove heavy metals from wastewater whose effectiveness depends on thermal values, pH values, specific resin property, adsorbent concentration used, ion's property and contact time (Table 2.6).

2.3 NANOMATERIALS AS AN EMERGING APPROACH IN WWTPs

Release of contaminants into water bodies and their interaction with natural ecology is a global burning issue due to unsustainable development, which, in turn, causes to pure water demand. Exponential population growth affects industrialisation and ecological system. Surging pure water call for drinking, agricultural application, domestic purpose, etc. have created regular water paucity. Architectural development of town, metro cities and developing or developed countries may face too much health-related hardships in coming days with respect to drinkable water as a basic amenity

TABLE 2.6
Study of Hybrid Treatment Methods for Toxic Removal

Hybrid Treatment Methods	Main Tasks	Findings	Reference
Membrane filtration & membrane bioreactor	Single-step contamination removal needs hydraulic pressure.	Approximately 100% bacterial & other smaller ions removal efficiency by reverse osmosis. Membrane bioreactor has high fouling propensity. Membrane pollution due to regular alteration.	[26]
Advanced filtration & advanced oxidation processes	Removing contaminants from pharmaceuticals wastewater treatment. Exhibit complement action during membrane fouling removal.	Improved membrane performance. Membrane filtration has higher removal efficiency than advanced oxidation processes.	[27]
MBR & modularised rolled piped system	Better effluent condition in WWT. Low membrane fouling ability.	Higher organic pollutants and nitrogen-removal efficiency.	[28]
Adsorption by powered activated carbon & ultrafiltration	Low dose of powered activated carbon removal effectiveness. Reduced ultrafiltration membrane fouling in textile industry.	Approximate 97% discoloration rate from acidic solution.	[29, 32]
Advanced oxidation & adsorption	Combination of ozone and hydrogen peroxide for advanced oxidation & adsorption has effective absorption in industrial or pharmaceutical wastewater treatment.	Its chemical oxidation demand is reduced to nearly 93%. Higher contamination degradation after ozone effect in the presence of free hydroxyl radical. After ozonation, adsorption eliminates metabolites as post-treatment effects.	[30]
Absorbents, membrane filtration and activated sludge process	Retrieving of resources from wastewater treatment.	High pollutant removal performance. Low energy depletion & sludge generation. Small reactor volume. Improved effluent ability.	[26]

because of inadvertent control of their wastes into local land and other water bodies such as lakes, ponds and rivers. Available wastewater treatment techniques are uneconomic, so they consume significant share of local economy. With the advent of biological, physical and chemical nature of nanomaterials, modern researchers are also planning to add nanotechnology concepts to wastes processing to reduce cost, time and ineffectiveness. Due to unaccountable social, religious and regional perspective, it is tough to regulate and identify all their contaminants.

Nanomaterials with high specific surface area have been explored with their enhanced adsorption, response capacity and resolution mobility [31]. For example, zerovalent metal nanomaterials, metal oxides nanoparticles, carbon nanomaterials, composites, etc.

Zerovalent metals such as Ag^+, Fe^{+2}, Zn^{+2}, Al^{+3} and Ni^{+2} have short size, colossal surface area and are highly sensitive. They are performing appropriate and compatible results for separating pollutants from wastes. Silver nanomaterials have high antimicrobial effects against microbes for the purpose of water disinfection. Iron nanomaterials remove cadmium ions, nitrate complexes, dyes and other antibiotics with the help of adsorption, redox and precipitate forming actions.

Metals are available in abundance. So high-quality metal oxides are economic. For example, Fe_2O_3, MnO_2, Al_2O_3, TiO_2, etc. They have high removing efficiency and sorption ability against arsenic, uranium, phosphates and other organics. Titania is the best one out of them. It is highly economic and photostable with catalytic application. It has reduced selective efficacy. That is why it is applicable for all types of contaminants, such as chlorine complexes, dyes, phenyls, aromatics, agricultural pesticides and heavy metals. Zinc oxide nanomaterials have high eco-friendly, oxidising ability, immense wavelength region and photocatalytic abilities. Zinc oxides could catch more lighter than most of other semiconducting metal oxides. Incorporation of tin oxides with zinc oxides at nanoscale show nearly more than 99% adsorption efficiency, which is better than other ones.

Allotropes of carbon nanomaterials such as carbon nanotubes, fullerenes, graphene, nanodiamonds have superb kinetics, aromatical characters, structural, electrochemical, electronic, biological and physiochemical properties. They are used for basic or toxic complexes and follow the sorption mechanisms for wastewater activity. Carbon nanotubes have been gaining attention as an alternate adsorbent against variety of emerging contaminants such as dichlorobenzenes, ethylbenzenes, Pb^{+2}, Zn^{+2}, Cd^{+2} and Cu^{+2}. Graphene, as another carboneous materials, has honeycomb-like structure with singular carbon layer. It is of effective removal capabilities when it is integrated with magnetic manganese ferrite particles against lead and arsenic. Functionalised carbon nanotubes have surprised phenol adsorption characteristics, the pristine carbon nanotubes due to pi–pi stacking, and hydrogen bonding between phenol and functionalised carbon nanotubes.

2.4 FUTURE SCOPE

Emerging contaminants, a type of synthetic chemicals, are in low concentration; even their non-toxicity treatment requires high investment. Even though, it is working with desired efficiency for every type of contaminants, the integration of nanotechnology with genetic engineering opens another door of new hope for more pollutant's elimination with reduced cost, time and energy. Adverse environmental effects due to chemical and biological wastes, greener methods need to be explored for the non-toxic techniques for nanoparticles synthetisation via biological protocol to abide by high energy costs and high pressure. The appropriate selection conscience treatment techniques depend on accuracy, source water character, utilisation efficiency, accuracy, resilience, operational and maintenance cost. Already existing physical wastewater technology uses ultrasounds with adsorption and gamma radiation bombardment of gamma radiation. But there is a need to search for degradation of acidic contaminants.

REFERENCES

1. H. Forbes, T. Quested, C. O'Connor (2021) Food waste index report 2021. United Nations Environment Programme, Nairobi, Kenya.
2. D.L. Heller (2012) National environmental policy act legal research: Overview. libraryguides.law.pace.edu.
3. A. Azimi, A. Azari, M. Rezakazemi, M. Ansarpour (2017) Removal of heavy metals from industrial wastewaters: A review. *Chem BioEng* 4(1), 37–59.
4. P.R. Rout, T.C. Zhang, P. Bhunia, R.Y. Surampalli (2021) Treatment technologies for emerging contaminants in wastewater treatment plants: A review. *Sci Total Environ* 753, Article 141990, doi: 10.1016/j.scitotenv.2020.141990.
5. K. Dhangar, M. Kumar (2020) Tricks and tracks in removal of emerging contaminants from the wastewater through hybrid treatment systems: A review. *Sci Total Environ*, Article 140320, doi: 10.1016/j.scitotenv.2020.140320.
6. M. Bilal, M. Adeel, T. Rasheed, Y. Zhao, H.M. Iqbal (2019) Emerging contaminants of high concern and their enzyme-assisted biodegradation–a review. *Environ Int* 124, 336–353, doi: 10.1016/j.envint.2019.01.011.
7. E.C. Lima (2018) Removal of emerging contaminants from the environment by adsorption. *Ecotoxicol Environ Saf* 150, 1–17, doi: 10.1016/j.ecoenv.2017.12.026.
8. Y. Cao, A. Pawlowski (2012) Sewage sludge-to-energy approaches based on anaerobic digestion and pyrolysis: Brief overview and energy efficiency assessment. *Renew Sustain Energy Rev* 16, 1657–1665, doi: 10.1016/j.rser.2011.12.014.
9. A. Bhargava (2016) Physio-chemical wastewater treatment technologies: An overview. *Int J Sci Res Educ* 4, 5308–5319, doi: 10.18535/ijsre/v4i05.05.
10. N.R. Esfahani, M.N. Mobarekeh, M. Hoodaji (2018) Effect of grit chamber configuration on particle removal: Using response surface method. *J Memb Sep Technol* 7, 12–16, doi: 10.6000/1929–6037.2018.07.02.
11. S. Mintenig, I. Int-Veen, M.G. Löder, S. Primpke, G. Gerdts (2017) Identification of microplastic in effluents of wastewater treatment plants using focal plane array-based micro-Fourier-transform infrared imaging. *Water Res* 108, 365–372, doi: 10.1016/j.watres.2016.11.015.
12. S.H. Ammar, A.S. Akbar (2018) Oilfield produced water treatment in internal-loop airlift reactor using electrocoagulation/flotation technique. *Chin J Chem Eng* 26, 879–885, doi: 10.1016/j.cjche.2017.07.020.
13. C. Iskurt, R. Keyikoglu, M. Kobya, A. Khataee (2020) Treatment of coking wastewater by aeration assisted electrochemical oxidation process at controlled and uncontrolled initial pH conditions. *Sep Purif Technol* 248, Article 117043, doi: 10.1016/j.seppur.2020.117043.

14. A.A. Yaqoob, A. Khatoon, S.H. Mohd Setapar, K. Umar, T. Parveen, M.N. Mohamad Ibrahim, A. Ahmad, M. Rafatullah (2020) Outlook on the role of microbial fuel cells in remediation of environmental pollutants with electricity generation. *Catalysts* 10, 819, 10.3390/catal10080819.

15. M.S.S.A. Saraswathi, A. Nagendran, D. Rana (2019). Tailored polymer nanocomposite membranes based on carbon, metal oxide and silicon nanomaterials: A review. *J Mater Chem A* 7, 8723–8745, doi: 10.1039/C8TA11460A.

16. A. Shah, M. Shah (2020) Characterisation and bioremediation of wastewater: A review exploring bioremediation as a sustainable technique for pharmaceutical wastewater. *Groundw Sustain Dev*, Article 100383, doi: 10.1016/j.gsd.2020.100383.

17. A. Hosseinzadeh, M. Baziar, H. Alidadi, J.L. Zhou, A. Altaee, A.A. Najafpoor, S. Jafarpour (2020) Application of artificial neural network and multiple linear regression in modeling nutrient recovery in vermicompost under different conditions. *Bioresour Technol* 303, Article 122926, doi: 10.1016/j.biortech.2020.122926.

18. A. Hosseinzadeh, M. Baziar, H. Alidadi, J.L. Zhou, A. Altaee, A.A. Najafpoor, S. Jafarpour (2020) Application of artificial neural network and multiple linear regression in modeling nutrient recovery in vermicompost under different conditions. *Bioresour Technol* 303, Article 122926, doi: 10.1016/j.biortech.2020.122926.

19. M. Harb, E. Lou, A.L. Smith, L.B. Stadler (2019) Perspectives on the fate of micropollutants in mainstream anaerobic wastewater treatment. *Curr Opin Biotechnol* 57, 94–100, doi: 10.1016/j.copbio.2019.02.022.

20. E. Daneshvar, M.J. Zarrinmehr, A.M. Hashtjin, O. Farhadian, A. Bhatnagar (2018) Versatile applications of freshwater and marine water microalgae in dairy wastewater treatment, lipid extraction and tetracycline biosorption. *Bioresour Technol* 268, 523–530, doi: 10.1016/j.biortech.2018.08.032.

21. J. Ji, A. Kakade, Z. Yu, A. Khan, P. Liu, X. Li (2020) Anaerobic membrane bioreactors for treatment of emerging contaminants: A review. *J Environ Manag* 270, Article 110913, doi: 10.1016/j.jenvman.2020.110913.

22. K.S. Khoo, W.Y. Chia, D.Y.Y. Tang, P.L. Show, K.W. Chew, W.-H. Chen (2020) Nanomaterials utilization in biomass for biofuel and bioenergy production. *Energies* 13, 892, 10.3390/en13040892.

23. E.M. Cuerda-Correa, M.F. Alexandre-Franco, C. Fernández-González (2020) Advanced oxidation processes for the removal of antibiotics from water. An overview. *Water* 12, 102, doi: 10.3390/w12010102.

24. J. Shen, T. Ding, M. Zhang. Analytical techniques and challenges for removal of pharmaceuticals and personal care products in water. M.N.V. Prasad, et al. (Eds.), *Pharmaceuticals and Personal Care Products: Waste Management and Treatment Technology*, Butterworth-Heinemann (2019), pp. 239–257.

25. P. Borah, M. Kumar, P. Devi. Recent trends in the detection and degradation of organic pollutants. P. Singh, et al. (Eds.), *Abatement of Environmental Pollutants*, Elsevier (2020), pp. 67–79.

26. E. Obotey Ezugbe, S. Rathilal (2020) Membrane technologies in wastewater treatment: A review. *Membranes* 10, 89, doi: 10.3390/membranes10050089.

27. N. Rosman, W.N.W. Salleh, M.A. Mohamed, J. Jaafar, A.F. Ismail, Z. Harun (2018) Hybrid membrane filtration-advanced oxidation processes for removal of pharmaceutical residue. *J Colloid Interface Sci* 532, 236–260, doi: 10.1016/j.jcis.2018.07.118.

28. Q.-V. Bach, V.T. Le, Y.S. Yoon, X.T. Bui, W. Chung, S.W. Chang, H.H. Ngo, W. Guo, D.D. Nguyen (2018) A new hybrid sewage treatment system combining a rolled pipe system and membrane bioreactor to improve the biological nitrogen removal efficiency: A pilot study. *J Clean Prod* 178, 937–946, doi: 10.1016/j.jclepro.2018.01.038.

29. A. Hammami, C. Charcosset, R.R.B. Amar (2017) Performances of continuous adsorption-ultrafiltration hybrid process for AO7 dye removal from aqueous solution and real textile wastewater treatment. *J Membr Sci Technol* 7, 2, doi: 10.4172/2155–9589.1000171.

30. S. Patel, S. Mondal, S.K. Majumder, P. Das, P. Ghosh (2020) Treatment of a pharmaceutical industrial effluent by a hybrid process of advanced oxidation and adsorption. *ACS Omega* 5, 32305–32317, doi: 10.1021/acsomega.0c04139.

31. M. Nasrollahzadeh, T. Baran, N.Y. Baran, M. Sajjadi, M.R. Tahsili, M. Shokouhimehr (2020) Pd nanocatalyst stabilized on amine-modified zeolite: Antibacterial and catalytic activities for environmental pollution remediation in aqueous medium. *Sep Purif Technol* 239, Article 116542, doi: 10.1016/j.seppur.2020.116542.

32. Y. Zhang, B. Wu, H. Xu, H. Liu, M. Wang, Y. He, B. Pan (2016) Nanomaterials-enabled water and wastewater treatment. *Nano Impact* 3–4, 22–39, doi: 10.1016/j.impact.2016.09.004.

3 Synthesis of Natural Polymer–Carbonaceous Filler Composites

Mokae F. Bambo, Teboho C. Mokhena, and Kgabo P. Matabola

DSI/Mintek-Nanotechnology Innovation Centre

CONTENTS

DOI: 10.1201/9781003328094-3

3.1 INTRODUCTION

Polymer nanocomposites are generally defined as the combination of a polymer matrix and additives that have at least one dimension in the nanometer range [1,2]. They have since drawn significant academic interest, because they allow the design of high-performance materials that show noteworthy improved properties with regard to the pristine polymer. The extent of improvement ordinarily hinges on several parameters, including the size of the particles, their aspect ratio, their state of dispersion, and their surface chemical characteristics that determine the interaction between the filler and the polymer chains, and thus the interface of the polymer–filler system [2,3].

The additives can be one dimensional (nanotubes and fibres) [4–9], two dimensional (clay and graphite) [10–15], or three dimensional (spherical particles) [16–18]. Adding such additives to polymers can effectively enhance the properties of composites such as strength, modulus, and thermal stability.

The development of nanocomposites based on conventional polymers has been an attractive topic of study for materials to be used in different applications [19]. However, the frequent problem of synthetic nanocomposites is their relatively high price as well as the possible environment pollution they can provoke [20,21]. Recently, scientific attention has been mostly on the attainment of advanced applications that meet the growing technological development by facing the challenges like ongoing environmental concern and energy crisis. The reduction of plastic pollution in order to promote awareness against the growing impact of synthetic polymers on environment and ecosystem is very important. As a result, as an effort of environmental preservation and to reduce the material formulation cost, nanostructured materials based on biodegradable, renewable natural polymers, and low-cost polymers became of interest [22]. The research on these new materials technology is attracting the attention of researchers all over the world. Advancements are being made to improve the properties of the materials and also to find alternative precursors that can confer desirable properties on the materials. Considerable interest has recently developed in the area of nanostructured carbon materials. Carbon nanostructures are becoming commercially important with interest growing rapidly over the decade or so since the discovery of buckminsterfullerene, carbon nanotubes, and carbon nanofibers [23–25].

Since the discovery of carbon-based nanostructures starting with fullerene, their unique structural and transport properties have captured interest of researchers. Graphene—another nanostructure of carbon—has been tested to be the strongest material on Earth, and carbon nanotubes (CNTs) do not trail behind much in their strength. From a simple rule of mixtures to more complicated models, it is clear that carbon nanofillers promise ultra-strong-conducting polymer composites.

This chapter provides an overview of the current status in the synthetic methods of natural polymer–carbonaceous filler composites. First and foremost, the synthetic methods for the natural polymer–carbonaceous filler composites and modification methods for the carbonaceous materials (CNTs, graphene/graphene derivatives, carbon black and biochar carbon) are reviewed followed by the resultant properties of the fabricated composites. In conclusion, the perspective corresponding to the resultant composites is highlighted.

3.2 POLYMER/CNTs-BASED NANOCOMPOSITES

Carbon nanotubes (CNTs) are nano-scaled tubular structures and considered as the ideal reinforcing nanofiller for polymer matrix, precisely owing to their unique mechanical property, thermal stability, high aspect ratio, and electrical conductivity [7,24,26]. They can be single, double, or multiwalled depending on their synthesis methods. Their nonreactive surface and strong aggregative properties have limited the effectiveness of CNT with the polymer matrix they reinforce. If the unique multifunctional properties of CNT are to be utilised for effective reinforcement, the good dispersion of CNTs as well as the good interfacial adhesion between CNTs and the polymer matrix is a prerequisite [27–31].

3.2.1 CELLULOSE/CNTs-BASED NANOCOMPOSITES

Li et al. [32] prepared the cellulose acetate (CA)/carbon nanotube (CNT) nano-composites by melt compounding of CA with pristine multiwall carbon nanotube (MWCNT) or acid-treated one (MWCT-COOH). The solid-state mixtures of CA and MWNT-COOH with different ratios were mixed mechanically for 3 minutes. Then the mixtures were melt-extruded through a co-rotating twin-screw extruder with screw length of 440 mm and screw diameter of 11 mm. The extruder's temperature profile from Zone 1 to Zone 6 was set to be 235°C, 240°C, 245°C, 250°C, 260°C, and 260°C, while the screw speed was fixed at 180 rpm. The water bath was used to cool the extruded strands. After subsequent cooling, they were chopped into pellets and dried under vacuum at 180°C for 24 hours.

The modification of the pristine MWCNT was performed by refluxing a mixture of sulphuric acid (95%)/nitric acid (60%) (3:1 by mol ratio) at 120°C for 2 hours. Following cooling to room temperature, the solution was diluted with distilled water and filtered through a 0.22 μm pore size membrane until the pH approximated 7.0. Ultimately, the filtrated solid was dried in a vacuum oven at 120°C for 24 hours to achieve acid-treated MWCNT (MWCNT-COOH). The analysis has shown that the nanocomposites with various loadings (0.1, 0.5, 1.0, 3.0, 5.0, and 7.0 wt%) of MWCNT or MWCNT-COOH saw an increase in the thermal stability and dynamic mechanical properties. Furthermore, the electrical volume resistivities of CA/MWCNT-COOH nanocomposites were also found to be somewhat higher than those of CA/MWCNT composites [32].

Mostafa et al. [33] prepared the carbon nanotube (CNT)/cellulose composite sheets by utilising a simple solution-mixing method. The cellulose pulp is attained by blending the shredded pages (A4 size) in distilled water and 25 mL of the cellulose matrix. The uniformly dispersed solution of different amount of MWCNTs (0.25–3 wt%) is added into the mixture and stirred for 30 minutes. Thereafter, the resultant solution is poured into a Petri dish and dried using a hot air oven at 90°C for 3 hours. Subsequently, the moist fibres of the nanocomposites are placed between two glass plates and pressed under pressure of ~2.5 kPa to obtain composite sheets having even and smooth surface. The dispersion of the multiwalled CNTs (MWCNTs) is attained with the help of anionic sodium dodecyl sulphate (SDS). The latter is dissolved in distilled water by sonication for 10 minutes, and a certain amount of

MWCNTs is dispersed in the SDS aqueous solution. The ration of the surfactant to MWCNTs is adjusted to 1.2/1 (w/w) and sonicated for 30 minutes in order to attain uniformly dispersed solution of MWCNTs.

The results have revealed that the composites with different loadings of MWCNTs (0.5, 1, 1.0, 1.5, 2.0, 2.5, and 3.0) saw an appreciable thermal stability up to 550 K. They further demonstrated improved flame retardancy as compared to the virgin cellulose. In addition, the nanocomposites' conductivity appeared to increase with the increase in the loading of MWCNTs incorporated in the cellulose matrix. Similarly, the electrical conductivity was seen to increase with the increase in temperature. However, the sheet resistance of the nanocomposites steadily decreased as the concentration of the MWCNTs increased. This is as a result of the increased difficulty of high MWCNTs loading dispersing in the composites, and, for that reason, agglomeration occurs, which interrupts the cohesion between MWCNTs and polymer matrix leading to the loss of bonding. With these interesting properties, the composite sheets can find applications in different energy storage devices.

The MWCNTs and single-walled carbon nanotubes (SWCNT)-filled cellulose nanocomposites films and aerogels were made by Gnanaseelan et al. [34] using an alkaline aqueous solution. In respect of the preparation of cellulose-based films with CNTs, an NaOH/urea/CNT aqueous system was prepared by blending 7 g NaOH, 12 g urea, 77 g distilled water and 4 g of the 1 wt% CNT aqueous dispersion. The resultant mixture was stirred for 30 minutes and then pre-cooled to −12.0°C. Thereafter, the appropriate amount of cellulose was added immediately into the mixture with vigorous stirring for 5 minutes to achieve the cellulose/CNT dispersion. Subsequent to degasification, the dispersion was cast on a glass plate to obtain a gel sheet (thickness ~400 μm), which was immersed into a coagulation bath of 5 wt% H_2SO_4 for 5 minutes at room temperature to coagulate and regenerate.

The resultant cellulose/CNT composite hydrogels were washed with excess deionised water to remove residual chemical reagents. Finally, hydrogels were frozen by using the flash freezing method in which the hydrogels were immersed in liquid nitrogen (−196°C) for about 1 minute. Lyophilisation of the frozen mixture was carried out at −52°C under vacuum for 48 hours to obtain the aerogel. Thus, aerogels with 2, 3, 5, 8, and 10 wt% CNTs loadings with thicknesses between 0.1 and 0.7 mm were made [34]. The CNTs modification was fabricated by adding CNTs to a polyvinyl alcohol solution to obtain 1 wt% CNTs solution with a weight ration of PVOH/CNT of 1.5/1. The solution was mixed at somewhat rigorous conditions using a horn sonicator for 30 minutes at a constant output power of 40 W [34].

The results have indicated that the electrical conductivity of the cellulose/CNT composites is seen to increase with the CNT loading. Contrasting the SWCNT- and MWCNT-based composites, the former was found to have higher electrical conductivity than the latter at all compositions. At 10 wt% loading, the SWCNT-based composite films achieved the conductivity of 5.0 S/cm, while the MWCNT-based ones attained 0.9 S/cm. Lyophilisation generally led to a slight decrease in the electrical conductivity when comparing the composite films with the aerogels. However, the Seebeck coefficients of all investigated cellulose/CNT composites are positive. For SWCNT-containing aerogels, higher Seebeck coefficients than for films were measured at 3 and 4 wt% but significantly lower values at higher loadings. In the context of the films, the CNTs addition had the thermal conductivity increased, while the lyophilisation notably lowered it for the aerogels. With that said, the materials are promising candidates for use in thermoelectrical applications [34].

3.2.2   Chitosan/Multiwalled Carbon Nanotubes (MWNTs)-Based Nanocomposites

Wang et al. [35] examined the biopolymer chitosan/multiwalled carbon nanotubes (MWNTs) prepared by a simple solution evaporation method. Firstly, the MWCNTs were swelled in 100 mL of distilled water and homogenised in an ultrasonic bath for 60 minutes. Thereafter, 1 mL of acetic acid and 1 g of chitosan were added, and the mixture was shaken for 1 hour to dissolve the chitosan. The solutions were then stirred using a mechanical homogeniser at 18,000 rpm for 30 minutes, succeeded by sonication for 20 minutes to remove the bubbles. Subsequently, the chitosan/MWCNTs solutions were transferred into a plastic dish and heated at 50°C to evaporate water. Then, the dried uniform thin films (average thickness of 0.08 mm) were easily taken off their supports. Prior to nanocomposites preparation, the MWCNTs were modified by refluxing them in a mixture of concentrated sulphuric acid (98%)/nitric acid (65%) = 1/2 (v/v) for increasing more carboxylic and hydroxyl groups.

The results for the nanocomposites with the different content of the MWCNTs (0.2, 0.4, 0.8 and 2 wt%) have demonstrated that the addition of MWCNTs significantly improves the mechanical properties with the increase of MWCNTs loading. Most notably, the tensile modulus and strength of the nanocomposites increased significantly by about 78% and 84%, respectively, with only 0.4 wt% of MWCNTs filler compared with those of its neat counterpart. However, as the MWCNTs content increased to 0.8 wt%, the tensile modulus and strength of the chitosan/MWCNTs increased by about 93% and 99%, respectively [35].

3.2.3   Starch/CNTs-Based Nanocomposites

Cheng et al. [36] reviewed the preparation methods for starch-based carbon nanotubes composites. It emerged that there are two main forms of compounding starch and CNTs, which included convenient process-assisted mixing and casting method [37]. With regard to the former method, special care should be exercised with regard to the sequence of addition of ingredients that might affect the nanofiller dispersion, the gelatinisation/plasticisation and thus the final structure and properties of the nano-biocomposites. The starch functionalised CNTs or CNTs modified generally need the pretreatment of starch and CNTs to functionalise the surface. After conferring surface of starch with functional groups, a tighter nanocomposite is formed between starch and CNTs. Cheng et al. prepared starch-based CNTs films hinged on plasticised starch and modified CNTs by a simple casting method [37]. The CNTs were oxidised to prepare CNT oxide (OCNT) by Hummers method, and OCNTs were reduced to glucose to obtain reduced CNT (RCNT). Hence, the plasticised starch-CNT, plasticised starch-OCNT and plasticised RCNT composites could be used to fabricate composites by casting process easily. It was noticed that the reinforcing effect of RCNTs was more prominent than CNTs. The moisture resistance of

CNTs was better than that of OCNTs and RCNTs. Moreover, the electrical conductivities of plasticised starch-CNT, plasticised starch-OCNT and plasticised starch-RCNT composites were of no obvious difference.

A simple solution casting method was also used to make nanocomposite films from plasticised starch/functionalised MWCNTs to make a gas barrier [38]. The MWCNTs were functionalised by the hydrogen peroxide and H_2SO_4, and then mixed with plasticised starch by casting method. It is noticed that the thermal and electrical conductivity of starch/functionalised MWCNTs are improved. Cao et al. [39] prepared nanocomposites from plasticised starch and MWCNTs by a simple method of solution casting and evaporation. The results illustrated that the tensile strength and Young's modulus of the nanocomposites increased dramatically from 2.85 to 4.73 MPa and from 20.74 to 39.18 MPa, respectively, with an increase in MWCNTs content ranging from 0 to 3.0 wt%, respectively. However, the value of elongation at break of the nanocomposites was greater than that of plasticised starch and attained the highest value at 10 wt% MWCNTs content.

The novel starch-based nanocomposites comprising of very small amounts of MWCNTs were prepared by Fama et al. [40]. The composites were fabricated by adding 0.002 and 0.004 g of CNTs in the solubilised aqueous starch–iodine complex. The mixtures were sonicated in an ultrasonic bath for 1 hour. Thereafter, 4.5 g of Tapioca starch and 2.5 g of glycerol as plasticiser were added to each system. The mixture was heated at ~3°C/min till complete gelatinisation. The resultant gel was degassed with a vacuum pump for 10 minutes, and then cast on a glass Petri dish and dried at 50°C for 24 hours. Samples were conditioned over NaBr (~0.575 at 25°C) for 4 weeks before characterisation. Prior to composites preparation, MWCNTs were modified by wrapping them with an aqueous solution of starch–iodine complex prepared according to the method proposed by Star et al. [41]. The complex was performed at 0.5% of the same Tapioca starch of the matrix in distilled water with saturated iodine, stirred for 5 minutes at room temperature and stabilised during one night. The solubilised fraction of the complex was extracted avoiding any residual iodine. Thus, 93 g of the aqueous solution of the starch–iodine complex containing ~0.3 g of starch was obtained.

A significant improvement in Young's modulus (E) as well as an enhancement in tensile strength values were noticed from the incorporation of marginal amount of CNTs to the starch matrix. The stiffness improved by almost 70% and ultimate strength by ~35% with only 0.5 wt% of MWCNTs. Moreover, all tensile parameters also increased with filler contents. Consequently, improved tensile toughness was displayed by the nanocomposites and with filler loading.

3.2.4 Alginate/CNTs-Based Nanocomposites

Kawaguchi et al. [42] described the preparation of an alginate-CNT nanocomposite gel (CNT-Alg gel). The latter composites were prepared by dissolving 2 wt% of sodium alginate and CNTs (0.05 or 0.1 g) in 30 mL of deionised water, and then ethylenediamine-di (hydroxysuccinimide) (EDS) and 1-ethyl-3-(3-dimethylaminopropyl)-carbodiimide hydrochloride were added to the solution. The solution was poured into a plastic mould (50×30×15 mm) and allowed to set for 24 hours at 25°C.

The CNTs were modified by adding them into a mixture of 300 mL of sulphuric acid and 100 mL of nitric acid, and the mixture was sonicated in a bath-type sonicator for 7 hours at 40°C. Subsequently, the product was vacuum-filtrated using a PTFE membrane with a pore size of 0.2 μm. The resultant solid was then washed with deionised water and dried at 80°C in a vacuum oven for 24 hours.

The elastic deformation results showed that the CNT-Alg gels indicated significantly lower deformation than the alginate gels of SAlg-1 and Alg-2 (controls without CNTs). Further, increase in CNT content improved the resistance against deformation of CNT-Alg gels. These results suggested that CNT-Al gels could be useful as a scaffold material in tissue engineering.

3.2.5   Gelatin/MWCNTs-Based Nanocomposites

Kavoosi et al. [43] fabricated gelatin/MWCNT nanocomposite films by mixing different concentrations of MWCNTs (0.5, 1, 1.5, and 2% w/w gelatin) with 10% (w/v) gelatin in deionised water and stirred them for 12 hours at 50°C. To avoid photochemical reactions, all the solutions were kept in the dark. The gelatin/MWCNT solutions were further mixed and sonicated again using a sonication bath at 40 W for 1 hour at 50°C, and transferred onto the polystyrene Petri dish for film casting. The films were placed at room temperature until films were dried [44]. After drying, films were peeled off from the plate surface and left to equilibrate at relative humidity of 75% at 4°C in a closed box containing saturated salt solution of NaCl. The films' preparation was preceded by modification of the MWCNTs. The latter were dissolved in ethanol solution (50%) and sonicated using a sonication bath at 40 W for 1 hour at 50°C [45].

The mechanical properties of the resulting composite films saw an increase in the tensile strength (MPa), decrease in elongation at break (%) and an increase in Young's modulus (MPa) as the MWCNTs loading increased up to 1.5%. However, the addition of 2% MWCNTs caused no significant decrease in tensile strength, a significant increase in elongation at break and a no significant decrease in Young's modulus. These improved mechanical properties imply that the gelatin/MWCNTs composite films could be probed as potential packaging materials [43].

3.2.6   Natural Rubber (NR)/MWCNTs-Based Nanocomposites

Fakhru'l-Razi et al. [46] prepared the MWCNTs/natural rubber (NR) nanocomposites by a solvent casting method using toluene as a solvent. The method of preparing the composites started with the dissolution of the rubber in a suitable organic solvent (toluene). Then, a specific quantity of rubber (10 g) was added to a quantity of organic solvent (500 mL of toluene), thereby keeping an appropriate rubber weight ratio. This mixture was stirred and kept for certain duration of time until the rubber gets uniformly dissolved in the solvent. The dissolution/dispersion of CNTs in that study is made by adding a certain amount of carbon nanotubes to a specific amount of toluene solution. The solution was then sonicated by means of a mechanical probe sonicator qualified to vibrating at ultrasonic frequencies with a view to inducing an efficient dispersion of nanotubes. Different CNT solutions were prepared (containing CNTs in various weight ratios), 1, 3, 5, 7 and 10 wt% CNTs in 10 mL of toluene solutions. The mixing of rubber with nanotube solution generally involves complete mixing of the solutions prepared in the first and second stages, culminate in a solution that consists of a good blend of nanotubes in the rubber.

Following assessments, it appeared that the tensile strength significantly increased as the amount of CNTs concentration increased. Similarly, the Young's modulus increased with increase in the amount of the CNTs. However, the increase of the modulus is not as high as that of the tensile strength at 1 and 3 wt% of CNTs. The same value of the modulus and the tensile strength were observed at 5 wt% of CNTs while at 7 and 10 wt%, the modulus was higher than the tensile strength.

The CNT/natural rubber (NR) nanocomposite was prepared via solvent mixing by Sui et al. [47]. A certain amount of CNTs and NR were added into the toluene solution. The solution of CNTs was sonicated for 2 hours, while the solution of NR was stirred and kept for certain duration until the rubber was uniformly dissolved in the toluene. The toluene solution of CNTs was then dispersed into the solution of NR with stirring and ultrasonication at the same time for 2 hours. Following that, the dried CNT/NR mixture was obtained via evaporating the solvent off at 80°C under vacuum. Then, the preparation of CNT/NR nanocomposites was achieved by adding other ingredients inclusive of vulcanising agent in the formulation of composites in an open two-roll mill at room temperature with the nip gap of about 1 mm.

The composite preparation is preceded by the creation of the functional groups on the surface of the CNTs. This is achieved by dipping the CNTs in a blended acid solution with a volume ratio between nitric acid and sulphuric acid of 1:3. The loading of CNTs was 1 g for 10 mL of blended acid solution. The mixing solution was boiled and refluxed for 0.5 hour, and then the CNTs were carefully washed and filtrated with de-ionised water until chemically neutral. Moreover, the interface between the CNTs and NR matrix is improved by using silica-containing bonding system named HRH ([hydrated silica, resorcinol and hexamethylene tetramine with a weight ratio 15:10:6) [30, 31]. The dry acid-treated CNTs were blended with the HRH bonding systems with a weight ratio of 25:3. To further untie the entanglement of CNTs, and facilitate the co-mixing of CNTs and HRH bonding systems, the mixture was milled for 0.5 hour in a ball-milling machine. The results have shown that the CNT/NR nanocomposites exhibited great enhancements in tensile modulus (MPa) and tensile strength (MPa), while the elongation at break (%) decreased as compared to the pure NR. Moreover, the thermal stability of NR was also evidently improved upon the addition of the treated CNTs.

3.3 POLYMER/GRAPHENE-BASED NANOCOMPOSITES

Graphene is a two-dimensional single layer of carbon atoms with exceptional properties. Due to these excellent properties, graphene and its derivatives, in particular graphene oxide (GO) and reduced graphene oxide (rGO), have found interesting applications in nanocomposites [48–50]. Graphene oxide is produced by exfoliation of graphite oxide, and rGO is produced by reduction of GO [51,52]. The fabrication of graphite oxide is achieved mainly by the Hummer's, Brodie's or Staudenmaier's method, or modified methods [53–56].

3.3.1 Cellulose/Graphene (Graphene Derivatives)-Based Nanocomposites

The abundance of oxygen-containing functional groups on the surface of cellulose, which is widely present in nature, gives it the advantages of being inexpensive, highly hydrophilic, and biocompatible [57,58]. Cellulose and its derivatives make up the majority of cellulose-based products. Enzymatic, chemical, and mechanical processes can easily convert cellulose into various forms, such as micro nanofibrillated cellulose (MNFC), dissolved polymers, macro, and nanofibres [57,58]. For a range of applications, it is critical and desirable to develop functional graphene nanocomposite materials based on renewable and natural polymers, not only for economic or environmental considerations but also for high productivity. Numerous studies have been carried out in an effort to create practical and intelligent cellulose/graphene composite materials that integrate the advantages of graphene and the features of cellulose. But it's still difficult to get graphene layers to disperse at the nanoscale in polymer matrices. Therefore, determining the best way to disperse graphene throughout the cellulose matrix is one of the most difficult tasks in the field of functional composites.

As long as the utilised solvents can adequately dissolve the cellulose and disperse the graphene, the solution-mixing approach can be used to make graphene and cellulose nanocomposites. Under simple and facile conditions, Yang et al. [59] synthesised GO/cellulose (GOC) composite. Figure 3.1 below includes a detailed illustration of the process for creating GOC composite. The modified Hummers approach was used to create GO [60]. The cellulose (3.7 mmol) was added to the GO suspension (6.3 g/L), while being vigorously stirred for 10 minutes and sonicated. Gradually, a dark precipitate was formed. The suspension was combined with a 5 mL mixture of urea (1.358 g) and NaOH (0.75 g), while being vigorously stirred for an additional 10 minutes.

The combined solution was then rapidly agitated for 5 minutes at room temperature after being frozen overnight. After standing for 10 minutes, the resultant solution was injected into 10 mL of HNO_3 containing NaCl (10 wt%). Before drying under vacuum at 50°C, the resultant black product was washed multiple times with deionised water. Finally, a mortar and pestle were used to ground the GOC into a powder.

A real wastewater sample containing uranium (VI) was treated using the GOC composite. GOC is a suitable adsorbent for the treatment of uranium-containing

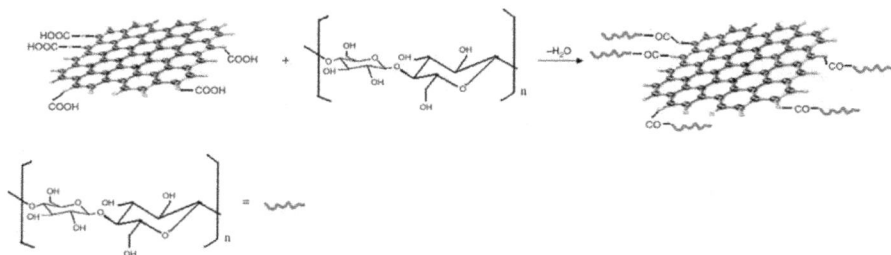

FIGURE 3.1 Procedure for the formation of GOC composite. (Reprinted from Ref. [59].)

industrial effluents because of its excellent adsorption ability for the uranium waste liquid and achieved a removal rate of more than 99%.

Another simple and green method was reported by Chen et al. [61] to distribute GO and dissolve cellulose in alkaline-urea aqueous solution followed by *in-situ* reduction. The solvent in this procedure was a mixture of NaOH, urea and water (7:12:81 ratio). The solvent was vigorously stirred while the specified amount of cellulose was dissolved in it. The cellulose solution was then vigorously stirred while a calculated amount of GO dispersion was added. The mixture was then cast onto a glass plate to create a gel sheet that was 400 mm thick. This gel sheet was then submerged in a coagulation bath containing 5 wt% H_2SO_4 at room temperature to coagulate and regenerate. As a result, rGO/cellulose composites were formed. Also using NaOH/urea as a solvent, Huang et al. [62] discovered that GO was completely exfoliated in cellulose. As a result, the cellulose/GO film's barrier qualities were enhanced, and the permeability coefficient of oxygen was decreased.

In order to generate the GO/Cellulose nanocomposite film, Wu et al. used an aqueous LiOH/urea solution, followed by ethanol coagulation [63]. As a result, the mechanical characteristics of GO/cellulose films were improved. As shown by Velusamy et al. [64], graphene and cellulose can be combined directly to create graphene–cellulose microfibers (GR/CMF) composite. GR (5 mg/L) was dissolved into the CMF solution using ultrasonication for around 30 minutes to create the GR–CMF composite. Further electrochemical research was conducted using the resulting Haemoglobin (Hb) immobilised GR–CMF (GR–CMF/Hb) composite-modified electrode. Figure 3.2 shows a schematic of the biosensor fabrication process.

Direct electron transport between Hb and electrode is made easier by the unique combination of GR/CMF composite qualities compared to GR- and CMF-modified electrodes. The prepared biosensor has a good electrocatalytic activity and analytical performance low LOD (10 nM), high sensitivity, quick response (3 seconds), and globally linear response range towards H_2O_2.

Modifying either the graphene or the cellulose is another way to make the composites of graphene and cellulose. In order to create graphene-loaded cellulose sheets, 1-butyl-3-methylimidazolium chloride (BMCl) was used [65]. In this study, graphene was employed to improve the mechanical, electrical and electroactive performance of composite actuators made of cellulose. A mass of microcrystalline cellulose

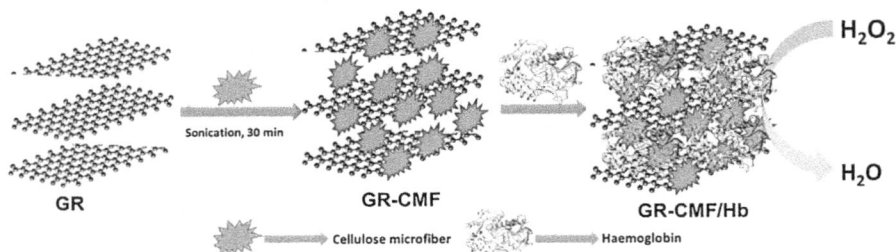

FIGURE 3.2 Formation of graphene–cellulose microfibers (GR/CMF) composite biosensor. (Reprinted from Ref. [64].)

powder (MCC) was added to the molten ionic liquid after first melting the BMCl. Until a uniform liquid solution was formed, the mixture was stirred. Then cellulose was dissolved, and DMA was added to the liquid. Using an ultrasonic homogeniser, Gr particle material was disseminated in the solution to create MCC/BMCl/Gr films. The mixture was then placed in a Petri dish and kept at room temperature in a vacuum oven for 12 hours. The films were removed from the Petri plates and divided into 80 mm² squares. It was observed that graphene loading did not affect the chemical structure of MCC/BMCl films. Graphene loading into the cellulosic film considerably increased electrical conductivity of MCC/BMCl film. The tensile strength and the Young's modulus of MCC–BMCl films increased with increasing graphene loading up to 0.25 wt%. Graphene nanoplatelets loading into the mixture of MCC, BMCl, and DMA resulted in better electroactive performance for cellulose-based composite actuator.

The sonochemical technique described by Shao et al. [66] embeds GO nanosheets uniformly and without functionalisation into bacterial cellulose (BC). The procedure for making BC is known and documented by Bakshia et al. [20]. To achieve good GO dispersion in the polymer matrix, GO aggregation was broken up by ultrasonication. The typical Hummers process was used to synthesise GO [6]. The BC homogenate was treated by ultrasonification at supersonic power of 500 W for 30 minutes while submerged in ice water, adding GO aqueous solution. The weight ratio of GO to BC was maintained at 2, 4, 6, and 8 wt%, respectively. After filtering the homogenous dispersions, BC/GO composite films were then freeze dried for 10 hours at 40°C. It was discovered that the BC/GO composites have good cytocompatibility. Additionally, the composites demonstrated remarkable anti-bacterial rates against *Staphylococcus aureus* and *Escherichia coli*, reaching 95.61% and 65.35%, respectively.

3.3.2 Chitosan/Graphene (Graphene Derivatives)-Based Nanocomposites

Chitosan (CS), a well-known compound of chitin *N*-deacetylation, shows many eco-friendly properties, such as biodegradation, good biocompatibility and antifungal activity. These characteristics make it a promising candidate in the fields of catalysis, materials, food, drugs, etc. [67–69]. Using solution-mixing technique, Fang et al. created graphene/chitosan composite sheets, and subsequently investigated their mechanical and biocompatibility aspects [70]. Graphene sheets were first ultrasonically mixed with acetic acid aqueous solution for 1 hour. Then, by agitating the suspension of graphene, chitosan was broken down. The solution casting technique was used to create the films containing graphene and chitosan. Composite films were created with varying amounts of graphene (0, 0.1, 0.2, 0.3, 0.6, and 2.3 wt%). To entirely eliminate the acetic acid, the composite films were first dried at ambient temperature and then placed in a vacuum chamber at 50°C overnight. At a relatively low concentration, the inclusion of graphene considerably enhanced the modulus of chitosan (between 0.1 and 0.3 wt%). When tested using vitro MTT assay, the composites demonstrated good biocompatibility with L929 cells.

Reduced graphene oxide/chitosan, (rGO/CS) was created by Hung et al. [71] using a hydrothermal reduction method, which is beneficial to the environment. When rGO and CS are mixed together, the occurrence of ionic complexation ($-COO-H^{3+} N-R$)

between the negatively charged carboxylate ions of GO and the positively charged protonated amines of CS create extreme membrane aggregation. This hydrothermal reduction method was useful for preventing undesired ionic complexation from mixing rGO and CS, because it can remove most carboxylate ions from GO. The GO was prepared from the modified Hummers' method [72]. The GO was dissolved in deionised water using ultrasound to create a GO suspension solution, which was then aliquoted and decreased using heat. After that a 50 wt% GO or rGO suspension was added to a CS solution. To make sure that GO or rGO was evenly distributed throughout the mixture, it was first homogenised and then ultrasonically processed. The hydrophilic CS molecular chain was inserted in between the rGO laminates, improving the dispersion and enabling the rGO/CS to stack-up and self-assemble into a lamellar structure (see Figure 3.3). Additionally, the GO/CS or rGO/CS mixture was cast on a PSf support layer. Selective water channels were formed between rGO/CS laminates, allowing only water molecules to pass through and resulting in an excellent methanol dehydration effect in the membrane; at a high temperature of 80°C, the permeation flux reached 690 g/m²h; and the water concentration on the permeate side was 96 wt%.

Graphene and chitosan must be chemically modified by adding new functional groups, crosslinking, or grafting with polymers and other moieties in order to improve its properties. By employing biopolymers as crosslinkers to join GO sheets together via noncovalent or covalent link, GO framework structures can be created. The carboxyl groups of GO can be amidated with CTS to create a homogeneous and evenly distributed GO composite. By crosslinking GO with CTS, Sabzevari et al. [73] created a simple approach for creating a GO composite that has superior physicochemical properties for sorption-based applications. In order to create the composite materials (GO/LCTS), a 3 mg/mL GO solution was created. 5 g of LCTS were dissolved with stirring in 500 mL of 1 v/v glacial acetic acid to create LCTS solution. Drop by drop, while stirring continuously for 4 hours, the resulting LCTS solution 1% (v/v) was added to the GO solution. After using 1 M NaOH to bring the pH down to 7, the liquid was stirred for 12 hours. In this study, a variety of crosslinked GO

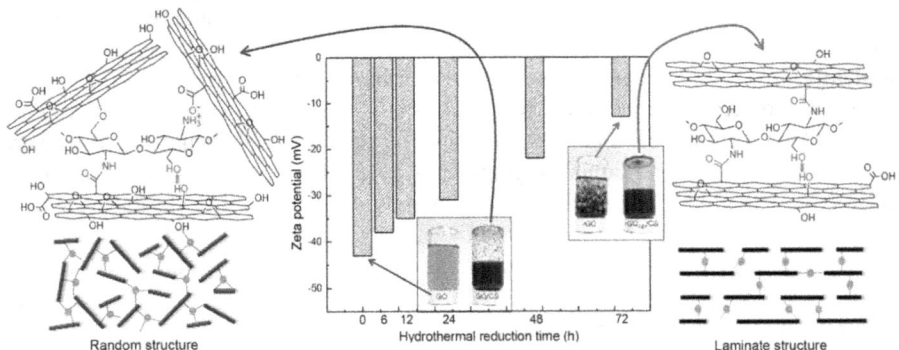

FIGURE 3.3 Proposed chemical interactions showing the random structure (GO/CS) and laminate structure (rGO/CS). (Reprinted from Ref. [71].)

samples were created using different precursor weight ratios. The GO/LCTS films, with an average thickness ranging from 20 to 60mm, were created by pouring the solution onto a glass surface, washing it with Millipore water, and letting it dry at room temperature for 48 hours. The biopolymer crosslinking approach used to create the developing GO-based composites exhibits better durability and enhanced adsorption characteristics for reuse in numerous adsorption–desorption processes.

According to Zhao et al.'s [74] procedure, GO-based composite hydrogels were made using an *in-situ* reduction method and self-assembly of CS and GO [27]. The CS molecules and GO sheets interact electrostatically and through hydrogen bonds to form the composite hydrogels. For the purpose of accelerating the gelation of GO sheets, CS molecules were added. GO-based composite hydrogels were created utilising CS, including GO/CS hydrogel and reduced GO (rGO)/CS hydrogel. In a typical experiment, 2.5% acetic acid (HAc) solution was used to prepare 30.0 mg/ mL of CS and 4.0 mL of an aqueous dispersion of GO (5 mg/mL) in a glass vial. The hydrogel was then subjected to a 2–3 minutes sonication process in order to achieve a homogenous gelation condition. To create the rGO/CS composite hydrogel, 4.0 mL of a GO aqueous dispersion (5 mg/mL) was combined with 1.0 mL of vitamin C solution (60 mg/mL), and the mixture was agitated for a while. The aforementioned solution was then combined with 0.4 mL of CS (30.0 mg/mL, produced in 2.5% HAc solution), and the resulting ternary combination was sonicated for approximately 10 minutes before being heated at 90°C for an additional 10 minutes. After heating, the gel's GO component changed to rGO, while the gel's condition somewhat shrank. According to the pseudo-second-order model, the adsorption capacity results show that the generated GO-based composite hydrogels can effectively remove the three tested dye molecules—Congo red, methylene blue, and Rhodamine B—from wastewater. The GO sheets are primarily responsible for the hydrogels' ability to absorb dye, while the chitosan molecule was added to speed up the GO sheets' gelation.

Microwave irradiation (MW) has been widely used in composite synthesis due to better conversion and quicker reaction times than those under conventional heating [75]. In the study by Ge and Ma, a novel chitosan-based composite was prepared using microwave technology [76]. Triethylenetetramine-modified graphene oxide/ chitosan composite (TGO/CS) was prepared by microwave in a customised microwave oven that had a stirrer and water that circulated at a consistent temperature. At room temperature, 4 g/L GO aqueous dispersion (50 mL) was mixed with 5 mL of triethylenetetramine, and the pH of the mixture was then brought down to 8 by adding diluted NaOH solution. The mixture was transferred to the microwave reaction system and radiated at 343 K under stirring for 15 minutes. The mixture was then exposed to radiation for 45 minutes, while being stirred at 343 K under, followed by the addition of epichlorohydrin (5 mL) and CS (1 g). The mixture was filtered and chilled following the reaction. The filter residue was thoroughly cleaned with distilled water, ethanol and 5% HCl before being dried in vacuum at 333 K to produce a composite known as TGOCS. The experimental results showed that the product made using MW had a greater yield and uptake than one made using a conventional method, and that TGO/CS had a higher absorption of Cr(VI) than recently published adsorbents.

A novel functionalised graphene oxide–chitosan (rGO/CS) structure was constructed by a facile microwave synthesis system that utilises yeast extract as the reductant by Shu et al. [77]. With an ultrasonic probe running at 800 W for 8 hours, the GO was exfoliated. Finally, deionised water was used to disseminate the exfoliated GO for subsequent use. Through the use of a microwave synthesis technique, GO was reduced with chitosan aqueous solution to create the partially reduced chitosan–GO nanosheets. This was used to stabilise GO nanosheets in physiological solution. The biosynthesis of rGO/CS was carried out using cell-free yeast extract and microwave-assisted reduction. The stored yeast cells were first given a kickstart by being inoculated into Luria–Bertani media and shaken for 18 hours at 25°C at 135 rpm. The active yeast cells were then added to a 2% sugar solution and shaken for an additional 6 hours at 25°C at 135 rpm. Next, centrifugation separation at 2000 rpm for 6 mL of cell-free yeast extract was obtained. About 25 mL of a 2% (v/v) acetic acid solution were used to dissolve 50 mg of chitosan, and yeast extract was added. A 5 mg GO solution was added, while being vigorously stirred with magnets. The produced solution was then put into the microwave synthesis apparatus for the microwave process. The microwave system's heating plan comprised heating to 80°C for 5 minutes and maintaining the temperature for an additional 5 minutes. The resulting rGO/CS was freeze dried for further usage after being purified with a 100-kDa MWCO filter. The drug encapsulation and delivery effectiveness of the nanocomposites as prepared was quite high. The rGO/CS/adriamycin nanosheets significantly and dose-dependently inhibited the growth of BT-474.

3.3.3 ALGINATE/GRAPHENE (GRAPHENE DERIVATIVES) NANOCOMPOSITES

Alginate–graphene oxide composites have been synthesised in a number of different ways, including wet spinning to create calcium alginate–GO fibres and sodium alginate–GO films [78–80]. For the creation of new GO/alginate (GO/Alg) nanocomposite films, Ionita et al. [81] used casting procedures. Using double distillate (DD) water and an ultrasonic process, graphene oxide aqueous suspensions were created. In a nutshell, 100 mL of DD water was combined with 0.1 g of GO. To achieve a uniform dispersion, the mixture was sonicated for 1 hour. The sodium alginate was dissolved in DD water as a 1% (w/v) solution to produce the alginate solution. By adding dropwise GO suspension to the sodium alginate solution, composite films with various GO contents (0, 0.5, 1, 2.5, and 6 wt%) were created. Using a magnetic stirrer, the resulting mixture was continuously stirred for 30 minutes. Each solution was poured into a clear glass Petri dish and allowed to stand at room temperature for 72 hours so that a thin film might form. The preparation showed that the mechanical and thermal stability of the nanocomposite films were considerably altered by hydrogen bonding and high interfacial adhesion between the GO filler and alginate matrix. Alginate films containing 6 wt% GO had an increase in tensile strength and Young's modulus, going from 71 MPa and 0.85 GPa to 113 MPa and 4.18 GPa, respectively. The Young's modulus of the GO/Alg composite films was improved by around 300% and their thermal stability was improved by about 25°C with the addition of 6 wt% GO.

Another procedure that involves turning the face of GO into modified GO (MGO), which increased the miscibility of fillers and polymers [82]. GO was ultrasonically suspended in 75 mL of water for 5 hours. Sodium alginate (SA) was then suspended as well. A variety of loading levels (0, 0.2, 0.5, 1.0, 1.5, and 2.0 wt%) of SA/GO mixture based on SA were made, then heated at 60°C for 4.5 hours with constant stirring, poured into a glass plate, and then heated at 50°C to obtain dry films after degassing under vacuum. Biocomposite films with the codes SA, SA/ GO-1, SA/GO-2, SA/GO-3, SA/GO-4, and SA/GO-5 were identified. The same process was used to create the SA/MGO-n biocomposite films, which were coded as SA/MGO-1, SA/MGO-2, SA/MGO-3, SA/MGO-4, and SA/MGO-5, where n denoted the various GO or MGO loadings. Prior to testing, dried films were conditioned at 40% relative humidity (RH) at room temperature. According to the method, increasing hydrogen bonding and electrostatic interactions can effectively improve the properties of SA biocomposite films by adding active groups on the basal planes.

Li et al. prepared the SA/GO composites by first creating various concentrations of GO solutions (0.1, 0.2, 0.4, and 0.6 mg/mL) by using ultrasonication [83]. To create homogeneous SA/GO solutions, 2 g SA was added to 100 mL of the prepared GO solutions and rapidly agitated for 6 hours. The SA/GO solutions were then dumped by the peristaltic pump into 100 mL of $CaCl_2$ (0.16 mol/L)-containing coagulation solutions, which were continuously stirred for 4 hours. The needle employed in this investigation has a diameter of 0.8 mm, and the flow rate of the alginate solution is 100 mL/h. The Ca-SA/GO created by GO solutions with a concentration of 0.1, 0.2, 0.4 and 0.6 mg/mL are represented by the series Ca-SA/GO0.1, Ca-SA/GO0.2, Ca-SA/GO0.4 and Ca-SA/GO0.6, respectively. In comparison to the pure Ca^{2+} crosslinked alginate gel beads (Ca-SA), the as-prepared Ca-SA/GO exhibits a reduced swelling degree, increased gel stability in salt solutions, and higher mechanical performance. All these can be explained by the uniform distribution of GO sheets in the Ca-SA matrix, the presence of hydrogen bonds, and the high interfacial adhesion between the GO filler and SA matrix [83].

By using a straightforward solution-mixing–evaporation technique, Yadav et al. [84] created composite blends of GO, carboxymethylcellulose (CMC), and alginate (GO/CMC/Alg). Briefly, 0.01 g of GO was dissolved in 50 mL of water and subjected to a 15 minutes sonication in an ultrasonic bath. Then, a solution containing 0.25 g of CMC, 0.75 g of sodium alginate (Alg) and 49 mL of deionised water was agitated for 5 hours using a magnetic stirrer to dissolve the CMC and Alg. The CMC/Alg solution was then mixed for 5 hours after the GO solution was introduced. The solution was then poured onto a glass plate at 80°C under vacuum until dry, after which it was degassed for 30 minutes to remove bubbles. Finally, with a thickness of 0.04 mm on average, the dried GO/CMC/Alg composite thin films were gently removed from the glass plate. The process for creating blended CMC/Alg films was the same as the one used to create GO/CMC/Alg films, with the exception that GO was not included. The addition of 1 wt% GO increased the tensile strength and Young's modulus by 40% and 1128%, respectively, over a CMC/Alg blend. It also demonstrated a greater storage modulus.

3.3.4 Natural Rubber/Graphene (Graphene Derivatives)-Based Nanocomposites

The preparation methods of graphene-based rubber composites, such as solution compounding melt/mechanical compounding, *in-situ* polymerisation, and latex co-coagulation, have been used to prepare the composites [85–88]. Solution processing method of filler–polymer composites involves the dispersion of filler in a polymer solution by energetic agitation, controlled evaporation of the solvent, and finally composite film casting [89–94]. This method illustrated by Rahman et al. [95] involved a liquid exfoliation technique to obtain the graphene dispersion. The mild sonication and simple mixture technique were used to prepare the natural rubber latex/graphene (NRL/G) composite. The infusion of graphene as filler into natural rubber latex matrices results in the performance of mechanical properties, as the tensile strength of NRL/G composite increased by 92%. In addition, the NRL/G composite exhibited the increase in conductivity value as 1.91×10^{-8} Sm^{-1} for pure NRL to 1.03×10^{-6} Sm^{-1}. The method is fully illustrated by the simple schematic diagram below (see Figure 3.4);

Li et al. [96] reported on the ammonium-assisted fabrication of G/NRL composite. In this method, removable ammonium was used to improve the dispersion of graphene in NRL, which possesses two advantages over the traditional stabilisers used in composite fabrication: (i) It is a green method, because no residual stabilisers are left in the final composites; and (ii) it has the potential for manufacturing of practical NRL material, because ammonium is a cheap stabiliser for NRL in NR industry. Ammonium solution is also used to improve the stability of graphene sheets. After the introduction of graphene, the glass transition temperature (T_g) and the crosslink density are improved to 68°C and 7%, respectively, with a slight decrease in overall free-volume number (3.5%) and fraction (1%). In another procedure, Wu et al. [97] used a silane coupling agent bis-(triethoxysilylpropyl)tetrasulfide (BTESPT) to graft onto the surface of GO; this is generally used in the case of poor adhesive qualities

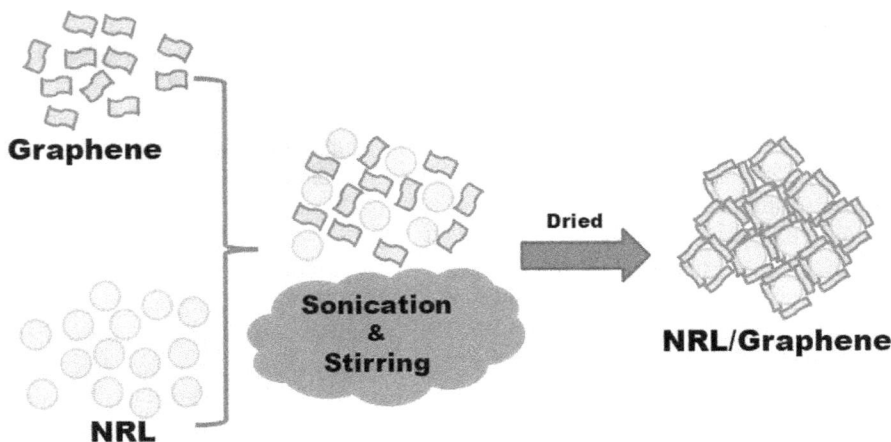

FIGURE 3.4   Preparation procedure of the NRL/G composite. (Reprinted from Ref. [95].)

of a filler for a polymer. This coupling agent is able to establish molecular bridges at the interface between the polymer matrix and the filler surface as in the case of a silica-filled hydrocarbon rubber. By this way, the hydrophilic character of GO, not compatible with the hydrophobic NR, is reduced, allowing a better dispersion of the filler in the host polymer. It is shown that the mechanical properties of NR are significantly improved at very low filler loading, since a 100% increase in the tensile strength and a 66% improvement in the tensile modulus are achieved with only 0.3 wt% of grafted GO.

Melt compounding seems to be the most economically attractive method to prepare graphene-filled elastomer composites. However, its disadvantages of poor dispersion and serious breakage of graphene sheets under high shearing force are inevitable, and should not be ignored [98]. Azira et al. [99] prepared natural rubber (NR) and functionalised graphene sheets (FGSs) nanocomposites by conventional two-roll mill mixing. To harness the exceptional mechanical properties of GO in polymer nanocomposites, the dispersion state and the interfacial interaction are crucial. The dispersion state strongly depends on the mixing method and the interfacial interaction. In this study, melt compounding method was used in that two sets of nanocomposite were prepared, namely, Epoxidised Natural Rubber (ENR)/GS (graphene in powder form) and ENR/GL (graphene in liquid) using internal mixer. For GL, dispersion of the graphene in a medium is achieved by tip sonication with optimum sonication conditions. An appropriate amount of graphene was dispersed into ethanol by sonicating the suspension for 30 minutes (three times separated by a rest of 3 minutes) using a Perkin–Elmenr Sonic Dismembranator FB 705 operating at 50% amplitude with on and off cycles, respectively, equal to 4 and 2 seconds. The compound formulation with calculation based on parts per hundred resin/rubber (phr) and the mixing of the compounds was carried out in Haake Banbury Rotor Mixer. ENR/GL nanocomposites with a conductive segregated network exhibiting good electrical conductivity, water vapour permeability and high mechanical strength are prepared by self-assembly in latex and static hot pressing. The composite exhibits a percolation threshold of 0.62 vol% and a conductivity of 0.03 S/m at a content of 1.78 vol%, which is five orders of magnitude higher than that of the composites made by conventional methods at the same loading fraction.

In another work by Wu and Yu [100], cornmeal graphene (CGE) was modified with coupling agents to improve the mechanical and thermal properties of the NR nanocomposites. The scheme below represents the formation of modified CGE/NR composite. The authors used the self-made CGE produced from corn flour so as to reduce costs of making graphene (Figure 3.5).

The CGE was placed in a beaker with distilled water at 60°C under ultrasonication for 30 minutes, then filtered and dried. After that, the CGE was coated with a modifier KH550 (KH590 or Si69), stirred until a homogeneous dispersion, and dried to prepare modified CGE. The NR was kneaded on a double-roll mill manufactured (SK-160 B) at a roller temperature of 45°C and a distance of 0.8 mm for 20 minutes. The thermal conductivity of the KH590-CGE/NR nanocomposites was improved by 6.0 mW/g, the decomposition temperature was increased by 37°C, its tensile strength was increased by 16%, elongation at break was increased by 14%, and also wear resistance was improved by 17%.

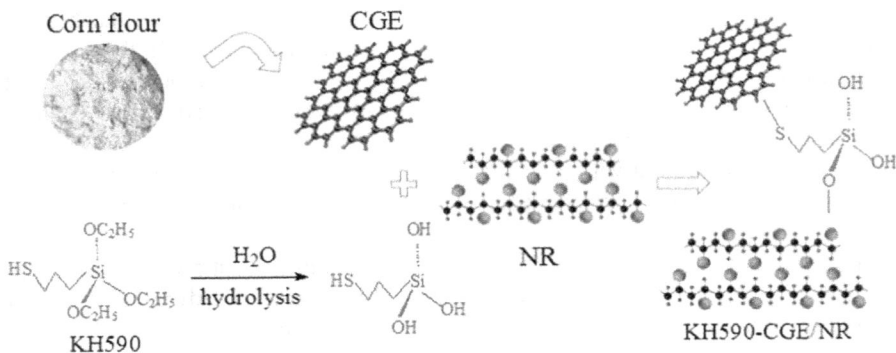

FIGURE 3.5 Preparation of modified CGE/NR composite. (Reprinted from Ref. [100].)

Hernadez et al. [88] prepared natural rubber (NR) and functionalised graphene sheets (FGSs) nanocomposites by conventional two-roll mill mixing. All the compounds were prepared in an open two-roll laboratory mill at room temperature. First, the mastication of the rubber took place. Afterwards, all the vulcanising additives except sulphur were added to the rubber prior to the incorporation of the filler; and, finally, sulphur was added. NR compounds were vulcanised in an electrically heated hydraulic press at 150°C and 200 MPa. The optimum cure time, t90 was derived from the curing curves previously determined by means of a Rubber Process Analyser. The interfacial interactions between FGS and the rubber matrix accelerate the cross-linking reaction, and increase the electrical conductivity and cause an important enhancement on the mechanical behaviour of the NR nanocomposites.

The dispersion of GO in rubber matrix is bottle-neck issue for fabrication of high-performance GO/rubber composites in the industry scale so far, because high viscosity of rubber matrix and GO sheet tends to restack and form aggregates. Therefore, great efforts have been contributed to obtain satisfactory dispersion of GO through the composites. Other authors used the most common fabrication method to compound graphene into rubbers directly, as reported above such as melt compounding [99,100]. However, they always yields weak dispersion quality. Others used the solution-blending method; but it also has several drawbacks, as reported previously by Rahman et al. [95]. Considering that GO can be readily exfoliated in water, and most of rubbers have their latex type that can easily blend with GO suspension, latex co-coagulation technology is a more effective method to fabricate GO/rubber composites. Latex co-coagulation appears to be a more environment-friendly and promising method [86]. Exfoliated graphene oxide (GO)–reinforced natural rubber (NR) composites were prepared by a simple and promising latex compounding and co-coagulation approach. The use of latex compounding and co-coagulation method realised the complete exfoliation and uniform dispersion of GO in NR matrix.

Zhan et al. [86] have proposed a new method, the ultrasonically assisted latex mixing and *in-situ* reduction (ULMR) process to prepare NR/graphene composites. GO platelets were prepared from natural flake graphite according to the known Hummers and Offeman method [53], and they were dispersed into NRL by ultrasonic irradiation due to the presence of oxygen-containing groups in the surface. The GO in the latex was then reduced *in situ* by hydrazine hydrate and followed by latex coagulation to get the well-dispersed and exfoliated GE/NR masterbatch. Through further dilution of the masterbatch with normal NR by a twin-roll mixing process, NR/GE composites with different GE contents were obtained. The results show that this method produces a better dispersion and exfoliation of graphene in the matrix, and contributes to an increase in the tensile strength compared to conventional direct mixing. The tensile strength, stress at 300% strain, and tear strength for NR/(2 wt%) GE composites are increased by ~47%, 175%, and 50%, respectively, which are much higher than those obtained by the conventional direct mixing.

Mao et al. [85] prepared the GO/NR composite by dispersing GO in natural rubber (NR) using latex co-coagulation technology. In this, graphene oxide (GO) sheets were obtained by sonicating graphite oxide aqueous dispersion at 25°C for 1 hour. Then the GO sheets suspension (0.1 wt%) and NR latex were mixed and vigorously stirred for 1 hour. The mixture was co-coagulated by pouring the mixture to HCl solution and washed with water until the pH of the filtered water was between 6 and 7. The dried compounds were mixed with other rubber ingredients on a two-roll mill. Then the compounds were vulcanised at 143°C in a standard mould for the optimum cure time (t90) determined by a rheometer. The GO sheets were finely dispersed with exfoliated structure into the NR matrix and had strong interface interaction with the NR matrix. The tensile strength and tear strength of GO/NR composite at GO content of 2 phr were efficiently enhanced to 22.8 and 47.5 MPa, respectively.

Dong et al. [94] used the latex method for preparation of GO/NR. An amount of GO aqueous suspension was added into NR latex. After being stirred vigorously for 20 minutes, the homogeneously mixed suspension was rapidly poured into $CaCl_2$ solution (1 wt%), accompanied with the rapid co-coagulation procedure. The obtained compound was cut up and washed with plenty of water for several times, and then dried in an oven at 50°C until a constant weight is achieved. The resultant composite showed that with increase in GO sheet content, the mechanical properties, fracture initiation and propagation resistance were all highly improved.

In another procedure that involves the modification of NR by Yin et al. [101], a facile and low-cost approach was devised, whereby vinyltriethoxysilane (VTES) was grafted onto natural rubber (NR) in latex form, using potassium persulfate (KPS) as an initiator. Then the GO was incorporated in the modified NR latex by a co-coagulation method. As a result, the GO particles were adsorbed or bonded to the hydroxyl groups derived from hydrolysis and condensation of the $-Si(OC_2H_5)_3$ groups in the NR-g-VTES macromolecular chains. Dynamic mechanical analysis showed that the interaction between GO and NR-g-VTES was better than that of the GO-reinforced NR.

3.4   POLYMER/CARBON BLACK NANOCOMPOSITES

Carbon black has been used as a reinforcement for polymers due to its attractive features, such as high surface area, excellent electrical conductivity, thermal conductivity and thermal stability [9,102,103]. In comparison with other carbon-based materials (e.g. graphene, carbon nanotubes (CNTs)), CB is the cheapest carbon-based filler, and hence it is being used to produce cheaper composite with excellent thermal and electrical conductivity [102,103]. A wide variety of polymers have been reinforced with CB to improve their overall properties. The choice of the polymer and the preparation methods are often chosen based on the intended application.

3.4.1 CELLULOSE/CB-BASED COMPOSITES

Cellulose and its derivatives have been reinforced with carbon black for a wide variety of applications, as summarised in Table 3.1 [104]. Cellulose is the most abundant natural polymer, making it cheaper. It is biocompatible, biodegradable, light weight, and has excellent mechanical performance and ease of functionalisation.

TABLE 3.1
Carbon Black–Natural Polymer Composites

Formulation	% CB	Preparation Method	Modification	Application	Highlights	Refs.
CA/CB	1.5	Electrospinning		Adsorbent for oil	1.5 wt% CB exhibited superior adsorption for both heavy machine oil (12.56 g/g) and light machine oil (12.01 g/g)	[104]
Alginate/CB	1a	Immobilisation on GCE		Paraquat detection and determination	Compared to other natural polymers, alginate showed superior potential as modifier for paraquat determination	[105]
CB/NR	0.09b	Melt mixing followed by open mill mixing and melt-pressing	–	Tyres	At 0.09 volume fraction there was more glassy fraction, indicating that the molecular chain segments of NR were completely immobilised	[106]
CMC/ GO-CNTs	3	*In-situ* incorporation followed by solution casting	–	Separation membrane	The membrane exhibited superior pervaporation separation performance higher than commercial PVA/ PAN composite membrane	[107]

a, mg; b, volume fraction.

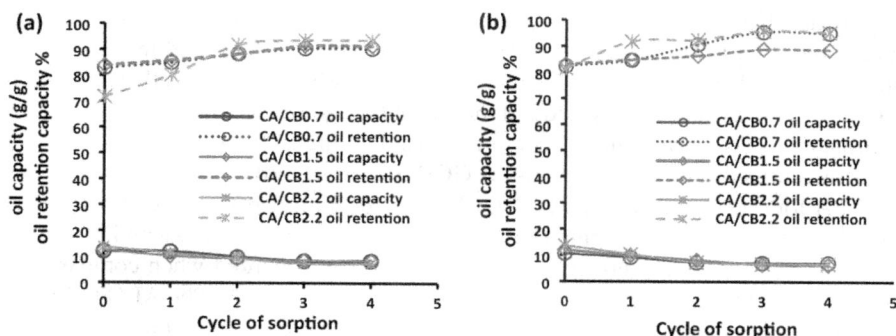

FIGURE 3.6   The reusability of CA/CB composite membrane for (a) heavy machine oil/water and (b) light machine oil/water for four adsorption/desorption cycles. (Reprinted from Ref. [104].)

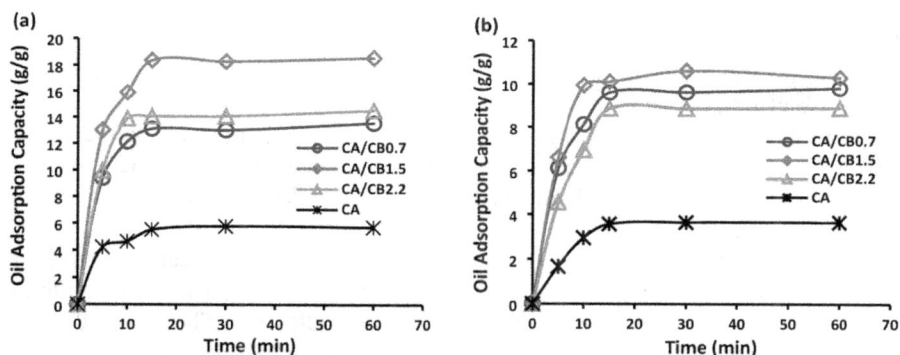

FIGURE 3.7   Oil adsorption capacity of (a) heavy machine oil and (b) light machine oil at room temperature. (Reprinted from Ref. [104].)

Cellulose acetate (CA)/CB composites were prepared using electrospinning technique to afford large surface area fibres for oil adsorption [104]. It was noticed that light machine oil was less adsorbed by the membrane when compared to heavy machine oil due to their difference in viscosity. The oil adsorption capacity of the membrane increased with an increase in CB content. The maximum oil retention for both the oils, however, was recorded for 1.5% CB-based composite membrane (Figure 3.6). The adsorption for the composites was faster within the first 5 minutes and reach plateau after 15 minutes due to oil molecules occupying the hollow cavities of the adsorbents, as shown in Figure 3.7.

3.4.2   Alginate/CB-Based Composites

Alginate is one of polysaccharides composed of D-mannuronic acid and L-guluronic acid linked together through 1,4 glycosidic bond [108–111]. It is polyelectrolyte copolymer that can be derived from algae. Pacheco et al. [105] decorated GCE with a mixture of alginate and CB for paraquat (PQ), herbicide often used in agricultural sector, determination. 1.0% alginate and 1 mg CB were mixed together and soni-cated for an hour followed by addition of 1% glutaraldehyde (GA) and sonication for 8 minutes to design an ink to decorate GCE. The highest current density of 37.4 μA was attained for alginate (Alg)-CB GCE, which slightly decreased to 23.4 and 26.8 μA for crosslinked alginate/CB GCE with epichlorohydrin and glutaraldehyde crosslinkers. ALG/CB performed better than alginate/graphite GCE with PQ current response of ~31 μA, but no significant difference was recorded when compared to functionalised CNTs with PQ current response of 33.2 ± 2.0 μA. The ALG/CB GCE exhibited suitable linearity in the concentration ranging between 0.4 and 2.0 mg/L, with calculated detection limit and limit of quantification values of 0.06 and 0.19 mL, respectively. In addition, the presence of other electroactive species, such as Zn, Cd, Cu, Pb, Cr, Hg, atrazine, glyphosate, and tebuconazole, have had no marked interfer-ence on the PQ current response.

3.4.3 CHITOSAN/CB-BASED COMPOSITES

Chitosan is often used for preparation of composites due to its surface functional-ities, ease of processing into different shapes, biocompatibility and non-conductive property. Several studies used chitosan as hot matrix, because of its insulating behav-iour to fabricate flexible and conductive wearable bioelectrode [103,112,113]. The presence of CB enhances mechanical performance and electrical conductivity of the resulting composite.

In the study, Buaki-Sogó and co-workers dissolved chitosan in acetic acid (0.3 M) followed by sonication for an hour and placed in ball-milling jar to afford addition of CB to achieve desired compositions of 25%–200% CB [103]. The obtained slurry after 30 minutes of ball milling was cast onto glass using casting knife and dried at room temperature. It was reported that less than 100% of CB resulted in compos-ites with suitable mechanical performance for wearable devices, with higher content above 100% the composites turn to be brittle and easily broken. Charge transfer resistance decreased with about 100 folds when 100% CB was embedded into chi-tosan and ten folds when 25% CB was used as a filler. It was demonstrated that the as-prepared bioelectrodes have a potential to be used for glucose biofuel cells and biosensors with limit of detection (LOD) and quantification for glucose of ~76 μM and of ~253 μM as well as energy harvesting of a glucose biofuel cell of ~21.3 μW. In most cases, chitosan is dissolved in a suitable solvent followed by inclusion of CB into the chitosan solution. However, depending on the intended application addi-tional step rather than traditional casting to allow solvent evaporation is employed. Elsewhere, chitosan–carbon black composites were used as a biosensing platform p53 detection [112]. The authors dissolved chitosan and binder (polyvinyldene fluo-ride) in a mixture of acetone and N-Methyl-2-pyrrolidone (NMP), and then CB was introduced into the mixture. The as-prepared solution was coated onto indium tin oxide (ITO) electrode using spin-coating technique. In this case, chitosan was used due to its poor conductive property and poor solubility, while CB can overcome the conductive issue as well as amplifying the electrochemical signal. Chitosan offered platform for interaction with antibodies due to the presence of surface functional amino groups in the presence of glutaraldehyde. The as-prepared immunosensor exhibited linear range of 0.01–2 pg/mL with LOD of 3 fg/mL, indicating that it can be used for as sensing device. It showed good selectivity, stability and reliability with acceptable recovery of 97.6%–102.0%.

The detection of ascorbic acid, dopamine and uranic acid using CB–chitosan ink was demonstrated by Dinesh and co-workers [114]. The authors prepared the ink within 15 minutes, followed by drop-casting on a glassy carbon electrode (GCE) and air dried at room temperature for ~10 minutes. In this case, 5 mg CB was dispersed in water and mixed with chitosan (0.5 wt%), followed by adjusting pH value to 3.5 using acetic acid. The presence of chitosan improved the dispersion of CB in water, which resulted in the ink being stable for 9 months without particles' settlement. CB–chitosan ink was drop coated onto glass carbon electrode (GCE) and air dried. CB–chitosan ink-modified electrode exhibited remarkable sensing and electrochemical oxidation of all analytes with current peak signals at defined peak potentials at linear concentrations of 25–1600, 0.1–1400 and 5–1800 μM, and reaching LOD of 0.1 μM (signal to ratio = 3). Using vitamin C tablets, dopamine hydrochloride injection and human urine, it was demonstrated that the as-prepared ink can be used for real-life applications for detecting all the three analytes without need for surface pretreatment and analyte adsorption, while maintaining 100% recovery values.

In order to fabricate strain sensor sponges composed of chitosan and CB, solution-mixing and freeze-drying techniques were employed, as demonstrated by Liu et al. [113]. In this case, chitosan:water:acetic acid (1:50:1 ratio) were mixed for 12 hours to obtain clear chitosan solution followed by addition of CB powder of desired percentages, i.e. 25%, 50%, 100% and 200%. The obtained mixture was then freeze dried to afford highly porous sponges. It was noticed that an increase in CB content significantly improved the density, electrical conductivity, thermal stability, compressive modulus and strength. The highly porous and lightweight composite had excellent strain detecting abilities. The composite detected muscle movements of the vocal cords and exhibited unique signal for movement depicted for pronunciation of each word with good repeatability. The gauge factor (GF) was estimated to be around 7.5, indicating that these sponges are promising strain detectors for sound monitoring and recognition. In the case of breathing test, the sensor was attached to buckle belt to record electrical response during exhaling and inhaling processes. A good linear response, reversibility and repeatability in sensing tiny compressive strain were attained. The sensor was also tested for bending strain from human activities, and the composite showed good sensitivity and stability for about 200 bending-recovery cycles.

3.4.4 Natural Rubber/CB-Based Composites

Natural rubber are usually used in the rubber industries because of their unique attributes such as excellent elasticity, high durability and low hysteresis [106,115]. CB has been used in almost all rubber materials as a reinforcing filler to improve physical properties and to strengthen vulcanisation. Such composites are used to produce commercially available products, which include gloves, tyres and shoe soles. The amount of the filler, filler dispersion, compounding methods and compounding conditions are essential to achieve desired composite product. In addition, the type of CB plays a major role on the properties of resulting composite product.

Despite the fact that CB is often embedded into NR to improve the properties of the resulting composite materials, there are few materials that are usually added to ensure processability without degradation as well as a complete vulcanisation of the product [6]. As mentioned above, the quantity of the CB plays a critical role on the properties of the resulting composite. Farida et al. [115] investigated the influence of CB content on the properties of CB/NR composites. The authors used open mill processing technique to mix all the required materials followed by vulcanisation process at 160°C and pressure of 135 MPa for 15 minutes. It was noticed that the fillers were homogeneously dispersed with NR matrix due to strong interaction between the CB and NR, as shown in Figure 3.8. There were cavities observed due to agglomeration of CB. In addition, the heat decreased with increase in CB content from 403.4 to 360.7 mJ. Kneader-type mixer was used to prepare NR/CB composite to study the effect of additional filler, i.e. silica [116]. It was demonstrated in this study that the features of the fillers play a major role on the resulting properties of the composites. The hardness decreased with increase in CB content, whereas abrasion value increased with CB content. CB have strong interaction with natural rubbers, which, in turn, reduces particle–particle interaction, thus improving its dispersion. Nonetheless, the optimal composite ratio of 40/20 for CB/silica displayed superior properties, *viz.* the shortest curing time and the maximum abrasion.

In their study, Nakaramontri et al. [117] demonstrated that the introduction of CB as secondary filler into CNT/NR composites using melt mixing resulted in enhanced electrical conductivity. The higher interfacial energy between CNT and NR contributed positively on the arrangement of the CNT within CNT/CB hybrid, which enhanced conductivity. This was ascribed to the CB bridging inter-particle gap between CNTs and thus serve as electron pathways for superior electrical conductivity. Li et al. [118] prepared continuous graphene/CNTs hybrid filler by mixing the fillers followed by sonication for 2 hours and then incorporated into NR latex followed by coagulation with saturated salt solution. Hydrazine hydrate was added into the as-prepared mixture for *in-situ* reduction of graphene. The hybrid composites were prepared using twin-roll milled followed by vulcanisation. The resulting hybrid composite exhibited superior reinforcing effect and toughness when compared to single-reinforced composite due to synergy between CNTs and graphene particles. The strong interaction between the fillers *via* π–π and homogeneous dispersion within the host matrix resulted in formation of sacrificial bonds that can dissipate energy before failure, thus resulting in improving both tensile strength and toughness.

Goa et al. [119] investigated the influence of CB on the properties of 80/20 NR/butadiene rubber (BR) blends. The authors prepared these composites using internal mixer followed by two-roll mill for ten times. The synergy between CNTs and CB resulted in enhanced abrasion resistance. Optimal abrasion was attained for composite containing 5 phr CNTs and 27.5 phr of CB. Hybridisation led to better thermal conductivity, tear strength and reinforcing effect when compared to single-reinforced composites.

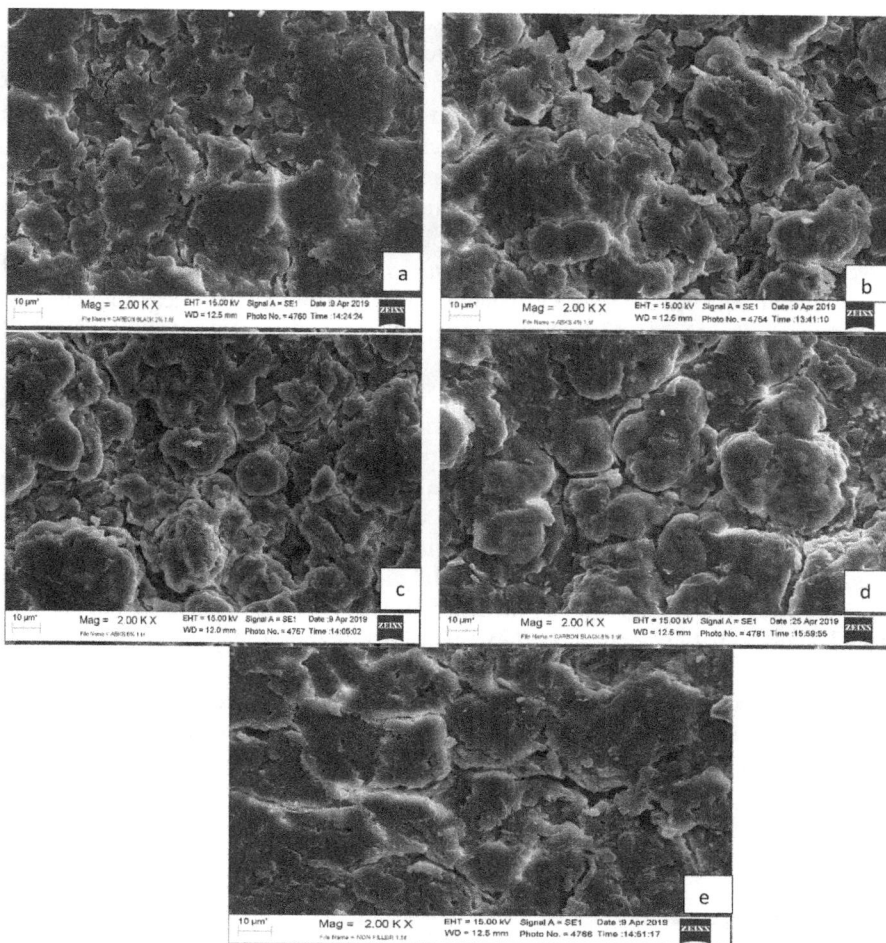

FIGURE 3.8 SEM images of CB/NR composites (a) 2 wt% CB, (b) 4 wt% CB, (c) 6 wt% CB, (d) 8 wt% CB, and (e) neat rubber. (Reprinted with permission from Ref. [115].) The combination of CNTs with CB can afford cheaper composite materials with superior mechanical, thermal and electrical conductivity.

3.5 OTHER CB-BASED COMPOSITES

Recent study based on curcumin, pigment of the turmeric, as a platform to fabricate electrode for electrocatalytic oxidation and electrochemical sensing application for sulphide was demonstrated by Dinesh et al. [120]. In this case, CB modified glassy carbon electrode (GCE) was decorated with a curcumin–quinone (Cur-Q) derivative *via in-situ* electrochemical oxidation. It was noticed that the CB-based electrode outperformed other carbon-based electrodes, *viz.* MWCNTs, SWCNTs, graphite nanopowder, graphitised mesoporous carbon, carbon nanofibers, graphene oxide and activated charcoal. The as-prepared CB-based electrode possessed excellent electro-chemical reaction and sensing for sulphide reaching a current sensitivity of ~37.0 μA/mM detection limit values of 7.12×10^{-6}M from flow injection analysis and detection values (signal to ratio of 3) of 2.6×10^{-6} from amperometric *i-t* analysis. The electrode showed no marked changes in the current signals in the presence of various electroac-tive species, i.e. arsenite (As^2), cysteine, dopamine, glucose, hydrogen peroxide (H_2O_2), NADH, nitrite, NO_2^- and uric acid, which often affect classic modified electrode.

In their study, Gandhi et al. [121] immobilised sesamol–quinone (ses-oxid) deriv-ative/CB on glassy carbon electrode *via in-situ* electrochemical oxidation method for efficient electrochemical, electroanalytical and bioelectroanalytical applications. Amongst all examined carbon-based materials, *viz.* graphene oxide, single-walled carbon nanotube (SWNT), double-walled carbon nanotube (DWCNT), MWCNT and carboxylic acid-functionalised MWCNT (f-MWCNT), CB showed excellent electrochemical reaction for *in-situ* formation of sesamol-quinone-based electrode (i.e., GCE/carbon@Ses-oxid). This was attributed to the additional formation of hydrogen bond between hydroxyl/functional from ses-oxid with –COOH and –OH surface functional groups of CB, besides π–π interaction on the electrode. The result-ing electrode showed capability of (i) quantification of sesamol in natural herb prod-ucts; and (ii) good transducer, bioelectrocatalytic reduction and sensing of hydrogen peroxide after decoration with hordedish peroxidase (HRP)-based enzymatic sys-tem. The as-prepared electrode had excellent specificity for hydrogen peroxide with current sensitivity of 0.1303 μA μM and detection limit values of 990 nM without interference from other chemicals such as ascorbic acid, cysteine, glucose, hydra-zine, hypoxanthine, oxygen and uric acid. Last, (iii) electrochemical immunosensing of white spot syndrome virus after relevant modifications (Figure 3.9).

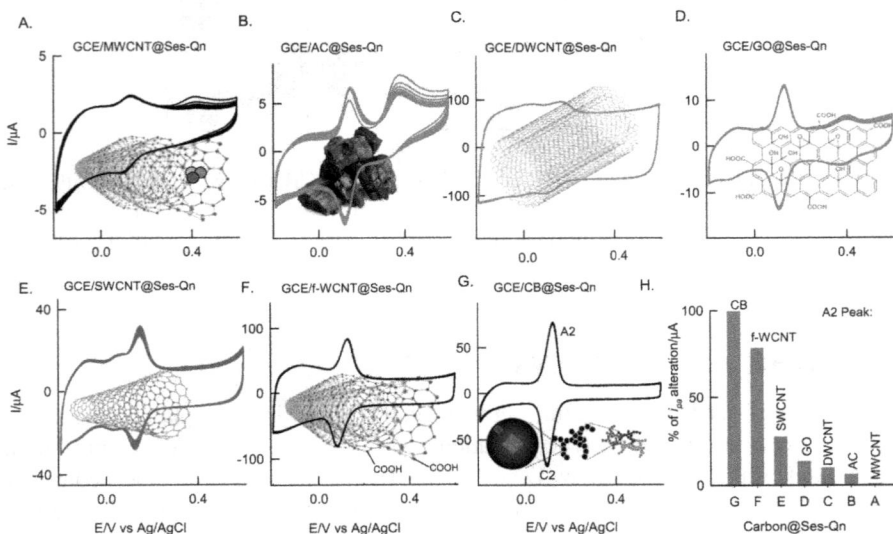

FIGURE 3.9 CV responses for ses-oxid (Ses-Qn) immobilised on carbon materials-modified GCEs in pH = 7 solution at $v = 50$ mV/s. (Reprinted with permission from Ref. [121].)

3.6 POLYMER/BIOMASS CARBON-BASED COMPOSITES

Pyrolysis of biomass can result in different carbon-based fillers with some contaminants (e.g. SiO_2) that can be used as cheaper filler for various applications [122–124]. The resulting carbonised biomass is usually ground and sieved to achieve desired particle size. These particles have been employed as reinforcing filler to enhance the overall properties of the ensuing composite materials (Table 3.2).

3.6.1 Chitosan/Biomass Carbon-Based Composites

Hydrogels are usually employed for wastewater treatment purposes because of their unique features, such as high specific area, high porosity, lightweight and excellent mechanical properties [122–124]. Hydrogels composed of chitosan and biomass carbon fillers were fabricated using grafting polymerisation in the presence acrylic acid to afford highly porous structure for dye adsorption [125]. It was reported that hydrogel composite exhibited high methylene blue (MB) adsorption capacity ranging between 1450 and 1950 mg/g, and maintained 91% removal efficiency after 5 adsorption/desorption cycles. Rodrigues et al. [126] investigated the effect of the rice husk on the swelling behaviour of composite hydrogel. It was reported that water uptake increased with increase in rice husk ash (RHA) up to 5 wt%. The water adsorbed increased from 141 g/g (i.e., for neat hydrogel) to 255 g/g, when 5% of RHA was embedded into the hydrogel. In addition, the incorporation of 5% RHA drastically reduced the swelling of the hydrogel because of increase in the number of crosslinking points within the polymer chains resulting from physico-chemical interaction of the filler and the host polymeric materials, thus reducing its elasticity.

In-situ grafting of poly(acrylic acid) onto chitosan and biochar (BC) was carried out by first mixing chitosan and BC followed by introduction of ammonium persulfate (APS) followed by acrylic acid (AA) monomer, as shown in Figure 3.10 [127]. BC was obtained by pyrolysing rice straw at 600°C for 2 hours under nitrogen atmosphere. The resulting BC had specific area of ~339 m^2/g and decreased to ~1.93 m^2/g for hydrogel composite material, meanwhile the average pore size increased from ~2.04 to ~18.4 nm. The adsorption capacities calculated from Langmuir model were 111, 115, 99.0, 476, 370, 139, 135 and 313 mg/g for Cu(II), Zn(II), Ni(II), Pb(II), Cd(II), Mn(II), Co(II) and Cr(III), respectively. The adsorption efficiency mechanism was attributed to the surface complexation with available –COOH, –OH, and –NH_2. Similar observations, where physical properties drastically changed when compared to as-prepared biochar [128]. The authors reported that as-prepared sewage sludge biochar (SSBC) surface area decreased from 41.0 to 2.35 m^2/g when used to prepare composite materials, and the pore width increased from 11.4 to 31.6 nm. The adsorption capacity of the composite for Pb(II) and Hg(II) increased with an increase in carboxymethyl chitosan (CMC) content because of the presence of a large number of exposed chelating active sites (carboxylic, hydroxyl and amino groups). The maximum adsorption capacity at 30°C and pH value of 5 for Hg(II) and Pb(II) within 60 minutes were ~210 and ~594 mg/g, respectively. The removal capacity increased to ~240 and ~769 mg/g after 4 hours at 30 pH value of 3 for Pb(II) and Hg(II), respectively. The composite displayed strong resilience with only 15% and 4% decrease for Pb(II) and Hg(II) after five sorption–desorption cycles.

TABLE 3.2

Summary on the Preparation and Application of Biomass-Based Composites

Formulation	%C	Preparation Method	Modification	Application	Highlights	Refs.
CS-grafted PAA (polyacrylic acid)/RHA	5	Solution blending	Grafting polymerisation	Dye adsorption	Optimal adsorption capacity of 1952 mg/g was attained with 5% of the filler	[125]
Starch/RHA	5	Solution blending	Grafting polymerisation		Optimal adsorption capacity of 1906.3 mg/g was attained with 5% of the filler	[130]
CS/CKSC/iron oxide nanoparticles	3a	Solution blending	Coagulation–GA solution	Metalloids adsorption	Approximately 48 mg/g Cr(VI) adsorption capacity was achieved	[129]
CS-g-PAA/biochar	0.5a	Solution blending	In-situ grafting of PAA onto CS/BC composite	Metalloids adsorption	Strongly selective for Cr^{3+} when compared to other ions, viz., Pb^{2+}, Cu^{2+}, Cd^{2+}, Ni^{2+}, Zn^{2+}, Co^{2+}, and Mn^{2+}	[127]
CMC/SSBC	2:1 CS:SSBC ratio	Solution blending	GA		Short equilibrium Pb(II) adsorption of 60 minutes and maximum Hg(II) adsorption capacity of 594.2 mg/g	[128]
CS/SSBC	0.45a	Solution blending	–		Equilibrium adsorption were reached within 60 minutes (~82%), 90 minutes (~97%), 45 minutes (~88%), and 30 minutes (~90%) for Cr, Cu, Se and Pb ions due to the presence of a large number of surface active sites	[123]
CS/SSBC	0.45a	Physical blending	–		Removal efficiency was poor than neat CS, SSBC, and CS/SSBC hydrogel	

a, gram (g).

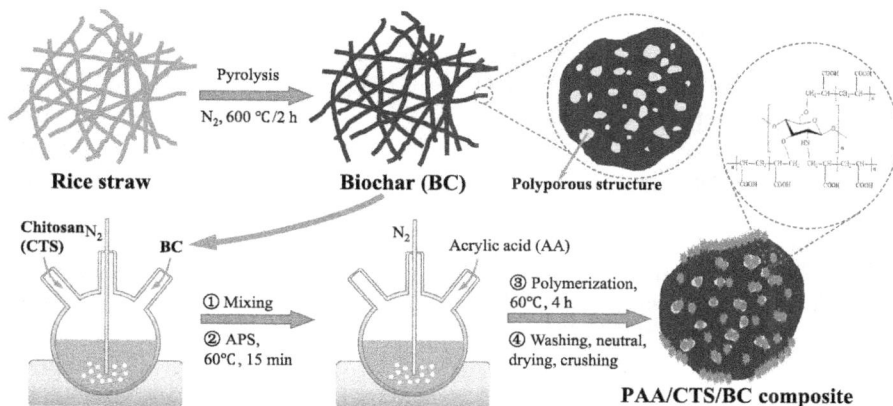

FIGURE 3.10 Schematic illustration of *in-situ* grafting of PAA onto CS/BC composite. (Reprinted with permission from Ref. [127].)

Coagulation method involves the introduction of the solution mixed composite into a coagulation bath to afford crosslinked hydrogel or beads. These materials are highly porous, physically and/or chemically crosslinked, and capable of holding large quantities of aqueous solutions without losing their integrity. Hydrogel beads containing chitosan, cherry kernel shell pyrolytic charcoal (CKSC) and iron oxide nanoparticle (Fe_2O_3) were prepared by dissolving chitosan and incorporating CKSC and nanoparticles into the system [129]. Beads were prepared by adding the solution drop wise into coagulation bath of sodium hydroxide solution that was kept overnight. To improve the stability of the beads, glutaraldehyde was employed as a crosslinking agent. The maximum adsorption capacity of 47.6 mg/g was achieved. It was noticed that the adsorption capacity and mechanical integrity was compromised after four sorption–desorption cycles. The decline in adsorption efficiency to 85.13% was ascribed to the disintegration of the membrane and iron leaching.

3.6.2 STARCH/BIOMASS CARBON-BASED COMPOSITES

Hydrogel composed of starch and RHA was designed using grafting polymerisation by de Azevedo et al. [130]. In this case, the presence of 5% RHA exhibited superior MB adsorption capacity. When the content filler increased beyond 5%, adsorption capacity declined due to –OH groups of RHA interacting with –COOH of PAA that are supposed to interact with MB molecules. The composite hydrogel maintained high adsorption capacity for MB dye, with 8% decline recorded after five adsorption/desorption cycles.

3.6.3 ALGINATE/BIOMASS CARBON-BASED COMPOSITES

A number of studies have shown that the presence of activated carbon in alginate-based adsorbents can improve the removal efficiency of the resulting membranes [131–134].

Benhouria et al. [131] prepared three adsorbents, i.e. bentonite–alginate beads, activated carbon–alginate beads and activated carbon–bentonite–alginate beads (ABA) for adsorption of MB. The beads were prepared by mixing desired components and then dropping into coagulation bath using burette to afford highly porous beads with large surface area, i.e. BET surface area of 185.28 m^2/g and pore volume of 0.159 cm^3/g. The Langmuir monolayer adsorption capacities of ~757 mg/g at 30°C, ~982 mg/g at 40°C and ~994 mg/gat 50°C were obtained for ABA beads. The membrane displayed resilience after six sorption–desorption, indicating their potentiality for rehabilitation of dye-contaminated wastewater. Using the same preparation method, Annudurai et al. [132] prepared alginate/activated carbon beads for Rhodamine 6G removal. They reported that the presence of activated carbon significantly increased the adsorption capacity of the beads. In the case of metalloids removal, activated carbon incorporated in sodium alginate beads was also studied [135]. It was reported that the maximum As (V) adsorption of ~67 mg/g was attained at 30°C. The removal of zinc and toluene using activated carbon/alginate/zeolite beads was demonstrated by Choi et al. [133]. The maximum adsorption of 4.3 and 13 g/kg for zinc and toluene was attained, respectively.

Biochar as the cheaper alternative when compared to activated carbon thus can be incorporated into alginate-based beads for removal of pollutants from wastewater streams [134,136–138]. Roh et al. [137] prepared beads by dropping the mixture of alginate and biochar into coagulation bath using a peristaltic pump. The maximum adsorption capacities of the beads for Cd, 2,4,6-trinitrotoluene (TNT) and 1,3,5-trinitro-1,3,5-triazacyclohexane (RDX) obtained from Langmuir adsorption isotherm were 9.7, 90.1 and 28.1 mg/g, respectively. A maximum Cd (II) adsorption capacity of 40.0 mg/g using balled milled char/alginate beads was reported by Wang et al. [139]. In this case, the authors used a needle for dropping the solution mixture. The maximum Cd (II) adsorption capacity ranging between 37 and 46 mg/g was attained using a composite of alginate and biochar from water hyacinth (*Eichhornia crassipes*). Encapsulation of biochar with alginate resulted in a capsule capable of removing lead ions (Pb(II)) from aqueous solution [140]. The maximum adsorption capacity of 263 mg/g was attained at pH value of 5. The maximum Pb (II) adsorption capacity of 254 mg/g was obtained at 35°C, 224 mg/g at 30°C and 209 mg/g at 25°C [141]. In this case, the composite was prepared in that biochar solution was added dropwise into alginate solution, and achieved surface area and pore volume of 91.6 m^2/g and 0.087 cm/g, respectively.

Zhixing et al. [138] reported on the modification of biochar using KMnO$_4$ to enhance adsorption efficiency of polyvinyl alcohol/alginate beads. The beads showed high adsorption of Cu (II) ions from aqueous solution reaching the maximum adsorption capacity of ~78 mg/g because of modification of char introducing MnO$_2$ into the system promoting the formation of complexes such as Mn–O–Cu and Cu–O. In addition, the beads displayed excellent Cu (II) removal efficiency values, which were greater than 95% in real water samples, i.e. deionised water, lake water, river water and sea water. The beads exhibited good resilience for ten sorption–desorption cycles with adsorption capacity decreasing from 89.7 to 82.2 mg/g, indicating potentiality that these beads can be used as cheaper adsorbents for treating polluted water.

3.7   CONCLUSIONS AND FUTURE PERSPECTIVES

The methods for the synthesis and properties of the natural polymer–carbonaceous material composites are reported in this chapter. The above techniques are not comprehensive methods for the synthesis of nanocomposites, but rather most widely used in the laboratory and industry to achieve uniform dispersion of different types of nanofillers in polymer matrices. From the reported studies, the most widely used carbonaceous materials for the preparation of these composites were carbon nanotubes (its derivatives), graphene (its derivatives, mainly graphene oxide and modified graphene oxide), carbon black and biochar carbon, and related materials. These were used as fillers to enhance the properties of the composites. The various natural polymers of cellulose, chitosan, alginate starch, and natural rubber were used as a polymer matrix.

The solution-based processing technique such as solution mixing was found to be the most common technique to form composites. The typical procedure starts with dispersing filler in a suitable solvent and combining them with that of the polymer solution. This is followed by the film cast and then the solvent is evaporated, and the result is a nanocomposite film or sheet. This technique tends to be limited to the polymers that are soluble in a particular solvent.

Melt mixing or blending is successfully exploited for dispersion of filler inside various polymer matrices. It has been discovered that, in comparison to solution-mixing approaches, melt-mixing techniques are less successful at breaking the agglomeration of specific fillers, such as CNTs and graphene. The synthesis of composites can also be done using other common and also less common procedures, such as wet spinning, microwave techniques, latex processing, and twin-screw extrusion methods. The production of composites is still a complex topic, because no one method can be considered perfect on all counts. As a result, several attempts have been made to combine different processes, such as solution processing with melt mixing, microwave synthesis with solvent processing, or in latex compounding and co-coagulation approach to produce a desired composite. The composite prepared can be produced in different forms, such as thin films, sheets, powder, or aerogels depending on the method used or the intended application.

The most crucial factor to be taken into consideration when selecting the type of synthesis method for the composites is the uniform dispersion of the filler. The majority of interfacial interactions between the filler and matrices determine the mechanical, chemical, and physical characteristics of polymer nanocomposites. The secret to creating a composite with better attributes is a solid grasp of the interaction mechanism between the components. Understanding the precise impact of each component in the composites (filler particle size, shape, orientation, dispersion, compatibility with matrix, and volume/weight fraction) is beneficial, and how to control them is essential for the synthesis of the composites with the desired qualities. Furthermore, a successful method of polymer nanocomposites requires an understanding of the connection between these elements. The primary considerations that should direct the synthesis of composites are largely dependent on various factors such as applications, necessary features, functionality, cost, environmental constraints, and processing. A key requirement for developing new composite materials with specific functionality is the capacity to forecast how changes in chemical and macromolecular properties, as well as changes in the crystalline structures of filler and polymers would impact the behaviour of the final composites as a whole.

Natural polymers incorporated with carbonaceous materials should promote innovative research works and contribute to industrial developments.

REFERENCES

1. Sperling LH. *Introduction to Physical Polymer Science*: John Wiley & Sons; 2005.
2. Bokobza L. Elastomeric composites. I. Silicone composites. *Journal of Applied Polymer Science* 2004;93:2095–104.
3. Bokobza L. The reinforcement of elastomeric networks by fillers. *Macromolecular Materials and Engineering* 2004;289:607–21.
4. Thostenson ET, Ren Z, Chou T-W. Advances in the science and technology of carbon nanotubes and their composites: A review. *Composites Science and Technology* 2001;61:1899–912.
5. Lozano K, Barrera EV. Nanofiber-reinforced thermoplastic composites. I. Thermoanalytical and mechanical analyses. *Journal of Applied Polymer Science* 2001;79:125–33.
6. Lozano K, Bonilla-Rios J, Barrera EV. A study on nanofiber-reinforced thermoplastic composites (II): Investigation of the mixing rheology and conduction properties. *Journal of Applied Polymer Science* 2001;80:1162–72.
7. Breuer O, Sundararaj U. Big returns from small fibers: A review of polymer/carbon nanotube composites. *Polymer Composites* 2004;25:630–45.
8. Macak JM, Gong BG, Hueppe M, Schmuki P. Filling of TiO_2 nanotubes by self-doping and electrodeposition. *Advanced Materials* 2007;19:3027–31.
9. Mokhena TC, Matabola KP, Mokhothu TH, Mtibe A, Mochane MJ, Ndlovu G, et al. Electrospun carbon nanofibres: Preparation, characterization and application for adsorption of pollutants from water and air. *Separation and Purification Technology* 2022;288:120666.
10. Giannelis EP. Polymer layered silicate nanocomposites. *Advanced Materials* 1996;8:29–35.
11. Iatrou H, Hadjichristidis N. Synthesis of a model 3-miktoarm star terpolymer. *Macromolecules* 1992;25:4649–51.
12. Sinha Ray S, Bousmina M. Biodegradable polymers and their layered silicate nanocomposites: In greening the 21st century materials world. *Progress in Materials Science* 2005;50:962–1079.
13. Chen B, Evans JRG, Greenwell HC, Boulet P, Coveney PV, Bowden AA, et al. A critical appraisal of polymer–clay nanocomposites. *Chemical Society Reviews* 2008;37:568–94.
14. Cote LJ, Cruz-Silva R, Huang J. Flash reduction and patterning of graphite oxide and its polymer composite. *Journal of the American Chemical Society* 2009;131:11027–32.
15. Mokhena TC, Sadiku ER, Ray SS, Mochane MJ, Motaung TE. The effect of expanded graphite/clay nanoparticles on thermal, rheological, and fire-retardant properties of poly(butylene succinate). *Polymer Composites* 2021;42:6370–82.
16. Gangopadhyay R, De A. Conducting polymer nanocomposites: A brief overview. *Chemistry of Materials* 2000;12:608–22.
17. Lund R, Willner L, Richter D, Dormidontova EE. Equilibrium chain exchange kinetics of diblock copolymer micelles: Tuning and logarithmic relaxation. *Macromolecules* 2006;39:4566–75.
18. Chang M-T, Chou L-J, Hsieh C-H, Chueh Y-L, Wang ZL, Murakami Y, et al. Magnetic and electrical characterizations of half-metallic Fe3O4 nanowires. *Advanced Materials* 2007;19:2290–4.

19. Averous L, Boquillon N. Biocomposites based on plasticized starch: Thermal and mechanical behaviours. *Carbohydrate Polymers* 2004;56:111–22.
20. Shogren RL, Lawton JW, Doane WM, Tiefenbacher KF. Structure and morphology of baked starch foams. *Polymer* 1998;39:6649–55.
21. Sorrentino A, Gorrasi G, Vittoria V. Potential perspectives of bio-nanocomposites for food packaging applications. *Trends in Food Science & Technology* 2007;18:84–95.
22. Saini P. *Fundamentals of Conjugated Polymer Blends, Copolymers and Composites: Synthesis, Properties, and Applications*: John Wiley & Sons; 2015.
23. Wong Eric W, Sheehan Paul E, Lieber Charles M. Nanobeam mechanics: Elasticity, strength, and toughness of nanorods and nanotubes. *Science* 1997;277:1971–5.
24. Treacy MMJ, Ebbesen TW, Gibson JM. Exceptionally high Young's modulus observed for individual carbon nanotubes. *Nature* 1996;381:678–80.
25. Ebbesen TW, Lezec HJ, Hiura H, Bennett JW, Ghaemi HF, Thio T. Electrical conductivity of individual carbon nanotubes. *Nature* 1996;382:54–6.
26. Uchida T, Kumar S. Single wall carbon nanotube dispersion and exfoliation in polymers. *Journal of Applied Polymer Science* 2005;98:985–9.
27. Coleman JN, Khan U, Blau WJ, Gun'ko YK. Small but strong: A review of the mechanical properties of carbon nanotube–polymer composites. *Carbon* 2006;44:1624–52.
28. Coleman JN, Khan U, Gun'ko YK. Mechanical reinforcement of polymers using carbon nanotubes. *Advanced Materials* 2006;18:689–706.
29. Xia H, Song M, Jin J, Chen L. Poly(propylene glycol)-grafted multi-walled carbon nanotube polyurethane. *Macromolecular Chemistry and Physics* 2006;207:1945–52.
30. Villmow T, Pötschke P, Pegel S, Häussler L, Kretzschmar B. Influence of twin-screw extrusion conditions on the dispersion of multi-walled carbon nanotubes in a poly(lactic acid) matrix. *Polymer* 2008;49:3500–9.
31. Novelli MMPC, Nitrini R, Caramelli P. Validation of the Brazilian version of the quality of life scale for patients with Alzheimer's disease and their caregivers (QOL-AD). *Aging & Mental Health* 2010;14:624–31.
32. Li M, Kim I-H, Jeong YG. Cellulose acetate/multiwalled carbon nanotube nanocomposites with improved mechanical, thermal, and electrical properties. *Journal of Applied Polymer Science* 2010;118:2475–81.
33. Mostafa U, Rahman MJ, Mieno T, Bhuiyan MAH. Carbon nanotube-incorporated cellulose nanocomposite sheet for flexible technology. *Bulletin of Materials Science* 2020;43:142.
34. Gnanaseelan M, Chen Y, Luo J, Krause B, Pionteck J, Pötschke P, et al. Cellulose-carbon nanotube composite aerogels as novel thermoelectric materials. *Composites Science and Technology* 2018;163:133–40.
35. Wang S-F, Shen L, Zhang W-D, Tong Y-J. Preparation and mechanical properties of chitosan/carbon nanotubes composites. *Biomacromolecules* 2005;6:3067–72.
36. Chen Y, Guo Z, Das R, Jiang Q. Starch-based carbon nanotubes and graphene: Preparation, properties and applications. *ES Food & Agroforestry* 2020;2:13–21.
37. Cheng J, Zheng P, Zhao F, Ma X. The composites based on plasticized starch and carbon nanotubes. *International Journal of Biological Macromolecules* 2013;59:13–9.
38. Swain SK, Pradhan AK, Sahu HS. Synthesis of gas barrier starch by dispersion of functionalized multiwalled carbon nanotubes. *Carbohydrate Polymers* 2013;94:663–8.
39. Cao X, Chen Y, Chang PR, Huneault MA. Preparation and properties of plasticized starch/multiwalled carbon nanotubes composites. *Journal of Applied Polymer Science* 2007;106:1431–7.
40. Famá LM, Pettarin V, Goyanes SN, Bernal CR. Starch/multi-walled carbon nanotubes composites with improved mechanical properties. *Carbohydrate Polymers* 2011;83:1226–31.

41. Star A, Liu Y, Grant K, Ridvan L, Stoddart JF, Steuerman DW, et al. Noncovalent side-wall functionalization of single-walled carbon nanotubes. *Macromolecules* 2003;36:553–60.
42. Kawaguchi M, Fukushima T, Hayakawa T, Nakashima N, Inoue Y, Takeda S, et al. Preparation of carbon nanotube-alginate nanocomposite gel for tissue engineering. *Dental Materials Journal* 2006;25:719–25.
43. Kavoosi G, Dadfar SMM, Dadfar SMA, Ahmadi F, Niakosari M. Investigation of gelatin/multi-walled carbon nanotube nanocomposite films as packaging materials. *Food Science & Nutrition* 2014;2:65–73.
44. Wan YZ, Wang YL, Yao KD, Cheng GX. Carbon fiber-reinforced gelatin composites. II. Swelling behavior. *Journal of Applied Polymer Science* 2000;75:994–8.
45. Voge CM, Johns J, Raghavan M, Morris MD, Stegemann JP. Wrapping and dispersion of multiwalled carbon nanotubes improves electrical conductivity of protein–nanotube composite biomaterials. *Journal of Biomedical Materials Research Part A* 2013;101A:231–8.
46. Fakhru'l-Razi A, Atieh MA, Girun N, Chuah TG, El-Sadig M, Biak DRA. Effect of multi-wall carbon nanotubes on the mechanical properties of natural rubber. *Composite Structures* 2006;75:496–500.
47. Sui G, Zhong WH, Yang XP, Yu YH, Zhao SH. Preparation and properties of natural rubber composites reinforced with pretreated carbon nanotubes. *Polymers for Advanced Technologies* 2008;19:1543–9.
48. Wang J, Jin X, Li C, Wang W, Wu H, Guo S. Graphene and graphene derivatives toughening polymers: Toward high toughness and strength. *Chemical Engineering Journal* 2019;370:831–54.
49. Toto E, Laurenzi S, Santonicola MG. Recent trends in graphene/polymer nanocomposites for sensing devices: Synthesis and applications in environmental and human health monitoring. *Polymers* 2022;14.
50. Asghar F, Shakoor B, Fatima S, Munir S, Razzaq H, Naheed S, et al. Fabrication and prospective applications of graphene oxide-modified nanocomposites for wastewater remediation. *RSC Advances* 2022;12:11750–68.
51. Srivastava M, Elias Uddin M, Singh J, Kim NH, Lee JH. Preparation and characterization of self-assembled layer by layer NiCo2O4–reduced graphene oxide nanocomposite with improved electrocatalytic properties. *Journal of Alloys and Compounds* 2014;590:266–76.
52. Dideikin AT, Vul' AY. Graphene oxide and derivatives: The place in graphene family. *Frontiers in Physics* 2019;6.
53. Hummers WS, Offeman RE. Preparation of graphitic oxide. *Journal of the American Chemical Society* 1958;80:1339.
54. Brodie BC. On the atomic weight of graphite. *Philosophical Transactions of the Royal Society of London* 1859;149:249–59.
55. Staudenmaier L. Verfahren zur Darstellung der Graphitsäure. *Berichte der deutschen chemischen Gesellschaft* 1898;31:1481–7.
56. Zaaba NI, Foo KL, Hashim U, Tan SJ, Liu W-W, Voon CH. Synthesis of graphene oxide using modified hummers method: Solvent influence. *Procedia Engineering* 2017;184:469–77.
57. Seddiqi H, Oliaei E, Honarkar H, Jin J, Geonzon LC, Bacabac RG, et al. Cellulose and its derivatives: Towards biomedical applications. *Cellulose* 2021;28:1893–931.
58. Lavoine N, Desloges I, Dufresne A, Bras J. Microfibrillated cellulose – Its barrier properties and applications in cellulosic materials: A review. *Carbohydrate Polymers* 2012;90:735–64.
59. Yang A, Wu J, Huang C. Graphene oxide-cellulose composite for the adsorption of uranium (VI) from dilute aqueous solutions. *Journal of Hazardous, Toxic, and Radioactive Waste* 2018;22.

60. Ma J, Liu C, Li R, Wang J. Properties and structural characterization of oxide starch/chitosan/graphene oxide biodegradable nanocomposites. *Journal of Applied Polymer Science* 2012;123:2933–44.
61. Chen Y, Pötschke P, Pionteck J, Voit B, Qi H. Smart cellulose/graphene composites fabricated by in situ chemical reduction of graphene oxide for multiple sensing applications. *Journal of Materials Chemistry A* 2018;6:7777–85.
62. Huang H-D, Liu C-Y, Li D, Chen Y-H, Zhong G-J, Li Z-M. Ultra-low gas permeability and efficient reinforcement of cellulose nanocomposite films by well-aligned graphene oxide nanosheets. *Journal of Materials Chemistry A* 2014;2:15853–63.
63. Wu Y, Li W, Zhang X, Li B, Luo X, Liu S. Clarification of GO acted as a barrier against the crack propagation of the cellulose composite films. *Composites Science and Technology* 2014;104:52–8.
64. Velusamy V, Palanisamy S, Chen S-M, Chen T-W, Selvam S, Ramaraj SK, et al. Graphene dispersed cellulose microfibers composite for efficient immobilization of hemoglobin and selective biosensor for detection of hydrogen peroxide. *Sensors and Actuators B: Chemical* 2017;252:175–82.
65. Sen I, Seki Y, Sarikanat M, Cetin L, Gurses BO, Ozdemir O, et al. Electroactive behavior of graphene nanoplatelets loaded cellulose composite actuators. *Composites Part B: Engineering* 2015;69:369–77.
66. Shao W, Liu H, Liu X, Wang S, Zhang R. Anti-bacterial performances and biocompatibility of bacterial cellulose/graphene oxide composites. *RSC Advances* 2015;5:4795–803.
67. Ge H-J, Du S-K, Lin D-H, Zhang J-N, Xiang J-L, Li Z-X. Gluconacetobacter hansenii subsp. nov., a high-yield bacterial cellulose producing strain induced by high hydrostatic pressure. *Applied Biochemistry and Biotechnology* 2011;165:1519–31.
68. Bakshi PS, Selvakumar D, Kadirvelu K, Kumar NS. Chitosan as an environment friendly biomaterial – A review on recent modifications and applications. *International Journal of Biological Macromolecules* 2020;150:1072–83.
69. Chadha U, Bhardwaj P, Selvaraj SK, Kumari K, Isaac TS, Panjwani M, et al. Advances in chitosan biopolymer composite materials: From bioengineering, wastewater treatment to agricultural applications. *Materials Research Express* 2022;9:052002.
70. Fan H, Wang L, Zhao K, Li N, Shi Z, Ge Z, et al. Fabrication, mechanical properties, and biocompatibility of graphene-reinforced chitosan composites. *Biomacromolecules* 2010;11:2345–51.
71. Hung W-S, Chang S-M, Lecaros RLG, Ji Y-L, An Q-F, Hu C-C, et al. Fabrication of hydrothermally reduced graphene oxide/chitosan composite membranes with a lamellar structure on methanol dehydration. *Carbon* 2017;117:112–9.
72. Tsou C-H, An Q-F, Lo S-C, De Guzman M, Hung W-S, Hu C-C, et al. Effect of microstructure of graphene oxide fabricated through different self-assembly techniques on 1-butanol dehydration. *Journal of Membrane Science* 2015;477:93–100.
73. Sabzevari M, Cree DE, Wilson LD. Graphene oxide–chitosan composite material for treatment of a model dye effluent. *ACS Omega* 2018;3:13045–54.
74. Zhao H, Jiao T, Zhang L, Zhou J, Zhang Q, Peng Q, et al. Preparation and adsorption capacity evaluation of graphene oxide-chitosan composite hydrogels. *Science China Materials* 2015;58:811–8.
75. Huacai G, Wan P, Dengke L. Graft copolymerization of chitosan with acrylic acid under microwave irradiation and its water absorbency. *Carbohydrate Polymers* 2006;66:372–8.
76. Ge H, Ma Z. Microwave preparation of triethylenetetramine modified graphene oxide/chitosan composite for adsorption of Cr(VI). *Carbohydrate Polymers* 2015;131:280–7.
77. Shu M, Gao F, Zeng M, Yu C, Wang X, Huang R, et al. Microwave-assisted chitosan-functionalized graphene oxide as controlled intracellular drug delivery nanosystem for synergistic antitumour activity. *Nanoscale Research Letters* 2021;16:75.

78. Ajeel SJ, Beddai AA, Almohaisen AMN. Preparation of alginate/graphene oxide composite for methylene blue removal. *Materials Today: Proceedings* 2022;51:289–97.

79. Zheng H, Yang J, Han S. The synthesis and characteristics of sodium alginate/graphene oxide composite films crosslinked with multivalent cations. *Journal of Applied Polymer Science* 2016;133.

80. He Y, Zhang N, Gong Q, Qiu H, Wang W, Liu Y, et al. Alginate/graphene oxide fibers with enhanced mechanical strength prepared by wet spinning. *Carbohydrate Polymers* 2012;88:1100–8.

81. Ionita M, Pandele MA, Iovu H. Sodium alginate/graphene oxide composite films with enhanced thermal and mechanical properties. *Carbohydrate Polymers* 2013;94:339–44.

82. Nie L, Liu C, Wang J, Shuai Y, Cui X, Liu L. Effects of surface functionalized graphene oxide on the behavior of sodium alginate. *Carbohydrate Polymers* 2015;117:616–23.

83. Li J, Ma J, Chen S, Huang Y, He J. Adsorption of lysozyme by alginate/graphene oxide composite beads with enhanced stability and mechanical property. *Materials Science and Engineering: C* 2018;89:25–32.

84. Yadav M, Rhee KY, Park SJ. Synthesis and characterization of graphene oxide/carboxymethylcellulose/alginate composite blend films. *Carbohydrate Polymers* 2014;110:18–25.

85. Mao Y, Wang C, Liu L. Preparation of graphene oxide/natural rubber composites by latex co-coagulation: Relationship between microstructure and reinforcement. *Chinese Journal of Chemical Engineering* 2020;28:1187–93.

86. Zhan Y, Wu J, Xia H, Yan N, Fei G, Yuan G. Dispersion and exfoliation of graphene in rubber by an ultrasonically-assisted latex mixing and in situ reduction process. *Macromolecular Materials and Engineering* 2011;296:590–602.

87. Ozbas B, O'Neill CD, Register RA, Aksay IA, Prud'homme RK, Adamson DH. Multifunctional elastomer nanocomposites with functionalized graphene single sheets. *Journal of Polymer Science Part B: Polymer Physics* 2012;50:910–6.

88. Hernández M, Bernal MdM, Verdejo R, Ezquerra TA, López-Manchado MA. Overall performance of natural rubber/graphene nanocomposites. *Composites Science and Technology* 2012;73:40–6.

89. Kim H, Miura Y, Macosko CW. Graphene/polyurethane nanocomposites for improved gas barrier and electrical conductivity. *Chemistry of Materials* 2010;22:3441–50.

90. Yang L, Phua SL, Toh CL, Zhang L, Ling H, Chang M, et al. Polydopamine-coated graphene as multifunctional nanofillers in polyurethane. *RSC Advances* 2013;3:6377–85.

91. Tang Z, Wu X, Guo B, Zhang L, Jia D. Preparation of butadiene–styrene–vinyl pyridine rubber–graphene oxide hybrids through co-coagulation process and in situ interface tailoring. *Journal of Materials Chemistry* 2012;22:7492–501.

92. Potts JR, Shankar O, Du L, Ruoff RS. Processing–morphology–property relationships and composite theory analysis of reduced graphene oxide/natural rubber nanocomposites. *Macromolecules* 2012;45:6045–55.

93. Wang J, Zhang K, Fei G, Salzano de Luna M, Lavorgna M, Xia H. High silica content graphene/natural rubber composites prepared by a wet compounding and latex mixing process. *Polymers* 2020;12.

94. Dong B, Liu C, Zhang L, Wu Y. Preparation, fracture, and fatigue of exfoliated graphene oxide/natural rubber composites. *RSC Advances* 2015;5:17140–8.

95. Rahman MA, Tong GB, Kamaruddin NH, Wahab FA, Hamizi NA, Chowdhury ZZ, et al. Effect of graphene infusion on morphology and performance of natural rubber latex/graphene composites. *Journal of Materials Science: Materials in Electronics* 2019;30:12888–94.

96. Li C, Feng C, Peng Z, Gong W, Kong L. Ammonium-assisted green fabrication of graphene/natural rubber latex composite. *Polymer Composites* 2013;34:88–95.

97. Wu J, Huang G, Li H, Wu S, Liu Y, Zheng J. Enhanced mechanical and gas barrier properties of rubber nanocomposites with surface functionalized graphene oxide at low content. *Polymer* 2013;54:1930–7.
98. Araby S, Zaman I, Meng Q, Kawashima N, Michelmore A, Kuan H-C, et al. Melt compounding with graphene to develop functional, high-performance elastomers. *Nanotechnology* 2013;24:165601.
99. Azira AA, Kamal MM, Rusop M. Reinforcement of graphene in natural rubber nano-composite. *AIP Conference Proceedings* 2016;1733:020003.
100. Wu W, Yu B. Cornmeal graphene/natural rubber nanocomposites: Effect of modified graphene on mechanical and thermal properties. *ACS Omega* 2020;5:8551–6.
101. Yin C, Zhang Q, Liu J, Gao Y, Sun Y, Zhang Q. Preparation and characterization of grafted natural rubber/graphene oxide nanocomposites. *Journal of Macromolecular Science, Part B* 2019;58:645–58.
102. Wu Y, Yan X, Meng P, Sun P, Cheng G, Zheng R. Carbon black/octadecane composites for room temperature electrical and thermal regulation. *Carbon* 2015;94:417–23.
103. Buaki-Sogó M, García-Carmona L, Gil-Agustí M, García-Pellicer M, Quijano-López A. Flexible and conductive bioelectrodes based on chitosan-carbon black membranes: Towards the development of wearable bioelectrodes. *Nanomaterials* 2021;11.
104. Elmaghraby NA, Omer AM, Kenawy E-R, Gaber M, El Nemr A. Electrospun composites nanofibers from cellulose acetate/carbon black as efficient adsorbents for heavy and light machine oil from aquatic environment. *Journal of the Iranian Chemical Society* 2022;19:3013–27.
105. Pacheco MR, Barbosa SC, Quadrado RFN, Fajardo AR, Dias D. Glassy carbon electrode modified with carbon black and cross-linked alginate film: A new voltammetric electrode for paraquat determination. *Analytical and Bioanalytical Chemistry* 2019;411:3269–80.
106. Shi X, Sun S, Zhao A, Zhang H, Zuo M, Song Y, et al. Influence of carbon black on the Payne effect of filled natural rubber compounds. *Composites Science and Technology* 2021;203:108586.
107. Wu J-K, Ye C-C, Liu T, An Q-F, Song Y-H, Lee K-R, et al. Synergistic effects of CNT and GO on enhancing mechanical properties and separation performance of polyelectrolyte complex membranes. *Materials & Design* 2017;119:38–46.
108. Mokhena TC, Jacobs NV, Luyt AS. Electrospun alginate nanofibres as potential bio-sorption agent of heavy metals in water treatment. *Express Polymer Letters* 2017;11:652–63.
109. Mokhena TC, Jacobs NV, Luyt AS. Nanofibrous alginate membrane coated with cellulose nanowhiskers for water purification. *Cellulose* 2018;25:417–27.
110. Mokhena TC, Jacobs V, Luyt A. A review on electrospun bio-based polymers for water treatment. 2015.
111. Mokhena TC, Mochane MJ, Mtibe A, John MJ, Sadiku ER, Sefadi JS. Electrospun alginate nanofibers toward various applications: A review. *Materials* 2020;13.
112. Aydın EB, Aydın M, Sezgintürk MK. Electrochemical immunosensor based on chitosan/conductive carbon black composite modified disposable ITO electrode: An analytical platform for p53 detection. *Biosensors and Bioelectronics* 2018;121:80–9.
113. Liu Y, Zheng H, Liu M. High performance strain sensors based on chitosan/carbon black composite sponges. *Materials & Design* 2018;141:276–85.
114. Dinesh B, Saraswathi R, Senthil Kumar A. Water based homogenous carbon ink modified electrode as an efficient sensor system for simultaneous detection of ascorbic acid, dopamine and uric acid. *Electrochimica Acta* 2017;233:92–104.
115. Farida E, Bukit N, Ginting EM, Bukit BF. The effect of carbon black composition in natural rubber compound. *Case Studies in Thermal Engineering* 2019;16:100566.

116. Ulfah IM, Fidyaningsih R, Rahayu S, Fitriani DA, Saputra DA, Winarto DA, et al. Influence of carbon black and silica filler on the rheological and mechanical properties of natural rubber compound. *Procedia Chemistry* 2015;16:258–64.

117. Nakaramontri Y, Pichaiyut S, Wisunthorn S, Nakason C. Hybrid carbon nanotubes and conductive carbon black in natural rubber composites to enhance electrical conductivity by reducing gaps separating carbon nanotube encapsulates. *European Polymer Journal* 2017;90:467–84.

118. Li H, Yang L, Weng G, Xing W, Wu J, Huang G. Toughening rubbers with a hybrid filler network of graphene and carbon nanotubes. *Journal of Materials Chemistry A* 2015;3:22385–92.

119. Gao J, He Y, Gong X, Xu J. The role of carbon nanotubes in promoting the properties of carbon black-filled natural rubber/butadiene rubber composites. *Results in Physics* 2017;7:4352–8.

120. Dinesh B, Shalini Devi KS, Kumar AS. Curcumin-quinone immobilised carbon black modified electrode prepared by in-situ electrochemical oxidation of curcumin-phytonutrient for mediated oxidation and flow injection analysis of sulfide. *Journal of Electroanalytical Chemistry* 2017;804:116–27.

121. Gandhi M, Rajagopal D, Parthasarathy S, Raja S, Huang S-T, Senthil Kumar A. In Situ immobilized sesamol-quinone/carbon nanoblack-based electrochemical redox platform for efficient bioelectrocatalytic and immunosensor applications. *ACS Omega* 2018;3:10823–35.

122. Yadav A, Bagotia N, Sharma AK, Kumar S. Advances in decontamination of wastewater using biomass-basedcomposites: A critical review. *Science of the Total Environment* 2021;784:147108.

123. Song J, Messele SA, Meng L, Huang Z, Gamal El-Din M. Adsorption of metals from oil sands process water (OSPW) under natural pH by sludge-based Biochar/Chitosan composite. *Water Research* 2021;194:116930.

124. Chin JF, Heng ZW, Teoh HC, Chong WC, Pang YL. Recent development of magnetic biochar crosslinked chitosan on heavy metal removal from wastewater – Modification, application and mechanism. *Chemosphere* 2022;291:133035.

125. Vaz MG, Pereira AGB, Fajardo AR, Azevedo ACN, Rodrigues FHA. Methylene blue adsorption on chitosan-g-poly(acrylic acid)/rice husk ash superabsorbent composite: Kinetics, equilibrium, and thermodynamics. *Water, Air, & Soil Pollution* 2016;228:14.

126. Rodrigues FHA, Fajardo AR, Pereira AGB, Ricardo NMPS, Feitosa JPA, Muniz EC. Chitosan-graft-poly(acrylic acid)/rice husk ash based superabsorbent hydrogel composite: Preparation and characterization. *Journal of Polymer Research* 2012;19:1.

127. Zhang L, Tang S, He F, Liu Y, Mao W, Guan Y. Highly efficient and selective capture of heavy metals by poly(acrylic acid) grafted chitosan and biochar composite for wastewater treatment. *Chemical Engineering Journal* 2019;378:122215.

128. Ifthikar J, Jiao X, Ngambia A, Wang T, Khan A, Jawad A, et al. Facile one-pot synthesis of sustainable carboxymethyl chitosan – Sewage sludge biochar for effective heavy metal chelation and regeneration. *Bioresource Technology* 2018;262:22–31.

129. Altun T, Ecevit H. Cr (VI) removal using Fe 2 O 3-chitosan-cherry kernel shell pyrolytic charcoal composite beads. *Environmental Engineering Research* 2020;25:426–38.

130. de Azevedo ACN, Vaz MG, Gomes RF, Pereira AGB, Fajardo AR, Rodrigues FHA. Starch/rice husk ash based superabsorbent composite: High methylene blue removal efficiency. *Iranian Polymer Journal* 2017;26:93–105.

131. Benhouria A, Islam MA, Zaghouane-Boudiaf H, Boutahala M, Hameed BH. Calcium alginate–bentonite–activated carbon composite beads as highly effective adsorbent for methylene blue. *Chemical Engineering Journal* 2015;270:621–30.

132. Annadurai G, Juang R-S, Lee D-J. Factorial design analysis for adsorption of dye on activated carbon beads incorporated with calcium alginate. *Advances in Environmental Research* 2002;6:191–8.
133. Choi J-W, Yang K-S, Kim D-J, Lee CE. Adsorption of zinc and toluene by alginate complex impregnated with zeolite and activated carbon. *Current Applied Physics* 2009;9:694–7.
134. Wang B, Wan Y, Zheng Y, Lee X, Liu T, Yu Z, et al. Alginate-based composites for environmental applications: A critical review. *Critical Reviews in Environmental Science and Technology* 2019;49:318–56.
135. Hassan AF, Abdel-Mohsen AM, Fouda MMG. Comparative study of calcium alginate, activated carbon, and their composite beads on methylene blue adsorption. *Carbohydrate Polymers* 2014;102:192–8.
136. Ding Z, Hu X, Wan Y, Wang S, Gao B. Removal of lead, copper, cadmium, zinc, and nickel from aqueous solutions by alkali-modified biochar: Batch and column tests. *Journal of Industrial and Engineering Chemistry* 2016;33:239–45.
137. Roh H, Yu M-R, Yakkala K, Koduru JR, Yang J-K, Chang Y-Y. Removal studies of Cd(II) and explosive compounds using buffalo weed biochar-alginate beads. *Journal of Industrial and Engineering Chemistry* 2015;26:226–33.
138. Xiao Z, Zhang L, Wu L, Chen D. Adsorptive removal of Cu(II) from aqueous solutions using a novel macroporous bead adsorbent based on poly(vinyl alcohol)/sodium alginate/KMnO4 modified biochar. *Journal of the Taiwan Institute of Chemical Engineers* 2019;102:110–7.
139. Wang B, Gao B, Wan Y. Entrapment of ball-milled biochar in Ca-alginate beads for the removal of aqueous Cd(II). *Journal of Industrial and Engineering Chemistry* 2018;61:161–8.
140. Do X-H, Lee B-K. Removal of Pb2+ using a biochar–alginate capsule in aqueous solution and capsule regeneration. *Journal of Environmental Management* 2013;131:375–82.
141. Deng J, Li X, Liu Y, Zeng G, Liang J, Song B, et al. Alginate-modified biochar derived from Ca(II)-impregnated biomass: Excellent anti-interference ability for Pb(II) removal. *Ecotoxicology and Environmental Safety* 2018;165:211–8.

4 Characterizations of Polymer–Carbonaceous Composites

M. Somaiah Chowdary
National Institute of Technology Warangal
Prasad V Potluri Siddhartha Institute of Technology

G. Raghavendra
National Institute of Technology Warangal

M. Anupama Ammulu and M.S.R. Niranjan Kumar
Prasad V Potluri Siddhartha Institute of technology

Shakuntala Ojha
Kakatiya Institute of Technology & Science (KITS)

Vemuri Praveen Kumar
Koneru Lakshmaiah University

CONTENTS

4.1 INTRODUCTION

Polymers have long been a very well material in high-end applications. They are adaptable materials that can be molded to fit any requirement. Polymers' composites are gaining popularity as capable constituents for a wide range of uses, relatively less in cost, ease of handling, and compact nature, as well as the vast amount of filler and

DOI: 10.1201/9781003328094-4

polymer mixtures that allow for a diverse range of uses [1]. Particluates can give or adjust exact qualities of the polymer in addition to their reinforcing role, especially when utilizing nanofillers that permit the degree of crystallinity and structural, electrical, or thermal properties of the polymer to be controlled. As a result, PMCs are becoming more popular as research and development subject when reinforced with nanofillers. Carbon and carbonaceous reinforcements have greatly impacted polymer composites among the various nanofillers. Because carbon is one of the glorious elements that has changed materials science, researchers have focused their attention on selecting different carbonaceous materials from multiple sources [2].

Excellent carbon-based substances were used in a range of manufacturing uses, such as the degradation of various contaminants from water, as a catalyst, but as one of the finest electrically insulating substances. They have a greater surface area and a well-developed pore structure. Most carbon black is used as reinforcing in the plastics industry for cutting-edge technologies and to boost matrix performance in comparison to other well-known materials. The origin, conditions of production, and chemical processes of the carbon black in composites heavily influence its properties. Such fillers with diverse forms, particle sizes, and sources added to the polymer matrix could produce various microstructures and affect the characteristics of polymer composites in multiple ways [3]. These bio-fillers have a high degree of stiffness, are affordable, low density, renewable, environmentally beneficial, and flexible during dispensation. Bio-fillers can be transformed to carbon, turning undesirable agricultural waste into valuable, usable materials [4]. On the other hand, when distributed in water, materials with a high amount of elemental carbon filler exhibit a very hydrophobic character. For instance, carbon-filled polyethene composites have been utilized as pipes in dangerous storms. In difference, carbon black-filled polypropylene geotextile composites are useful for many construction applications [5], including soil reinforcements.

Considering the functionality and adaptability of these composites in various applications and equipments, the amount of carbonaceous-nanofiller-reinforced polymeric materials has augmented over the previous two decades [6]. Due to their functional aspects, such as the total area of graphene (G), the more significant aspect ratio, and greater mechanical properties of carbon nanofibers (CNF), or the electrical and mechanical properties of carbon nanotubes(CNTs), these nanomaterials are appealing in the creation of multifunctional composites. In addition to mechanical reinforcement [7], piezo-resistive composites for power and distortion detecting can be created using carbonaceous nanofillers with customized electrical characteristics. The advantages of carbonaceous materials are their high surface area, excellent conductivity, and inexpensive cost. Because of this, their typical specific capacitance is significantly lower than that of transition-metal oxides, sulfides, and other materials, limiting super capacitors' overall efficiency. The capacitance of carbonaceous materials is heavily dependent on their structure, because it is mainly caused by charge accumulation at the interface of the electrolyte and electrode materials. This is, therefore, crucial for developing carbonaceous material with an appropriate structure. However, due to their lower cost and well-proven electrochemical characteristics [8], other types of carbonaceous materials, such as carbon aerogels and activated carbon, are still appealing.

This chapter mainly deals with the characterization of various carbonaceous fillers reinforced with polymers (thermosets and thermoplastics). Different characterizations include SEM [9,10], TEM [11–13], and XRD [14] on the fractured samples of polymer–carbonaceous composites.

4.2 VARIOUS CARBONACEOUS FILLERS

Carbon nanotubes, fullerenes, graphene, nanodiamonds, and carbon fibers are a few examples of significant carbonaceous nanofiller forms, which are commercially available. Figure 4.1 displays a few essential forms. The diameter of nanodiamonds (NDs), carbon nanoparticles with a truncated octahedral morphology, ranges from 2 to 8 nm. The NDs were first created using the detonation method in the USSR in the 1960s, but it wasn't discovered until the 1990s. NDs are a great option for biological applications because of their outstanding visual and mechanical qualities and nontoxic nature. Fullerenes are carbon nanoparticles with zero dimensions. It was initially created in 1985 after discovering a steady molecule called Buckminsterfullerene C60, which was then regarded as the third allotrope of carbon after graphite and diamond [14]. Fullerene has significance possible in various technical areas, such as power augmentation. Carbon materials, such as amorphous carbon with a zero-dimensional (0D) structure, carbon nanotubes (CNTs), and carbon nanofibers are unique from

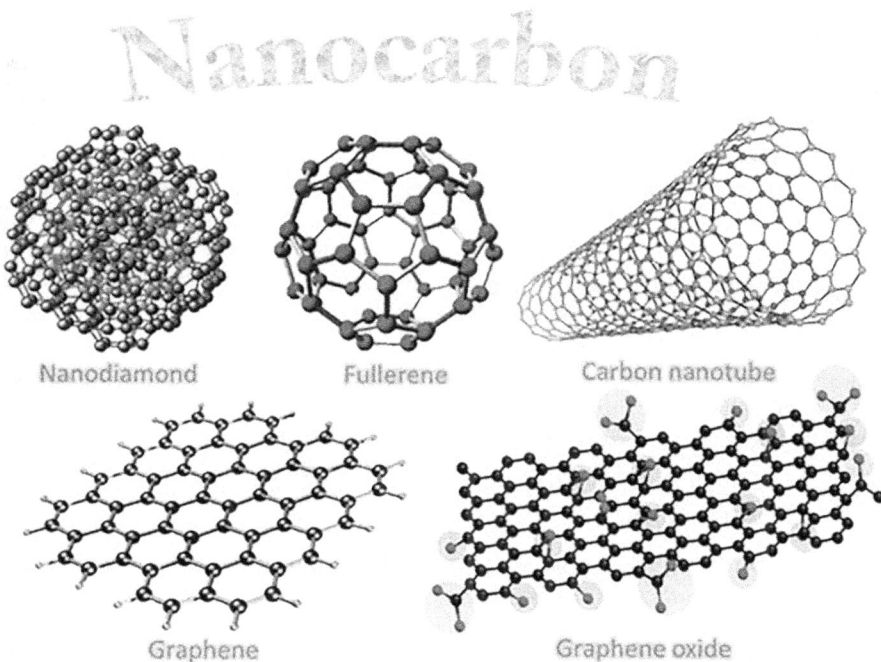

FIGURE 4.1 Various carbonaceous structures [20].

TABLE 4.1
Some Properties of Carbonaceous Materials [16]

	Fullerene	Nanodiamonds	Carbon Onion	Carbon Black	Reference
Density (g/cm³)	1.65	3.18	1.9	1.76–1.90	[17]
Specific area (m²/g)	–	–300	420	600	[18]
Hardness (GPa)	0.23 (nano-hardness)	120–420 (micro-hardness)	10	–	[19]

metallic materials (CNFs). Both the graphenes (GR) with a two-dimensional (2D) structure and a one-dimensional (1D) structure exhibit exceptional qualities like great toughness, high specific area, strong thermal conductivity, and outstanding electrical mobility [15]. These carbon materials are frequently used as efficient additions to enhance the magnetic, optical, electrochromic, conductive, mechanical, and dielectric properties of nanocomposites (Table 4.1).

Carbon black (CB) production, which is caused by partial combustion of oil feedstock, has historically been relatively expensive [21]. Agricultural left-over products are now good bases of raw materials for synthesizing CB among the numerous carbonaceous materials. This CB is created from the pyrolysis of lignocellulose biomass-based materials, which include wood, coal, coconut, oil palm, jute, banana, bamboo, and other carbonaceous and rich in organic materials. These agricultural byproducts are typically cheap, and their efficient use has been desired. Ojha et al. [22] studied the method to utilizing waste carbon as reinforcement in thermoset composite. CB wood apple shell particles were created by the carbonization process and concluded that the carbonization process increased the carbon percentage. The comparison of raw wood apple shell particles with the resulting CBs using scanning electron microscopy revealed the alterations brought on by the carbonization process, as illustrated in Figure 4.2. The raw wood apple shell particles initially had no discernible pores, and they also fell between the range of 1 and 212 mm, as illustrated in Figure 4.2a. Figure 4.2b shows that the shell particles are under the nano range after pyrolytic decomposition. The carbonization process causes the wood apple shell particulates to develop micropores on their surface. The CB particles from the wood apple shell had a lot of regular-sized micropores. After each carbonization stage, the pores in the particles' porous structure are larger than those that already exist.

4.2.1 FUNCTIONALIZATION OF CARBON

Surface functionalization is a method that modifies the chemistry of the solid surface to produce exact qualities. Unlike functionalization processes were used on various carbon species [23]. The shift in surface energy is one of the critical outcomes of this process. This modification allows for more vital molecular interactions with other materials by bonding specialized small compounds such as medicines, genetic materials, luminous agents, metallic nanostructures for catalysis or plasmonics, etc.

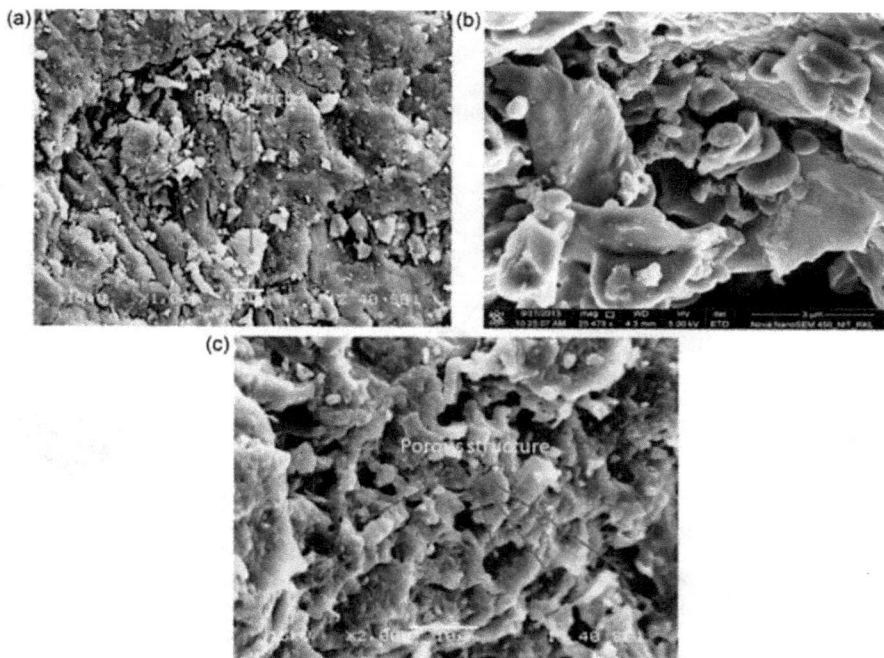

FIGURE 4.2 (a) Raw wood apple shell particles; (b) carbon black wood apple shell particles (400°C); (c) carbon black wood apple shell particles (600°C) [22].

Furthermore, by connecting polar/nonpolar groups to a surface, it's likely to adjust hydrophilicity, which leads to improved interaction with solvents or other materials at the macroscopic level, as in composites.

To adjust the surface's reactivity toward particular chemical species, the functionalization of the material's surface also involves attaching distinct molecules. This includes electrochemical characteristics customized for specific uses, including catalysis, batteries, supercapacitors, fuel cells, and organic photovoltaics. Additionally, the functionalization gives carbon-based nanostructures their luminous features or permits the bonding of molecular luminescent species to the material surface. As shown in Figure 4.1, functionalization is ultimately an essential stage in synthesizing carbon compounds for various uses (Figure 4.3).

4.3 REINFORCEMENT OF CARBONACEOUS PARTICLES WITH DIFFERENT POLYMERS

Polymers have long been used as well-known materials in cutting-edge applications. They are adaptable materials that are simple to shape for any needed purpose. One polymer cannot satisfy the demands of advanced applications; there are a few factors in the world of polymers to consider. As a result, polymer composites caught

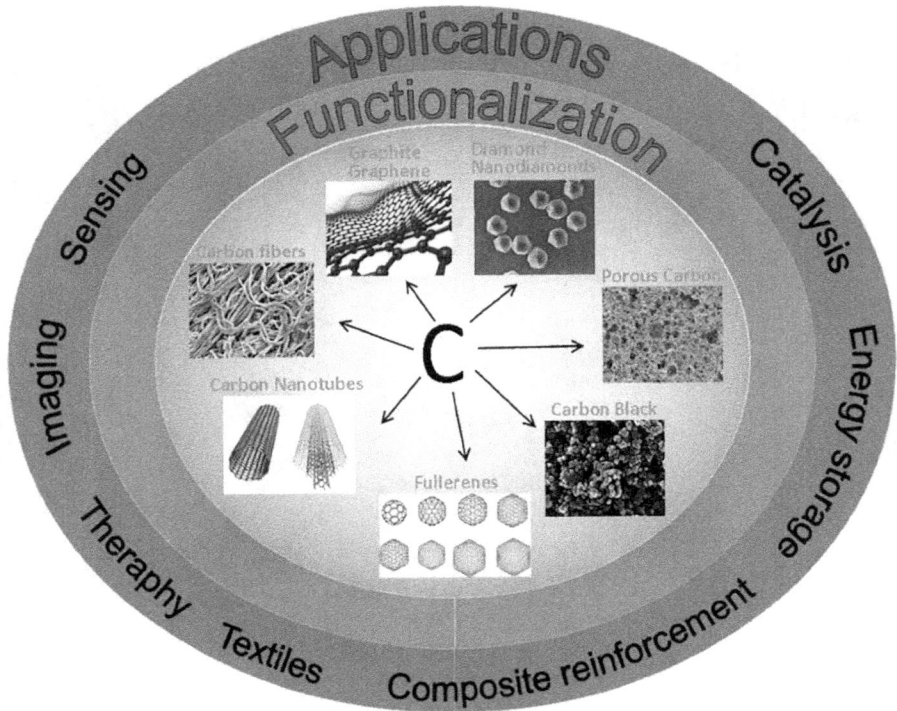

FIGURE 4.3   Various carbon-based compounds in their multiple forms [24].

the attention of everyone. At least two components make up a composite: the matrix and the reinforcement. Metals, ceramics, and other polymers may be used in the composite as a matrix and reinforcement. The matrix for polymer composites has traditionally been a mixture of thermosetting and thermoplastic resins. Thermosets have a lower density, whereas thermoplastics can be recycled and reused. Practically, every polymer that is significant for commerce has cutting-edge uses [25]. So polymers cannot alone satisfy the requirements. Some form of reinforcement makes them more suitable for specific applications.

4.3.1 VARIOUS CHARACTERIZATIONS OF THERMOSETS-CARBONACEOUS COMPOSITES

To learn about materials practically, characterization investigations are crucial. They can also be applied to link structure to characteristics. To identify substances, a variety of characterization procedures are used. They comprise NMR (nuclear magnetic resonance), scanning electron microscopy, transmission electron microscopy, X-ray diffraction, TEM, XRD, X-ray fluorescence, and synchrotron techniques [26]. The substance cannot, however, be thoroughly analyzed by a single process. In real-world research projects, results must be reached by combining two

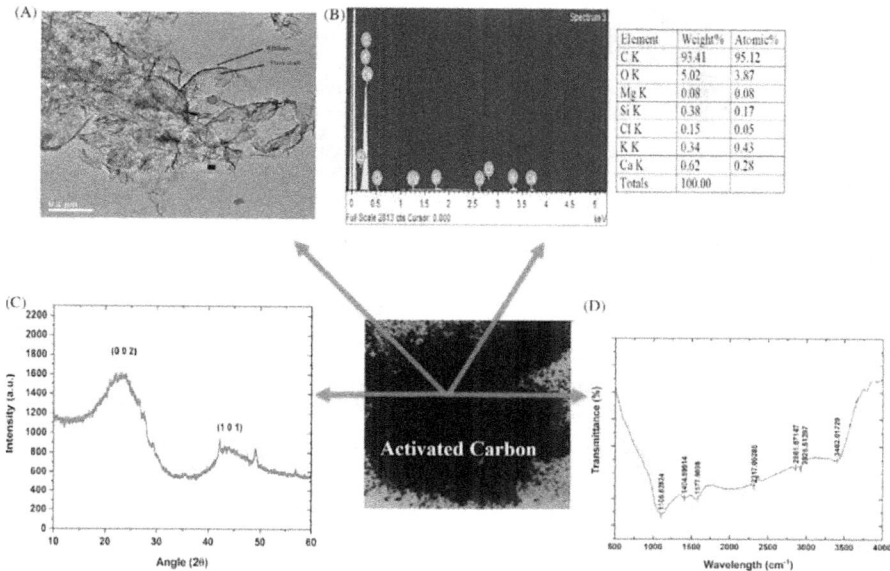

FIGURE 4.4 (a) TEM, (b) EDX, (c) XRD and (d) FTIR of images of activated carbon [28].

or three characterization methodologies [27]. Manoj et al. [28] studied environmental behavior and its impact on the highly porous, activated carbon epoxy composites. and concluded the following results from the images. The following two morphological traits are visible in the TEM picture in Figure 4.4. The surface of synthetic carbon material has a high porosity with a particular topographic characteristic and graphite threads. A thin wall connects the three-dimensional structure. The first two characteristics make activated carbon stand out. Figure 4.4b displays the EDX analysis's findings. The main component discovered by EDX analysis is carbon, which makes up over 90% of the entire composition. Figure 4.4c shows the material under XRD examination. Amorphous carbon is indicated by two large peaks in the XRD analysis: one at roughly $23°$ and the other at $43°$. These two peaks suggest the existence of graphene crystals, which are at $23°$ and $43°$, accordingly, and indicate the two planes (002) and (101). Figure 4.4d shows functional groups in the activated carbon at cm1 3402, 2926, 2867, 2317, 1577, 1404, and 1106. The peak authorizes the occurrence of the OH group at 3402 cm1. Indicated by the two peaks at 2850 and 2950 cm1, CH-stretching occurs. Whereas the peak at 1404 cm1 demonstrates the existence of nitro compound, the peak at 1577 cm1 approves the presence of amine groups (N–H).

They also calculated the erosion wear behaviors of activated carbon reinforced in epoxy under various environments. Figure 4.5 shows the fractured samples under erosion tests with 2 wt% reinforced epoxy. Although the particles in epoxy resin are distributed properly, it might be challenging to distinguish between particles and epoxy. The uniform dissemination of particles over the volume of the entire composite might have been a factor in the improvement of the friction coefficient.

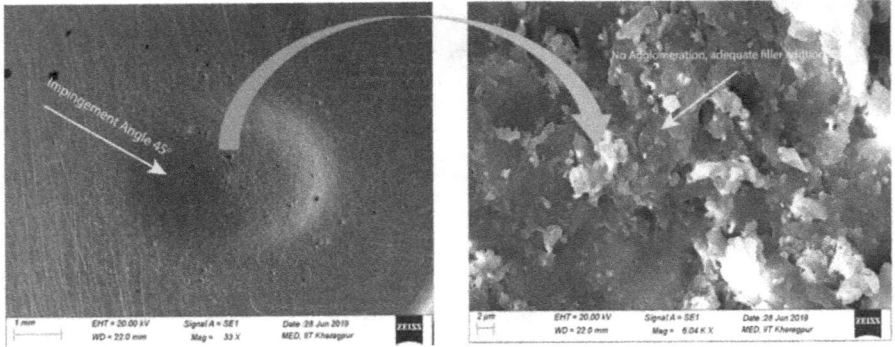

FIGURE 4.5   Activated carbon wear sample reinforced at 2 wt% at 45° impact angle [28].

Om Prakash et al. [29] examined TEM images of activated carbon prepared from Arhar stalk biochar synthesis. Figure 4.6 depicts the structure of AC, which has a honeycomb-like appearance and exhibits a continuous porous structure divided by pore walls. Most apertures are microspores, with a small number of the mesoporous present. The particles of the activated carbon are nano sized.

They also prepared samples with 1, 2, 3 wt% activated carbon reinforced with epoxy with various activation temperatures and concluded that activation temperatures show an important role in the properties of composites, and 800°C was found to be a better activation temperature. The surface morphology of the 2% nanoporous activated carbon epoxy composites is displayed in Figure 4.15. It is obvious that there are no visible cracks, yet the fragmented surface does have specific patterns that resemble rivers. The surface's wave- or river-like patterns demonstrate the composites' brittle breakdown (Figure 4.7).

Ojha et al. [30] analyzed that wood apple shell was used to create carbon black, which was then used as a filler in polymer composites after being pyrolyzed at 400°C. An experimental investigation has been done to compare the erosional wear behavior of epoxy resin matrix composites filled with raw and carbon black wood apple shell particles. On the erosion rate of the composite, the impact of the concentration of wood apple shell particles at various impingement angles (30°, 45°, 60°, and 90°) and a continuous impact velocity of 48 m/s has been studied. However, it is discovered that when paralleled to raw particulate composite, the carbon black particulates composite displays the least amount of wear. Additionally, it exhibits semi-ductile-type failure, and a 60° impingement angle results in the highest rate of erosion, and SEM results validated the above results. From SEM analysis, Figure 4.8a depicts the results of an SEM study of composites made of raw materials (10 wt%) and carbon black (20 wt%) filler materials with continual impact velocities of 48 m/s and 45° (b). Figure 4.8a shows that some materials have been lost from the 10 wt% filler composite's degraded surface. It demonstrates the somewhat positive interfacial bonding between the matrix and filler. Due to an rise in the carbon, surface of composite hardens when black carbon particulates are added with epoxy. As a result,

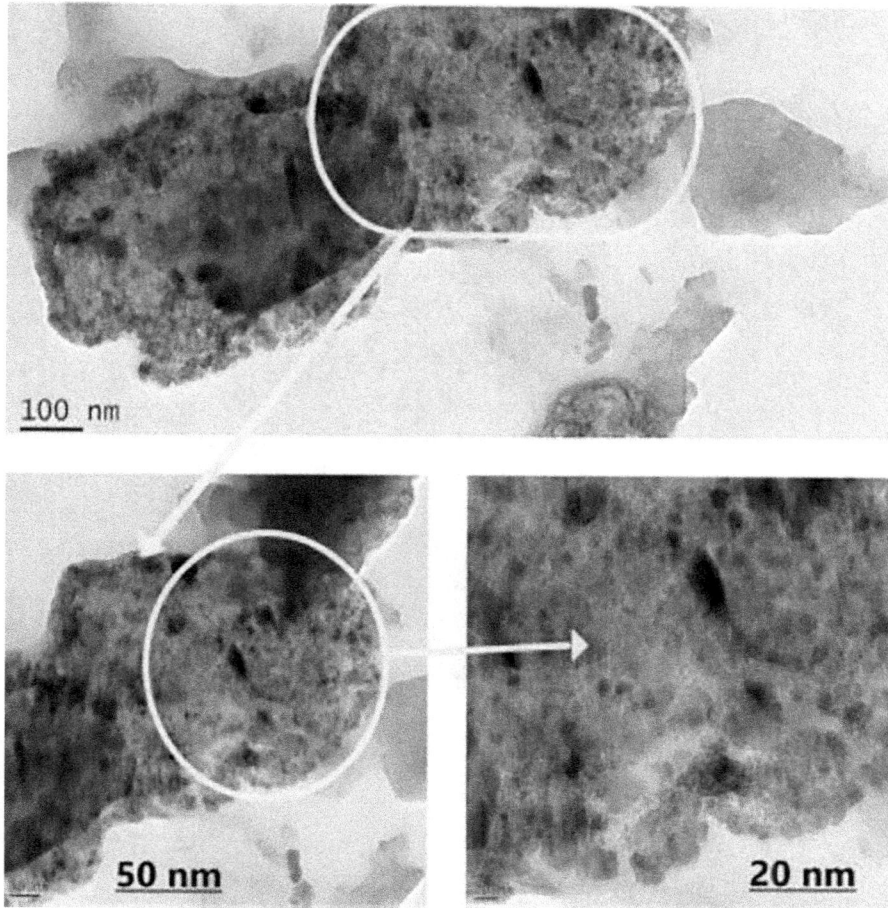

FIGURE 4.6 TEM of activated carbon [29].

the eroded composite surface depicted in Figure 4.8b has no voids or cracks, and there is a decent interfacial bonding among them.

Prakash et al. [31] studied recyclable porous nano-activated carbon made out of biomass waste effectively to produce composites. They calculated erosion wear and abrasion wear rates of the fabricated samples. They came to the conclusion that the reinforcement of porous nano-activated carbon in epoxy resin significantly increased the erosion resistance of composites. According to the results of the erosion wear tests, process variables like applied load and impact angle have an effect on how the activated carbon composites wear. In the current study, all of the composites showed semi-ductile eroding wearing characteristics, with the highest eroding wear occurring at a 45° impact angles, independent of the filler percentage. The present experiment's percent pore active carbon epoxy composites had the highest level of wear and erosion resistance of the composites. Regardless of the quantity of reinforcement, the

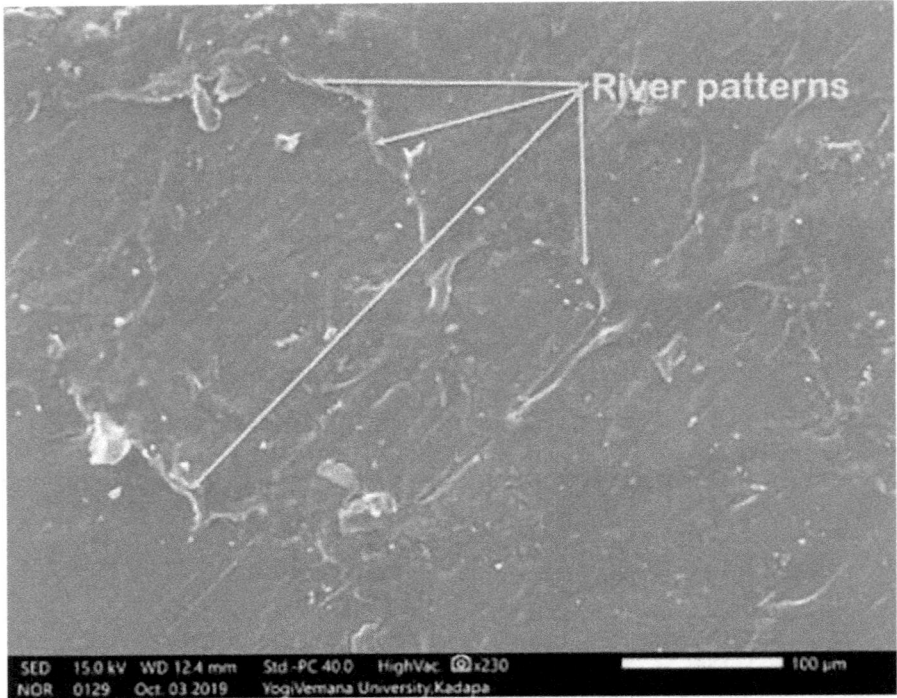

FIGURE 4.7   SEM images of 2 wt% nanocarbon reinforced epoxy composites [29].

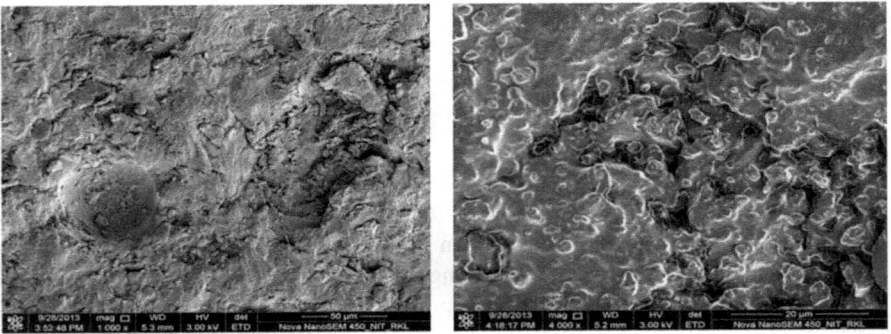

FIGURE 4.8   SEM showing an eroded composite surface with an impact at 48 m/s and a 45° angle for (a) 10% filler and (b) 20% filler [30].

load applied and countering area (abrasive surface) had a significant impact on the abrasive wear performance of the activated carbon composite.

Figure 4.9 demonstrates the degraded surfaces of samples of 2% and 3% activated carbon epoxy composites at a 45° impact angle. Both the surfaces have similar

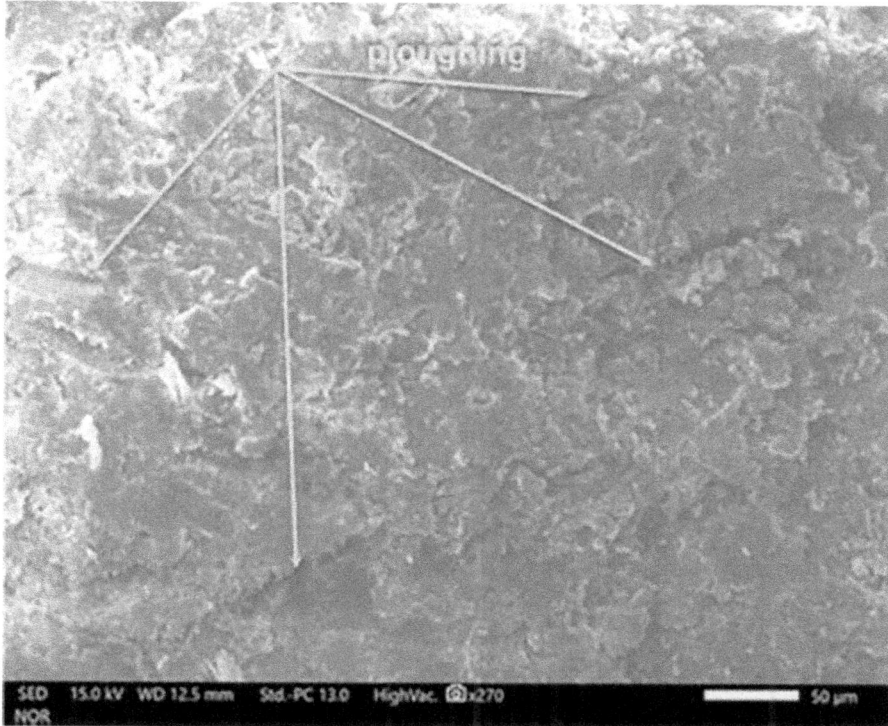

FIGURE 4.9 Erosion of samples with (a) 2 wt% (b) 3 wt% at 45° impact angle [31].

microcuts and channels, showing plastic material movement by erodent particles striking them. The loss of materials occurs when erodent penetrates the composite while hitting it at a reduced impact angle. Parallel pressures and vertical forces both have an impact on cutting and plowing at lower impact angles. As a result of the high influence of similar strength and high material loss, plastic deformation, microcuts are shown at a 45°.

Qin et al. [32] demonstrated the external morphologies of uncoated CF, epoxy-coated CF, and CF coated in GnPs, CNTs, and GnPs/CNTs hybrid in Figure 4.10. On noncoated CF, comparatively flat surfaces were seen. Noncoated CFs that developed throughout the spinning of the PAN precursor displayed continuous ridges and grooves in the fiber axis. Epoxy-coated CFs had a morphology that was comparable to that of uncoated CF. At the same time, the alteration of carbon nanoparticles altered the topographies of CF surfaces. At the very same modifying circumstances, the GnPs were effectively and evenly disseminated throughout CF surfaces. The CNTs were dropped on CF surfaces, as shown in the figure. On CF surfaces, little CNT aggregation was seen. The GnPs/CNTs hybrid covering coating on CF surfaces was evenly dispersed.

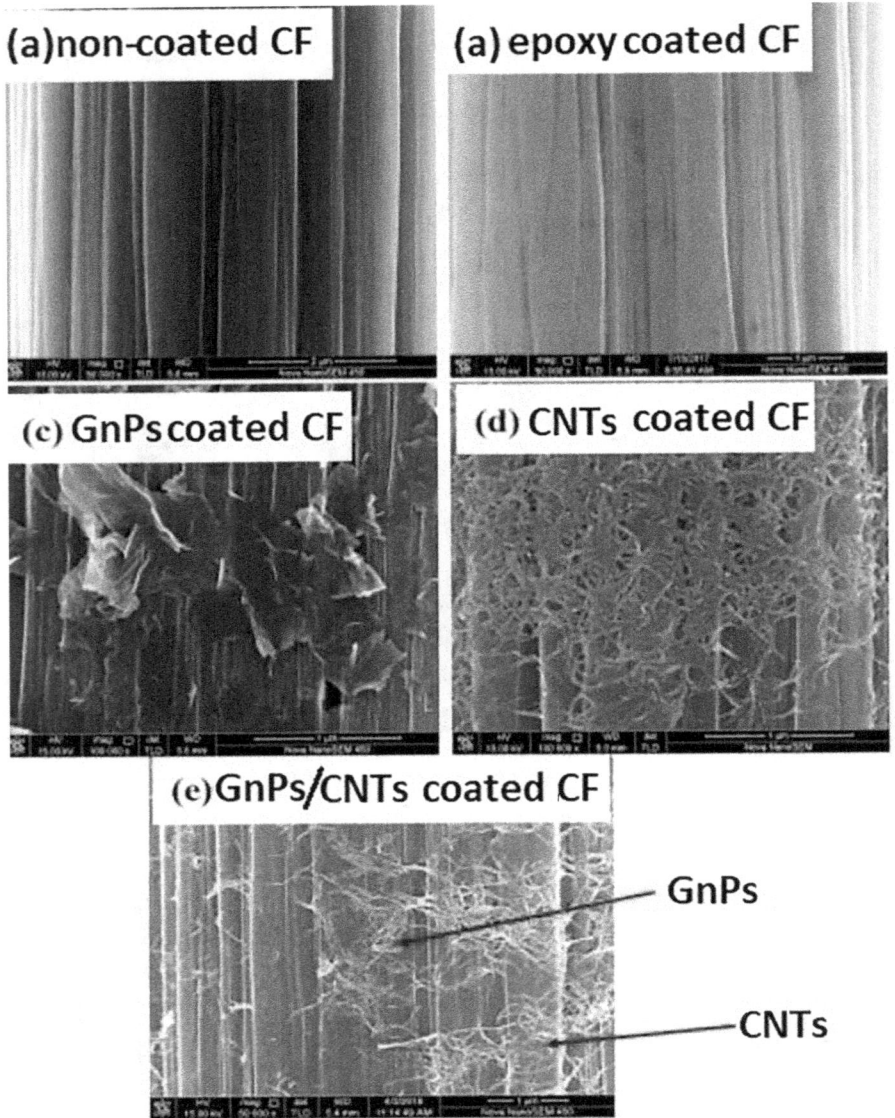

FIGURE 4.10   Surface morphology of CF in its different forms like uncoated, epoxy-coated, GnPs-coated, CNTs-coated, and GnPs/CNTs-coated [32].

4.3.2   Various Characterizations of Thermoplastics– Carbonaceous Composites

Xu et al. used a SEM to analyze the morphologies of carbon fiber/epoxy composites coated with a polyphosphazene layer of the Cyclomatrix type. Platinum was sputter

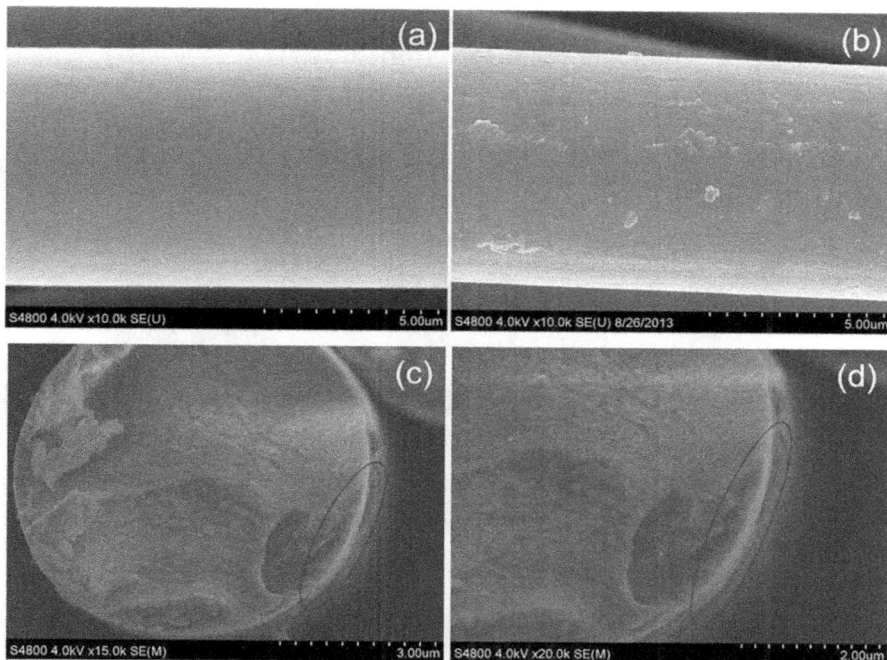

FIGURE 4.11 SEM images of (a) CFO, (b) CF-ACP, and (c) and (d) the transverse cross-section of CF-ACP [33].

dropped onto the samples. Figure 4.11 depicts the surface morphologies of coated and oxidized fibers. Sample oxidized carbon fibers (CFO) appear to have a clean, smooth surface (Figure 4.11a). Sample cyclomatrix-type polyphosphazene coated fibers (CF-ACP) exhibit some protrusions and a slightly rougher surface after treatment (Figure 4.11b). The attachment of the polyphosphazene microparticles may be the cause of the little protrusions. The SEM of the cross-section of covered fibers is depicted in Figure 4.11c and d. It is evident that a incessant, homogeneous, and crack-free coating firmly adheres to the surface of the fibers and completely wraps around the fiber surface. Affording to the diameter variance among CFO and CF-ACP fibers, the SEM measurement of the polymeric coating revealed an average thickness of roughly 200 nm [33].

Kim et al. [34] Epoxy/nylon 6 composites with carbon fiber reinforcement were created by VCC. The surface morphology was examined using the SEM. The microstructures of the composites are shown in Figure 4.12. The findings showed that the matrix and carbon fiber had a more difficult time adhering to the cracked substrate. A three-point bend test and a microbond test were also used to investigate the interface toughness. It was discovered that the reinforced carbon fiber had a diameter of 6–8 mm and a length of 50–200 mm [35].

The initial step was to create and characterize a new polymer reinforced with 3D graphene foam and 1D CBF in terms of its mechanical and thermal properties.

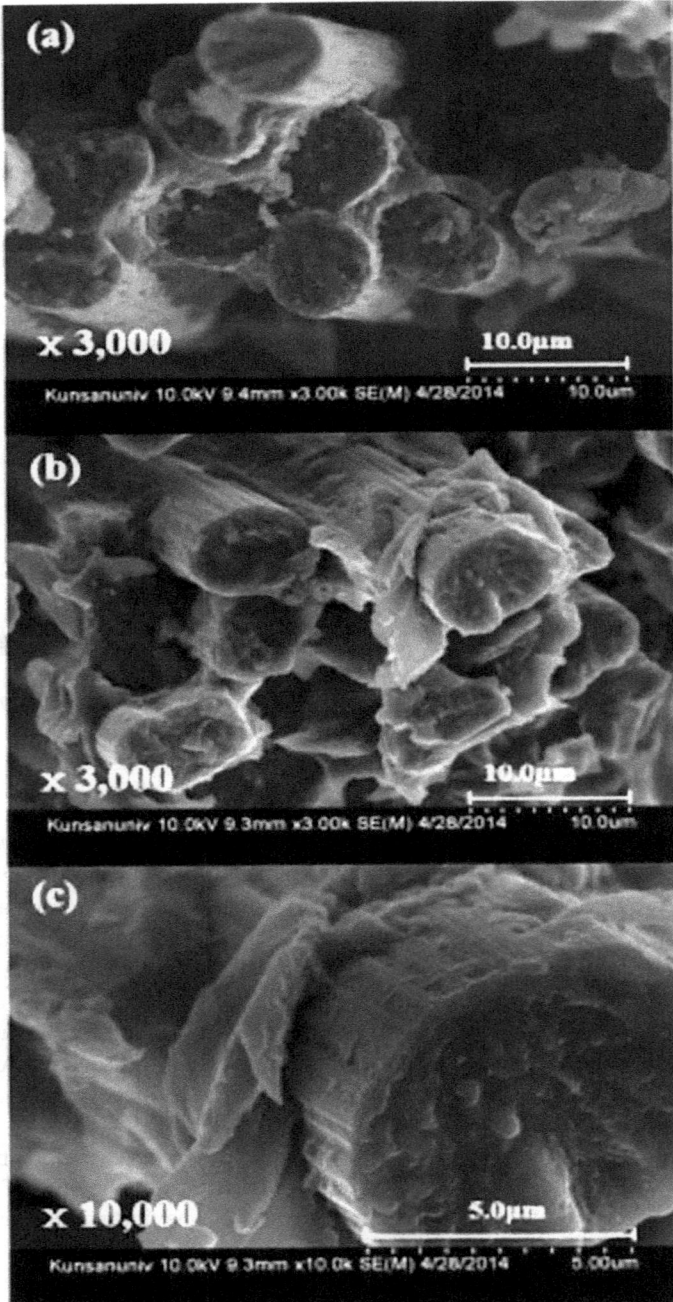

FIGURE 4.12   Epoxy composite materials (a), epoxy/nylon 6 matrix composites (6:4), and an expanded composite image (b) all display images of the fiber-reinforced epoxy/nylon 6 matrix composites with broken surfaces [34].

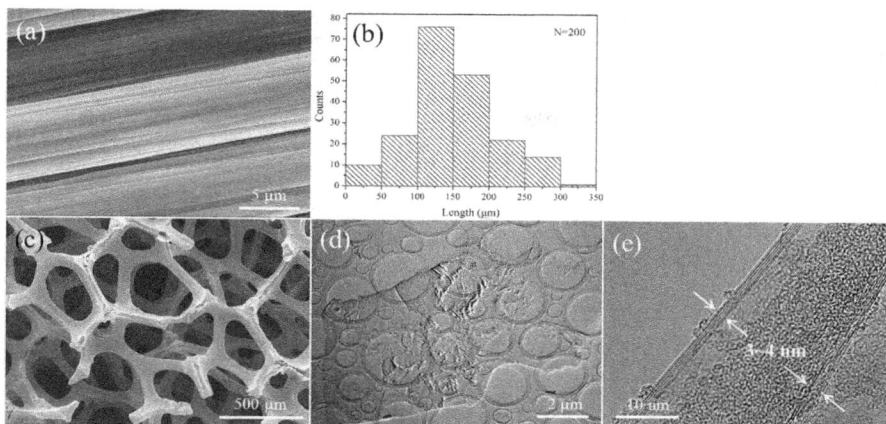

FIGURE 4.13 Surface morphology and the statistical length variation of CF are shown in (a). GF morphology is (c). (d, e) TEM pictures of the GS creating the GF [36].

SEM was used to examine the morphology of carbon fiber (CF), graphene foam (GF), and composite materials (SEM, S-4800, HITACHI). Figure 4.13 displays SEM images of the composite surfaces made of freeze-fractured graphene foam (GF)/polydimethylsiloxane (GF/PDMS) and CF/GF/PDMS at various CF loadings. Figure 4.13a shows that the liquid PDMS penetrated the GF arms without leaving any visible bubbles after filling the GF pores and entering the GF arms. Figure 4.13b–f demonstrates that the CF only takes place in the linked grid's interplanetary. Its length, which was greater than the thickness of the GF arms, prevented it from entering the inner of the GF arms, as seen in Figure 4.13c. Similar but with varying CF contents are in Figure 4.13d–f. The PDMS matrix's uniform and random CF distribution has been observed, and this can be linked to the effects of high-speed clipping and rousing. Additionally, the contact among CF and PDMS is sound without any obvious freeze fracture pulling out of CF [36].

Mohamed H. Gabr et al. examined the thermal and mechanical performance of carbon fiber/polypropylene (CF/PPc) composites filled with organoclay. By piling the preimpregnated carbon fabric with the created PPc/organoclay and the presence of reinforcement/matrix volume fraction of 52/48 2 (v/v), carbon fiber laminates were created. SEM was used on fractured surfaces that were acquired from fracture toughness testing utilizing JSM7001FD technology.

Prior to SEM examination, all specimens were spray coated with a layer of gold to prevent electrical charge. A micrograph for filled CF/PPc also demonstrates a gradual morphology at the conclusion of the insertion film and the fiber/matrix interface debonding (Figure 4.14a). The PPc resin appears to be coated across large areas, which indicates that the debonding resistance has diminished. The hybrid CFRP composite samples with organoclay in the matrix frequently had rougher matrix surfaces than those with plain PPc because of the pin and shatter tip splitting mentioned in Figure 4.14b–d. The high magnification in Figure 4.14c shows two distinctive

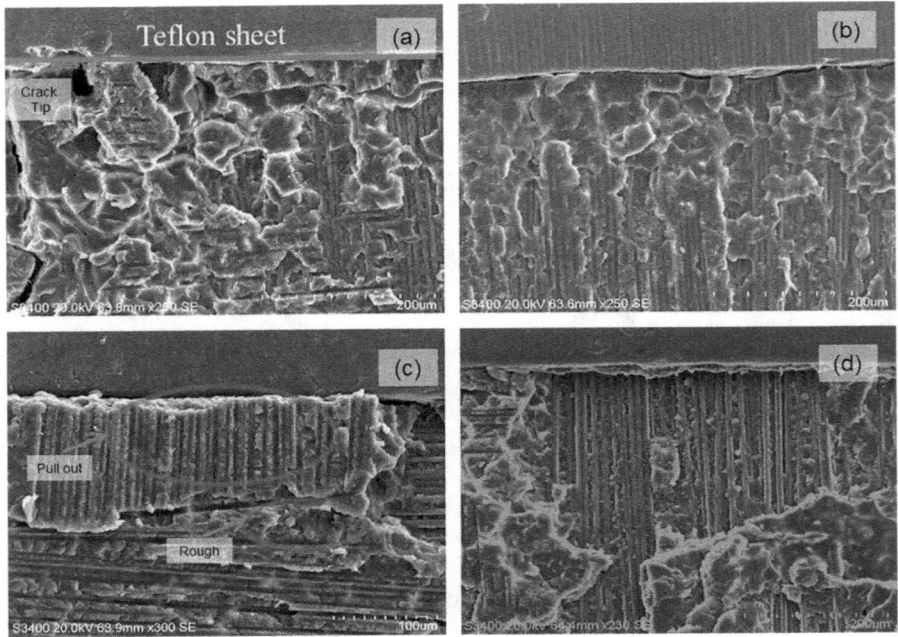

FIGURE 4.14   SEM images for CF/PPc composites filled with different ratios of organoclay at fracture surface of fracture toughness testing specimens: (a) unfilled, (b) 1%, (c) 3% and (d) 5% organoclay [37].

features of the CF/PPc-filled, 3 wt% organoclay: the first is the rough region showing reinforced adhesion, and the moment is the very front of the tip of the film spacer where some fibers can be seen of been drawn out or cracked without matrix coating. The greatest fracture toughness value is present in this area. The organoclay aggregates tend to grow larger with increased clay content, although the rough area and matrix distortion among fibers were more wide and profounder at great clay absorptions, resulting in relatively small rises in quasi-static fracture toughness (Figure 4.14d) [37].

To create the unidirectional carbon nanotube reinforced poly(biphenyl dianhydride-p-phenylenediamine) (BPDA-PDA) polyimide composites (Figure 4.15), we use the spray winding technique. Our findings demonstrated that multiwalled carbon nanotubes (MWNTs) sheets were well infiltrated by polyimide, and that long, aligned, and high volume fractions of MWNTs were realized in the composites. On a JEOL 6400F microscope with a 5 kV, SEM study of the MWNTs network and composite fracture surface was performed. The preponderance of the MWNTs are orthogonal to each other and matched in Figure 4.15, which is a SEM view of the sheet of MWNTs as it was drawn. To make sure the polyimide has broken, the composite sample was heated in TGA before the image was taken. The van der Waals interactions allow for continuous joining of carbon nanotubes. When the polymer matrix was coated whereas the MWNT sheets were wound onto the spindle,

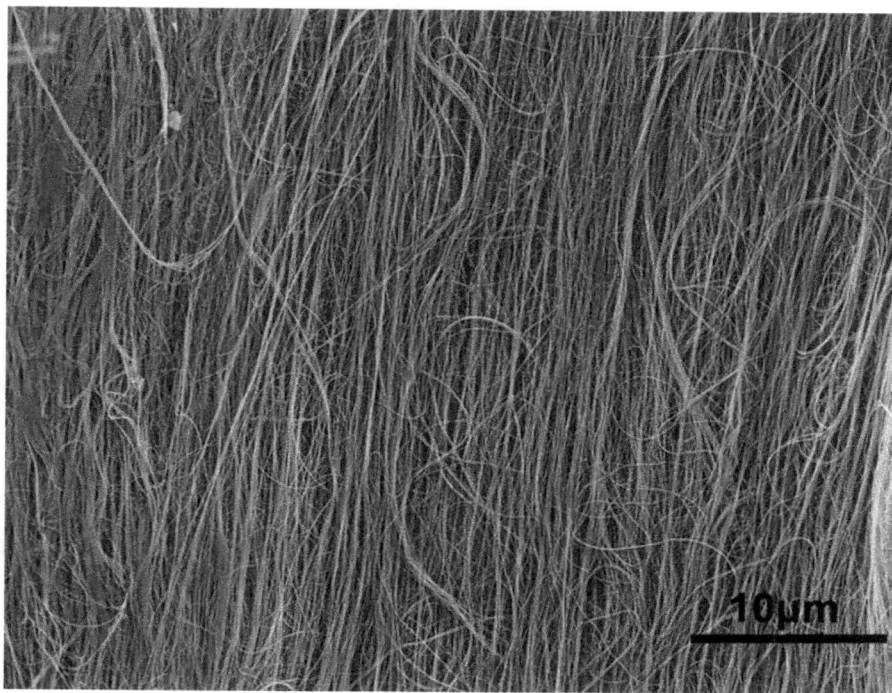

FIGURE 4.15 Unidirectional MWNTs composite's EM structure with polyimide degraded in TGA [38].

the greater standard of MWNT alignments was kept. Additionally, the additional zigzagging process compressed the MWNT assemblage due to the typical forces (Figure 4.16) [38].

A significant nanocarbon nanomaterial is ND [39]. The focus has been on adding ND to thermoplastic and thermosetting polymers. The mechanical and tribological qualities of the materials have been known to improve with the use of ND nanofiller. An epoxy/polyamide blend loaded with aminated ND was suggested by Neitzel et al. [10]. Images taken using a transmission electron microscope (TEM) revealed discrete ND-NH2 nanoparticles scattered throughout the matrix (Figure 4.17). The nanoparticle is 100 nm in diameter [35].

Multiwalled carbon nanotubes (MWCNTs) were assorted with an epoxy–polyamide mixture in this study after being dispersed in a hardener. The effects of various MWCNT weight ratio percentages were investigated. The samples were created using an axillary ultrasonic process and mechanical molding. Known commercially as Nitocote EP (405), the polymer epoxy resin type (DGEBA) (diglycidyl ether of bisphenol A) is a viscid liquid with a silver hue that is supplied by Fosroc Jorden. It has a medium viscosity and good adherence. In this investigation, the weight ratio of resin to hardener is 5:1. To create a mixture matrix with an epoxy-to-polyamide weight ratio, polyamide polymer is combined with epoxy resin (4:1).

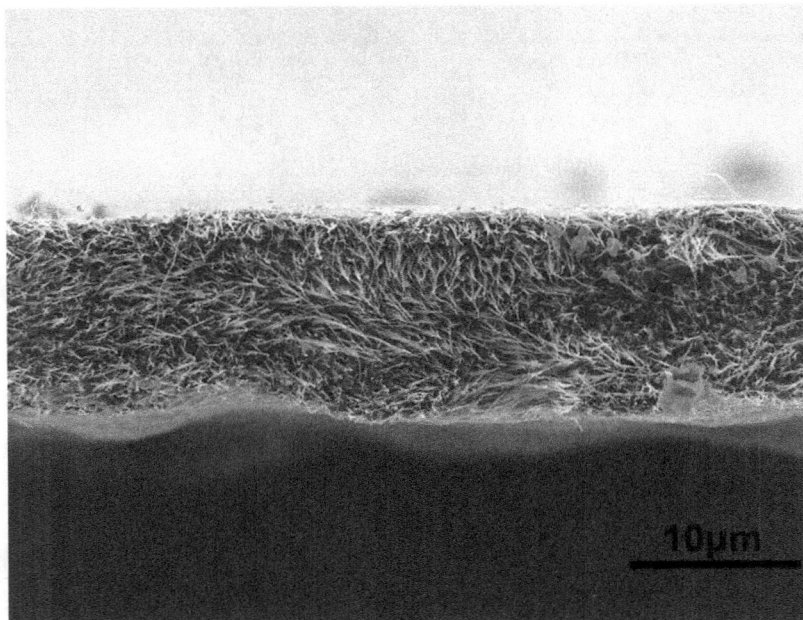

FIGURE 4.16   SEM photos of a MWNT/BPDA-PDA polyimide composite's surface morphology [38].

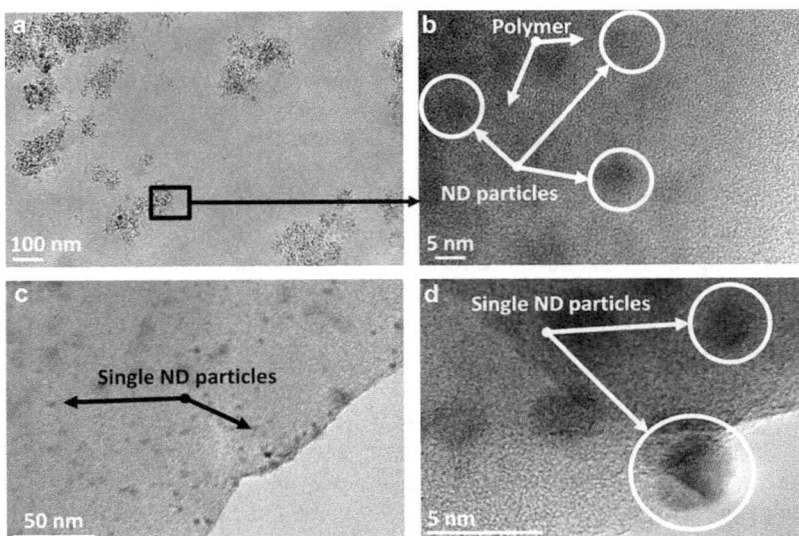

FIGURE 4.17   TEM images of epoxyeNDeNH2 composites produced with NDeNH2 contained in THF, each image being incrementally magnified: low magnification; close-up of loose or fractured aggregates of NDeNH2; a single NDeNH2 particles distributed in the epoxy; high resolution image of a single NDeNH2 particles spread in the epoxy [40].

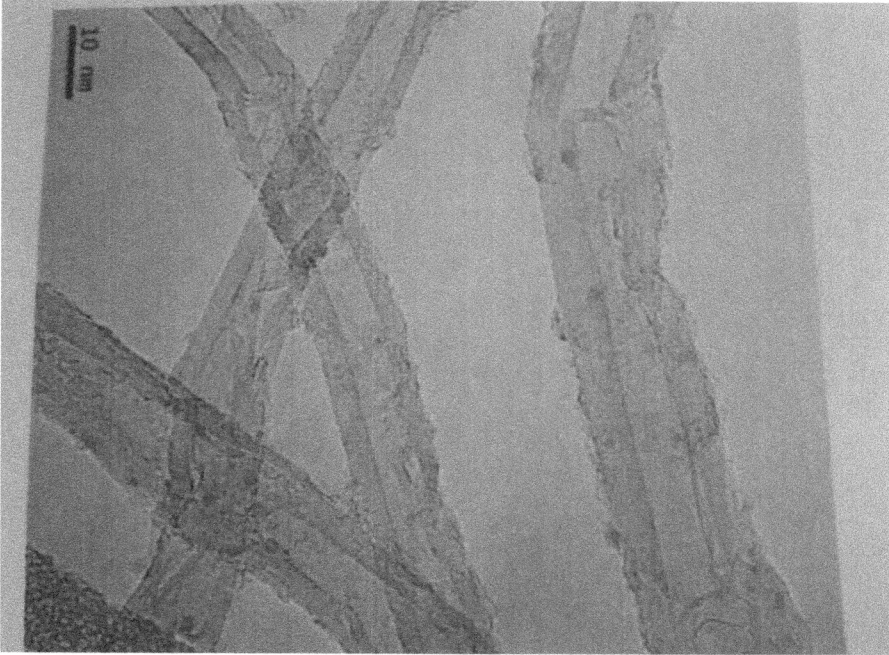

FIGURE 4.18   TEM of MWCNT [41].

Zhengzhou Dongyao Material Co., LTD in China provides the MWCNT that is utilized as a nano filler. The MWCNT tube's outer diameter is 13 nm, its length is equivalent to 3–12 nm, its wall thickness is 4.1 nm, and its layer count ranges from 8 to 15. Figure 4.18 shows the MWCNT TEM as captured by the production business. 0.06 g/cm^3 is the bulk density. There are 200 m^2 of surface per g. MWCNTs have a purity level of more than 90%. Figure 4.18 displays the TEM of MWCNTs produced by the company [41].

Utilizing XRD, the nanocomposites' structure was investigated. Cu Ka lines ($k = 1.5406$) were utilized to create the X-ray diffraction patterns using a Philips PW1050 diffractometer. The diffractrograms were scanned at a rate of 2/min from 2.1 to 35 (2h). On PPc/organoclay nanocomposites and on organoclay particles, X-ray diffractrograms were recorded. The PPc–organoclay series' XRD patterns are displayed in Figure 4.19. According to Bragg's diffraction rule, which states that $2d\sin h = nk$, the organonanoclay exhibits a diffusion ultimate of 2h at 4.12, which is equivalent to an interlayer space of nanoclay (d-spacing) of 21.13.

The absence of a diffraction peak at 1 wt% of organoclay in the PPc polymer either indicates that the matrix has interacted with the organoclay's interlayer space and improved the unique organoclay arrangement above the level of 7.5 nm (in which Bragg's law cannot satisfy) or that the organoclay's nanolayers have been arbitrarily discrete within the PPc polymer. In light of this, it can be said that the organoclays in the PPc polymer at a concentration of 1 wt% either generated an well-ordered

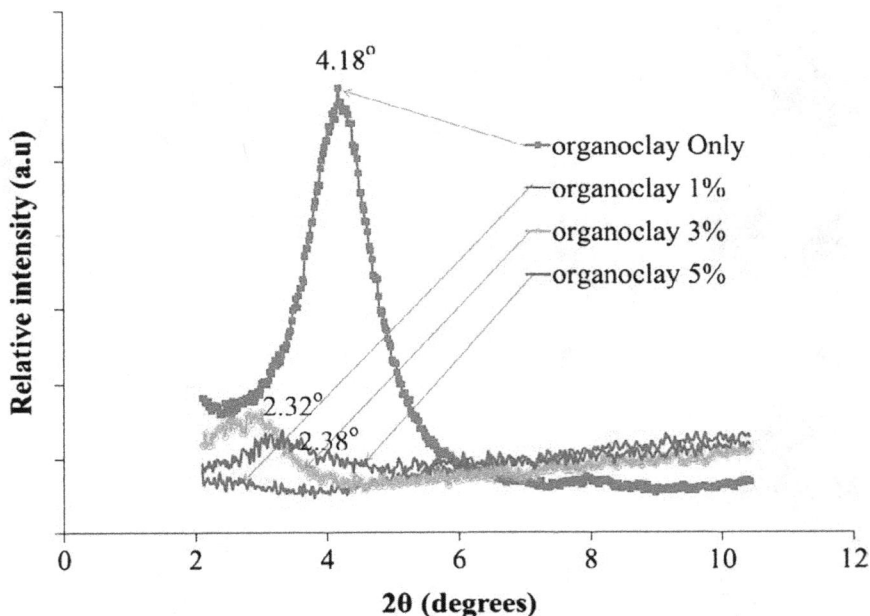

FIGURE 4.19   XRD pattern at 2.1–10.5_ of compatibilized PPc/organoclay composite [37].

exfoliated construction or an arbitrarily discrete clay exfoliated structure. They are at 3 and 5 wt% and above organoclay. According to XRD data (Figure 4.19) for organoclay concentrations of 3 and 5 wt%, the composites produced intercalated nanocomposites structures [37].

In a Perkin Elmer Pyris 1 machine, a thermogravimetric analysis (TGA) was conceded out using nitrogen (99.999%) and a heating rate of 10 C/min. To confirm the composite's tolerance to high temperatures, thermal gravimetric analysis was done. The TGA and DTG curves demonstrate the behavior of MWNT/BPDA-PDA polyimide from ambient temperature to 900°C, as illustrated in Figure 4.20. It took until 400°C for the composite to disintegrate. Only 18% of the composite's weight was lost at 900°C after a sluggish disintegration, demonstrating the material's outstanding thermal resilience at such high temperatures [38].

The mechanical characteristics of carbon fiber composites are significantly influenced by the interfacial qualities amid the carbon fiber (CF) and matrix. Carbon black (CB) was added to the surface of CFs by CVD in order to improve properties of fibers/epoxy composites without reducing the tensile strength of basic fibers (CVD). Using an atomic force microscope (AFM), the dispersal of CBs on the fiber surface and changes in the surface irregularity were examined.

AFM scans demonstrated the surface topographies of CFs in three dimensions and at the microscopic level more accurately. According to Figure 4.21a and d, the production procedure left the untreated CFs with a lot of relatively clean, shallow grooves that had a roughness of only 43.6 nm. Due to the homogeneous diffusion of

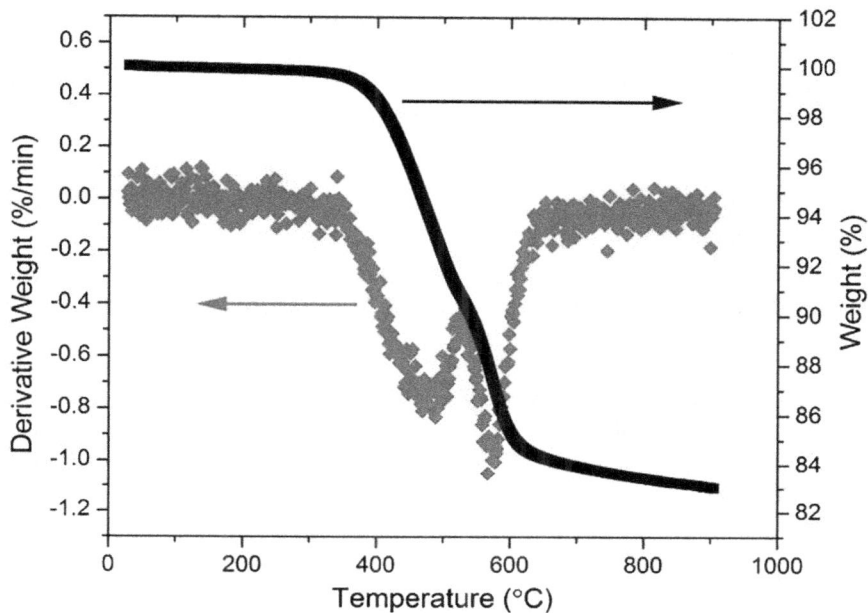

FIGURE 4.20 The TGA and DTG curves of MWNT/BPDA-PDA polyimide composite at 900°C [38].

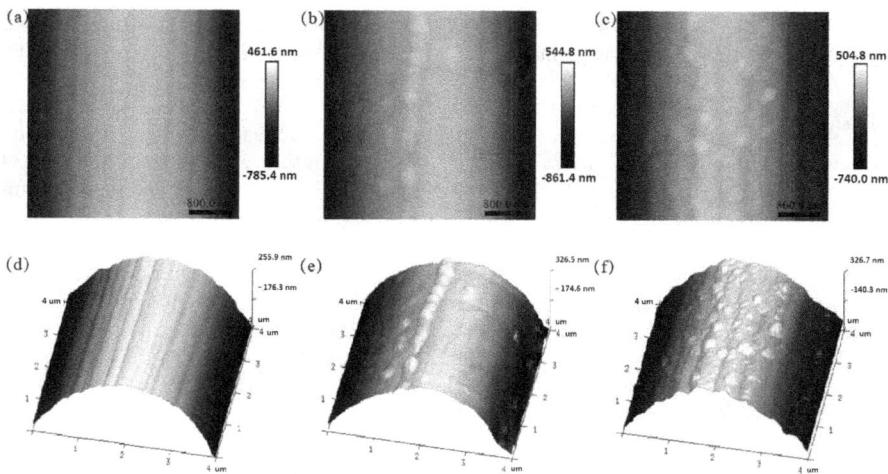

FIGURE 4.21 AFM images of CF surfaces: (a, d) untreated CF, (b, e) CF-5 min and (c, f) CF-10 min [42].

CBs over the surface, the CB-modified CFs (Figure 4.21b and e) develop significantly coarser (54.8 nm). Because CB accumulated and produced stuffing constructions with longer development times, the CF surface (Figure 4.21c and f) grew rougher (60.2 nm) than that of CF-5 min at the 10 minute mark [42].

4.4 CONCLUSIONS

Characterizations of various polymer–carbonecous filler composites have been studied in this chapter. Characterizations are very important aspects for determining the failure and fracture of the composite. SEM, TEM, XRD are some of the important characterizations.

1. This chapter mainly deals with SEM, TEM, and XRD characterizations on the fractured samples of polymer–carbonaceous composites.
2. Distributions of nanoparticles in the polymer and the bonds between the matrix and particulates have been thoroughly studied by using various characterization techniques.

REFERENCES

1. Friedrich K (2018) Polymer composites for tribological applications. *Adv Ind Eng Polym Res* 1:3–39. https://doi.org/10.1016/J.AIEPR.2018.05.001
2. Salahuddin B, Faisal SN, Baigh TA, et al. (2021) Carbonaceous materials coated carbon fibre reinforced polymer matrix composites. *Polymers (Basel)* 13. https://doi.org/10.3390/POLYM13162771
3. Zhang W, Zhang X, Liang M, Lu C (2008) Mechanochemical preparation of surface-acetylated cellulose powder to enhance mechanical properties of cellulose-filler-reinforced NR vulcanizates. *Compos Sci Technol* 12:2479–2484. https://doi.org/10.1016/J.COMPSCITECH.2008.05.005
4. Ojha S, Acharya SK, Raghavendra G (2015) Mechanical properties of natural carbon black reinforced polymer composites. *J Appl Polym Sci* 132. https://doi.org/10.1002/APP.41211
5. Abdul Khalil HPS, Noriman NZ, Ahmad MN, et al. (2007) The effect of biological studies of polyester composites filled carbon black and activated carbon from bamboo (Gigantochloa scortechinii). *Polym Compos* 28:6–14. https://doi.org/10.1002/PC.20239
6. Yan DX, Pang H, Li B, et al. (2015) Structured reduced graphene oxide/polymer composites for ultra-efficient electromagnetic interference shielding. *Adv Funct Mater* 25:559–566. https://doi.org/10.1002/ADFM.201403809
7. González C, Vilatela JJ, Molina-Aldareguía JM, et al. (2017) Structural composites for multifunctional applications: Current challenges and future trends. *Prog Mater Sci* 89:194–251. https://doi.org/10.1016/j.pmatsci.2017.04.005
8. Qin F, Brosseau C (2012) A review and analysis of microwave absorption in polymer composites filled with carbonaceous particles. *J Appl Phys* 111. https://doi.org/10.1063/1.3688435
9. Marturi N (2014) Vision and visual servoing for nanomanipulation and nanocharacterization in scanning electron microscope. Micro and nanotechnologies/Microelectronics. Université de FrancheComté, 2013. English.
10. Inkson BJ (2016) Scanning Electron Microscopy (SEM) and Transmission Electron Microscopy (TEM) for Materials Characterization. Elsevier Ltd.

11. Kumar, Challa SSR, ed. (2013) Transmission electron microscopy characterization of nanomaterials. Springer Science & Business Media.
12. Carter BA, Williams DB, Carter CB, Williams, DB (1996) Transmission electron microscopy: a textbook for materials science. Diffraction. II (Vol. 2). Springer Science & Business Media.
13. Ayache J, Beaunier L, Boumendil J, Ehret G, Laub D (2010) Sample preparation handbook for transmission electron microscopy: techniques (Vol. 2). Springer Science & Business Media.
14. Sherwood PMA (2001) Carbons and graphites: Surface properties of. *Encycl Mater Sci Technol*:985–995. https://doi.org/10.1016/b0-08-043152-6/00183–2
15. Wu H, Huang X, Qian L (2018) Recent progress on the metacompoistes with carbonaceous fillers. *Eng Sci* 2:17–25. https://doi.org/10.30919/es8d656
16. Kausar A (2019) Review of fundamentals and applications of polyester nanocomposites filled with carbonaceous nanofillers. *J Plast Film Sheeting* 35:22–44. https://doi.org/10.1177/8756087918783827
17. Kovářík T, Bělský P, Rieger D, et al. (2020) Particle size analysis and characterization of nanodiamond dispersions in water and dimethylformamide by various scattering and diffraction methods. *J Nanoparticle Res* 22:1–17. https://doi.org/10.1007/s11051-020-4755-3
18. Puzyr AP, Bondar VS, Bukayemsky AA, et al. (2005) Physical and chemical properties of modified nanodiamonds. *Synth Prop Appl Ultrananocrystalline Diam*:261–270. https://doi.org/10.1007/1-4020-3322-2_20
19. Richter A, Ries R, Smith R, et al. (2000) Nanoindentation of diamond, graphite and fullerene films. *Diam Relat Mater* 9:170–184. https://doi.org/10.1016/S0925–9635(00)00188–6
20. Liu Z, Liang XJ (2012) Nano-carbons as theranostics. *Theranostics* 2:235–237. https://doi.org/10.7150/thno.4156
21. He Y, Zhang GL (2009) Historical record of black carbon in urban soils and its environmental implications. *Environ Pollut* 157:2684–2688. https://doi.org/10.1016/J.ENVPOL.2009.05.019
22. Ojha S, Acharya SK, Raghavendra G (2016) A novel approach to utilize waste carbon as reinforcement in thermoset composite. *Proc Inst Mech Eng Part E J Process Mech Eng* 230:263–273. https://doi.org/10.1177/0954408914547118
23. Iqbal M, Dinh DK, Abbas Q, et al. (2019) Controlled surface wettability by plasma polymer surface modification. *Surfaces* 2:349–371. https://doi.org/10.3390/SURFACES2020026
24. Speranza G (2019) The role of functionalization in the applications of carbon materials: An overview. *C — J Carbon Res* 5:84. https://doi.org/10.3390/c5040084
25. Afzal A, Nawab Y (2021) Polymer composites. *Compos Solut Ballist*:139–152. https://doi.org/10.1016/B978–0–12–821984–3.00003–6
26. Ananthapadmanaban D (2020) Summary of some selected characterization methods of geopolymers. *Geopolymers Other Geosynth*:1–16. https://doi.org/10.5772/intechopen.82208
27. Panwar AS, Singh A, Sehgal S (2020) Material characterization techniques in engineering applications: A review. *Mater Today Proc* 28:1932–1937. https://doi.org/10.1016/J.MATPR.2020.05.337
28. Panchal M, Minugu OP, Gujjala R, et al. (2022) Study of environmental behavior and its effect on solid particle erosion behavior of hierarchical porous activated carbon-epoxy composite. *Polym Compos* 43:2276–2287. https://doi.org/10.1002/pc.26539
29. Om Prakash M, Gujjala R, Panchal M, Ojha S (2020) Mechanical characterization of arhar biomass based porous nano activated carbon polymer composites. *Polym Compos* 41:3113–3123. https://doi.org/10.1002/pc.25602

30. Ojha S, Acharya SK, Gujjala R (2014) Characterization and wear behavior of carbon black filled polymer composites. *Procedia Mater Sci* 6:468–475. https://doi.org/10.1016/j.mspro.2014.07.060

31. Prakash MO, Raghavendra G, Ojha S, et al. (2021) Investigation of tribological properties of biomass developed porous nano activated carbon composites. *Wear* 466–467:203523. https://doi.org/10.1016/j.wear.2020.203523

32. Qin W, Chen C, Zhou J, Meng J (2020) Synergistic effects of graphene/carbon nanotubes hybrid coating on the interfacial and mechanical properties of fiber composites. *Materials (Basel)* 13. https://doi.org/10.3390/ma13061457

33. Xu H, Zhang X, Liu D, et al. (2016) Cyclomatrix-type polyphosphazene coating : Improving interfacial property of carbon fi ber / epoxy composites and preserving fi ber tensile strength. *Compos Part B* 93:244–251. https://doi.org/10.1016/j.compositesb.2016.03.033

34. Kim KW, Kim DK, Kim BS, et al. (2017) Cure behaviors and mechanical properties of carbon fiber-reinforced nylon6/epoxy blended matrix composites. *Compos Part B Eng* 112:15–21. https://doi.org/10.1016/j.compositesb.2016.12.009

35. Kausar A (2020) Nanocarbon and macrocarbonaceous filler–reinforced epoxy/polyamide: A review. *J Thermoplast Compos Mater.* https://doi.org/10.1177/0892705720930810

36. Zhao YH, Zhang YF, Bai SL, Yuan XW (2016) Carbon fibre/graphene foam/polymer composites with enhanced mechanical and thermal properties. *Compos Part B Eng* 94:102–108. https://doi.org/10.1016/j.compositesb.2016.03.056

37. Gabr MH, Okumura W, Ueda H, et al. (2015) Mechanical and thermal properties of carbon fiber/polypropylene composite filled with nano-clay. *Compos Part B Eng* 69:94–100. https://doi.org/10.1016/j.compositesb.2014.09.033

38. Jiang Q, Wang X, Zhu Y, et al. (2014) Mechanical, electrical and thermal properties of aligned carbon nanotube/polyimide composites. *Compos Part B Eng* 56:408–412. https://doi.org/10.1016/j.compositesb.2013.08.064

39. Song SH, Park KH, Kim BH, et al. (2013) Enhanced thermal conductivity of epoxy-graphene composites by using non-oxidized graphene flakes with non-covalent functionalization. *Adv Mater* 25:732–737. https://doi.org/10.1002/adma.201202736

40. Neitzel I, Mochalin VN, Niu J, et al. (2012) Maximizing Young's modulus of aminated nanodiamond-epoxy composites measured in compression. *Polymer (Guildf)* 53:5965–5971. https://doi.org/10.1016/j.polymer.2012.10.037

41. Al Shaabania YA (2019) Wear and friction properties of epoxy-Polyamide blend nanocomposites reinforced by MWCNTs. *Energy Procedia* 157:1561–1567. https://doi.org/10.1016/j.egypro.2018.11.322

42. Dong J, Jia C, Wang M, et al. (2017) Improved mechanical properties of carbon fiber-reinforced epoxy composites by growing carbon black on carbon fiber surface. *Compos Sci Technol* 149:75–80. https://doi.org/10.1016/j.compscitech.2017.06.002

5 Advancements in Materials and Technologies for Wastewater Treatment

Pratibha Mishra, Arunesh K. Mishra,
Megha Das, and Ratnesh Das
Dr. Harisingh Gour Central University

CONTENTS

5.1   INTRODUCTION

The quantity of usable water is under intense demand due to the growing human population, lifestyle, and activities. At the same time, the quality of the water is also declining. Agricultural and industrial activity result in the generation of wastewater. Due to the constantly shifting needs and wants, new types of businesses are developing, which result in a large number of new contaminants being dumped into wastewater, necessitating the development of cutting-edge treatment methods. Advanced techniques are required to regulate wastewater flows that are always changing, and there is always a relationship between water and energy. Although it is impossible to entirely stop the formation of wastewater because no industry is 100% efficient, it is possible to create new and improved wastewater treatment and reuse techniques in order to meet water demand. Additionally, water reuse has a huge potential for refilling inventories of water resources that are currently overstretched. Wastewater treatment and reuse are crucial, since they are related to public health. Where contact, inhalation, or ingestion of a material or microbiological component of health concern occurs, the presence of pathogenic organisms and contaminated compounds

DOI: 10.1201/9781003328094-5

in wastewater poses the risk of detrimental health impacts. Acceptable levels have already been established for the effects of a number of parameters (including pH, temperature, colour, and particle size) and chemical elements (including cations, anions, and heavy metals) on human health. However, if industrial emissions make up a sizable fraction of the wastewater, it is necessary to investigate the impact of organic constituents in treated water used for non-potable purposes [1]. Water reuse has a huge potential to supplement the portfolios of water resources that are currently under stress, but using and disposing of biosolids is still difficult, especially in densely populated areas. The biggest issue continues to be public perception in both biosolids applications to land and water reuse. The challenge of perception may be far more difficult to overcome, even when improved technologies can help to reduce energy footprint and increase reliability. It can be particularly challenging to educate the public about emerging pollutants like medications and microbes resistant to antibiotics. Public worries about the safety of water reuse are greatly influenced by both historical and more current instances of disease carried by water (such as anaemia, typhoid, malaria, and dengue fever, respectively). A deeper understanding of how developed reused water compares to extant source waters can be very persuasive, even though modern technologies like sensors, membranes, and enhanced oxidation might ease perception [2]. Wastewater treatment protects the ecosystem by releasing less contamination, uses sustainable resources, provides the opportunity for unused products to be recycled, and handles residual wastes in a more biologically acceptable way. These factors make wastewater treatment an environmentally friendly process. The characteristics and types of contaminants present in the water, as well as the intended use of the treated water, determine the treatment method selection. Desalination of water, water filtration, wastewater treatment, groundwater treatment, and other nano-remediation processes are applications for purification and environmental clean-up. The ability to access clean, uncontaminated water is crucial for human survival. Human health, living things, and the ecosystem are all seriously threatened by the rising concentrations of dangerous toxins that are released into the environment as a result of industrial waste. The water sources used to provide water throughout recorded history were not always clean, thus water was treated in some way to enhance flavour, clarity, and smell or to get rid of microorganisms that cause sickness [3]. Water contamination with radionucleides such caesium, strontium, lanthanides, and actinides is currently the main issue. The operation of nuclear power reactors, research institutions, and the usage of radioisotopes in industry and diagnostic medicine are the principal sources of these radionuclides. A growing concern to civilian populations is radionuclide pollution of drinking water [4]. This need for water can be met in a way that is economical, effective, and diverse by using nanocomposites as potential adsorbent materials in the nano range (Figure 5.1).

The aquatic environment and recreational activities are in danger because of this type of ability to manage dye and heavy metal contamination of wastewater. In order to do this, low-cost, environmentally friendly monitoring and eradication techniques must be developed. One of the most crucial requirements for growth, boosting the economy and maintaining health is wastewater and drinking water quality, and heavy metal adsorption on nanomaterials appears to be a potential method for removing them from contaminated water [5].

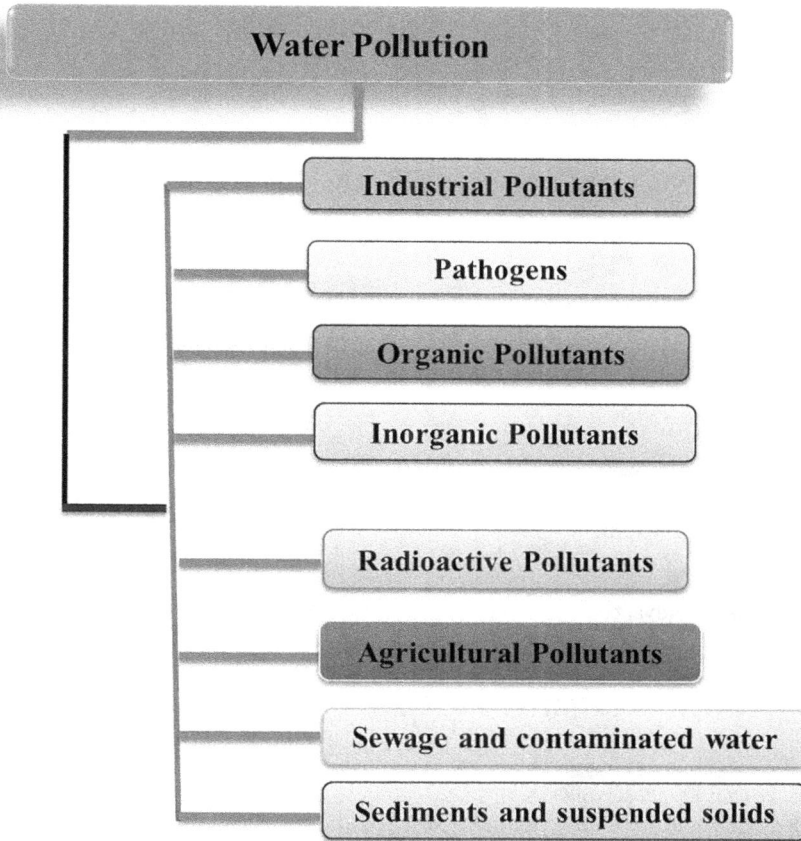

FIGURE 5.1 Different types of water pollutants.

When their tolerance thresholds are exceeded, heavy metals have negative effects on human health, physiology, and other biological systems. Water contains heavy metals from a number of industrial operations. These pollutants are dangerous to human health and natural ecosystems even at low doses [6]. Metal ion removal techniques have been used and improved over time, including chemical precipitation [7], ion exchange [8], reverse osmosis [9], electrodialysis [10], adsorption [11], membrane filtering [12], and flotation and sorption [13]. Both organic and mineral materials have been studied due to their ease of usage and affordability [14]. Heavy metal-induced wastewater contamination puts the aquatic ecosystem and recreational activities at danger. This calls for the creation of inexpensive and environmentally friendly monitoring and eradication methods. The removal of heavy metals from contaminated water by adsorption on nanomaterials appears to be a promising method [15]. For the elimination of these metal ions, numerous adsorbents have been reported, including natural and modified zeolites [16], bentonite [17], titanium oxide-clay [18], and non-modified and modified CNTs (Figure 5.2) [19].

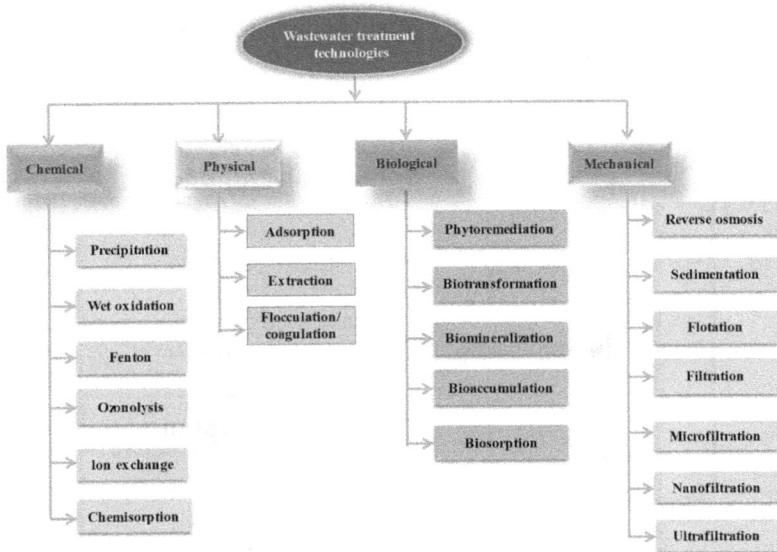

FIGURE 5.2   Advanced treatment technology for waste water.

5.2   WATER TREATMENT PROCESSES IN SEWAGE AND CONTAMINATED WATER

Processes for primary, secondary, and tertiary treatments are typically found in wastewater treatment facilities. Technologies from the physical, chemical, biological, and mechanical realms are frequently used in sewage treatment operations. Wastewater treatment techniques use chemicals that are intended to affect things through chemical reactions. They are always used in conjunction with physical and biological techniques. Since chemical processes are cumulative, they have a disadvantage over physical ones. In other words, wastewater typically has more dissolved components. This is a crucial factor to take into account if the effluent is to be recycled. The chemical processes include ion exchange, precipitation, wet oxidation, ozonolysis, and chemisorption. One of the earliest wastewater treatment technologies employed was physical methods, which remove impurities using physical forces. The majority of wastewater treatment process flow systems still employ them. When water is seriously polluted, several techniques are frequently used. Adsorption, extraction, and flocculation/coagulation are the three physical wastewater treatment techniques that are most frequently used. A wastewater treatment strategy must include biological water treatment technologies, because they are used to create safe drinking water. The strategies used for this are aerobic, anaerobic, and bioremediation processes. There are several biological treatment approaches, such as phytoremediation, biotransformation, biomineralisation, biosorption, and bioaccumulation. Technologies for mechanical water treatment that are currently being developed include reverse osmosis, sedimentation, flotation, filtration, microfiltration, nanofiltration, and ultrafiltration [20].

5.3 BIOPOLYMER-BASED NANOCOMPOSITES FOR HEAVY METAL REMOVAL

Heavy metal water contamination has developed into a global environmental risk, necessitating the treatment and disposal of contaminated industrial wastewater. There are numerous biopolymer-based nanocomposites with the capacity to remove heavy metals. Significant problems with heavy metal contamination exist in both developed and developing nations. It has always been difficult to select an effective adsorbent for extracting contaminants from an aqueous media. Investigations into the adsorption of heavy metals and subsequent release from reduced adsorbents are ongoing [21]. As prospective replacements for current, expensive heavy metal adsorbents, biopolymer-based nanocomposites are being researched in the field. Biopolymer-based nanocomposites have attracted the attention of researchers and decision-makers due to their distinctive qualities, such as their abundance, cost effectiveness, outstanding adsorption capacity, biocompatibility, biodegradability, and ease of structural modification [22,23]. One of the most popular nanocomposites products ever created for heavy metal removal from wastewater uses nanoparticles as an integrated material in cellulose, alginate, and chitosan nano-based biopolymer composites. High selectivity for heavy metals has been demonstrated for biopolymer-based nanocomposites in the presence of other heavy metals [24]. With the development of biopolymer-based nanocomposites, there is now a great deal of opportunity for generating different kinds of adsorbents and evaluating their applicability to wastewater treatment. The presence of harmful metal ions in the environment is concerning due to both acute and long-term toxicity [25]. Hazardous metal ions are present in water and are produced through the production of pesticides, mining, paper, textiles, medicines, and fossil fuels. Industrial effluents include large concentrations of metal ions as a result, constituting a serious environmental risk [26,27]. Chemicals, the environment, microbial assaults, and Hg, Pb, and Cd are all resistant to them. Standard biological treatments struggle to eliminate dangerous metal ions from effluents as a result of these properties. In the textile and chemical industries, for instance, Cr, Cu, Pb, Cd, Hg, and Ca are frequently utilised [10,28]. The properties of toxicity, mutagenicity, and carcinogenicity are all well recognised. Finding efficient methods to remove dangerous metals from wastewater is essential (Figure 5.3).

The removal of heavy metals from contaminated water by adsorption on nanomaterials appears to be a promising method [14].

5.4 REMOVAL OF TOXIC METAL IONS BY A VARIETY OF ADSORBENTS

Human health depends on clean water. Nanoscale science and engineering advances are creating new prospects for developing more cost-effective and environmentally friendly water purification processes. Because of their small size, nanomaterials have substantially bigger surface areas than bulk materials and display unique characteristics. Researchers are utilising the unique features of nanomaterials to build more effective sorbents and improve metal ion removal. Carbon nanotubes, zeolites, and other nanoparticles have been studied in the recent years for their capability to eliminate metal ions [29]. Heavy metals such as copper (Cu), zinc (Zn), manganese (Mn), iron

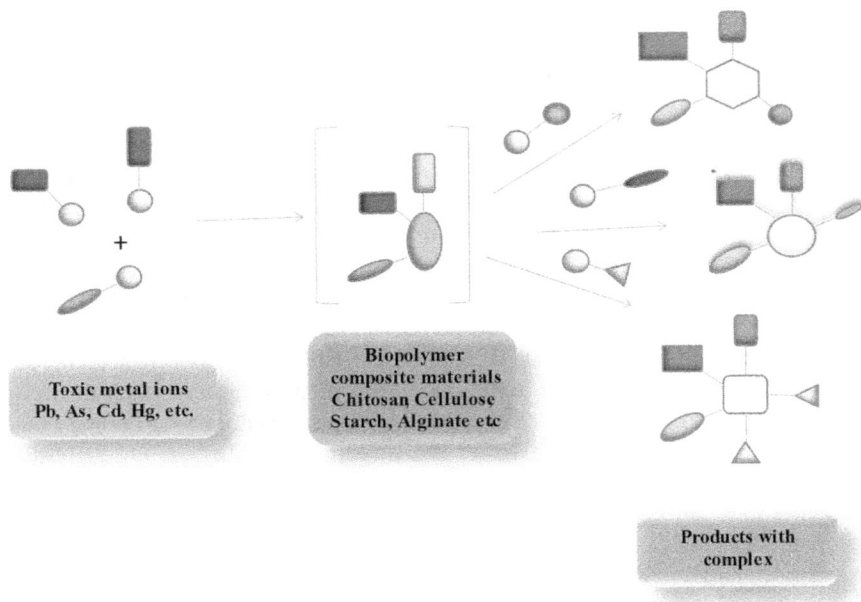

FIGURE 5.3   Removal of heavy metals from contaminated water by biopolymers.

(Fe), and cobalt (Co) play critical roles in biochemical processes in the human body. Excessive exposure to metal ions, on the other hand, can have harmful implications. Other heavy metals such as arsenic (As), cadmium (Cd), lead (Pb), mercury (Hg), and chromium (Cr) are harmful even at low concentrations (parts per billion, ppb), since they are non-degradable and can bioaccumulate in the basic systems of the human body [30].

5.5   REMOVAL OF COPPER, CADMIUM, LEAD, AND FLUORIDE FROM WASTEWATER BY ADSORBENT

Copper is a vital metal for plants, humans, and other animals on land and in water. In the human body, excessive Cu^{2+} ion uptake causes depression, anorexia, premenstrual syndrome, learning difficulties, liver damage, and allergies. Cu^{2+} ions are released into the aquatic environment through paints, electroplating, dyes, fossil fuel burning, the iron and steel industries, and pesticides [31]. In humans, low quantities of Cd^{2+} ions cause hypertension, pulmonary illness, hepatic damage, osteomalacia, and cardiac failure. Liquid waste from the alloying, metal polishing, mining, and ceramic industries releases them into the natural aquatic system [32]. Lead (Pb^{2+}) ions are the most harmful and common heavy metal ions, and they are released into the environment through leaded gasoline, smelting, battery recycling, and lead-containing pipes, among other things. It gradually deteriorates the central and peripheral nervous systems, resulting in renal, joint, and reproductive issues. It is also a carcinogen with the potential to

cause mental retardation and behavioural abnormalities in children [33]. Process effluents containing Cu^{2+}, Cd^{2+}, and Pb^{2+} are encountered in the chemical process industries. Massive volumes of heavy metals released into the environment have generated
a plethora of problems. Polypyrrole/ZnO (Polypy/ZnO) nanocomposite exhibits highly
enhanced adsorption of divalent copper, lead, and cadmium ions from aqueous solutions
[34]. N-carboxymethyl chitosan hydrogel (NCS-hydrogel) is utilised to treat Cu^{2+}, Cd^{2+},
and Pb^{2+} metal ions pollution in waste water, as well as the parameters that influence
adsorption efficacy [35]. Acid-pretreated alkali lignin acts as a hierarchical pore-form-
ing agent, boosting the simultaneous adsorption capacities of Pb^{2+}, Cu^{2+}, and Cd^{2+} metal
ions from wastewater [36]. Min Wang et al. created a unique calcium alginate–disodium
ethylenediaminetetraacetate dihydrate hybrid aerogel (Alg-EDTA) adsorbent with high
potential for heavy metal ions such as Cd^{2+}, Pb^{2+}, Cu^{2+}, Cr^{3+}, and Co^{2+} [37]. The focus
is on developing sensor materials for detecting heavy metal ions that are rapid, simple,
and usable by non-experts. The chemical interactions of harmful metal ions such as
Pb^{2+}, Cu^{2+}, and Cd^{2+} with nanomaterials, as well as the effect of metal ion concentration
and medium pH on their interaction. Excessive fluoride concentrations in surface and
groundwater, the primary source of drinking water, are a big concern today. Fluoride
concentrations in drinking water greater than 1.5 mg/L are hazardous to human health.
As a result, removing fluoride from water is critical both scientifically and healthwise.
Recently, fluoride was eliminated from water using nanostructured materials with
distinct properties. Magnetic nanocomposite of functionalised polypyrrole (PPy) and
Fe_3O_4 is widely used in the removal of fluoride from aqueous solution [38].

5.6 CONCLUSION

The creation and treatment of wastewater has become an increasingly important
issue in the 21st century, as a result of increased urbanisation and industry. The long-
term sustainability of the ecosystem is ensured via wastewater treatment. Physical,
chemical, and biological (primary to tertiary treatment) technologies are among the
wastewater treatment alternatives used to address the issue of increasing environmental contamination. Using certain treatment methods could result in the production of secondary pollutants. Water resource management requires planning, activity,
design, storage, and operation in order to implement wastewater treatment solutions
effectively. The ability to create water of almost any quality has been made possible
by advancements in wastewater recycling. To lessen the threats to the environment
posed by diverse reuse applications, water recovery systems contain a number of
safety measures. Both the fundamental science underlying water treatment techniques and the innovation employed in the procedure have undergone continuous
improvement. To treat significant amounts of wastewater with a single treatment
technique, nevertheless, is challenging based on the currently used treatment procedures. To assure high-quality water, decrease chemical and biological pollutants, and
improve industrial production processes, enhanced or integrated wastewater treatment technologies are urgently needed. Integrated strategies appear to be promising solutions for effective wastewater remediation, since they may go beyond the
limitations of single treatment procedures. Unfortunately, the majority of effective
treatment methods are limited in scope and incapable of being used commercially.

ACKNOWLEDGEMENTS

Author is grateful to University Grants Commission, New Delhi, India and Department of Chemistry, Dr. Harisingh Gour Central University, Sagar for financial support.

REFERENCES

1. Ahuja S. Overview of global water challenges and solutions. *ACS Symp. Ser.* 1206, 1–25 (2015).
2. Angelakis A. N. & Snyder S. A. Wastewater treatment and reuse: Past, present, and future. *Water.* 7(9), 4887–4895 (2015).
3. Zinicovscaia I. & Cepoi L. Cyanobacteria for bioremediation of wastewaters. *Cyanobacteria for Bioremediation of Wastewaters* (2016). doi: 10.1007/978-3-319-26751-7.
4. Lytle D. A., Sorg T., Wang L. & Chen A. The accumulation of radioactive contaminants in drinking water distribution systems. *Water Res.* 50, 396–407 (2014).
5. Dil E.A., Ghaedi M. & Asfaram A. The performance of nanorods material as adsorbent for removal of azo dyes and heavy metal ions: Application of ultrasound wave, optimization and modeling. *Ultrason. Sonochem.* 34, 792–802 (2017).
6. Jaishankar M., Tseten T., Anbalagan N., Mathew B. B. & Beeregowda K. N. Toxicity, mechanism and health effects of some heavy metals. *Interdiscip. Toxicol.* 7, 60–72 (2014).
7. Bose P., Aparna Bose M. & Kumar S. Critical evaluation of treatment strategies involving adsorption and chelation for wastewater containing copper, zinc and cyanide. *Adv. Environ. Res.* 7, 179–195 (2002).
8. Bhattacharyya K. G. & Gupta S. Adsorption of a few heavy metals on natural and modified kaolinite and montmorillonite: A review. *Adv. Colloid Interface Sci.* 140, 114–131 (2008).
9. Qdais H. A. & Moussa H. Removal of heavy metals from wastewater by membrane processes: A comparative study. *Desalination.* 164, 105–110 (2004).
10. Barakat M. A. New trends in removing heavy metals from industrial wastewater. *Arab. J. Chem.* 4, 361–377 (2011).
11. Di Natale F., Lancia A., Molino A. & Musmarra D. Removal of chromium ions form aqueous solutions by adsorption on activated carbon and char. *J. Hazard. Mater.* 145, 381–390 (2007).
12. Borbély G. & Nagy E. Removal of zinc and nickel ions by complexation-membrane filtration process from industrial wastewater. *Desalination.* 240, 218–226 (2009).
13. Zamboulis D., Peleka E. N., Lazaridis N. K. & Matis K. A. Metal ion separation and recovery from environmental sources using various flotation and sorption techniques. *J. Chem. Technol. Biotechnol.* 86, 335–344 (2011).
14. Fu F. & Wang Q. Removal of heavy metal ions from wastewaters: A review. *J. Environ. Manage.* 92, 407–418 (2011).
15. Ihsanullah et al. Heavy metal removal from aqueous solution by advanced carbon nanotubes: Critical review of adsorption applications. *Sep. Purif. Technol.* 157, 141–161 (2016).
16. Shi J. et al. Preparation and application of modified zeolites as adsorbents in wastewater treatment. *Water Sci. Technol.* 2017, 621–635 (2017).
17. Pandey S. A comprehensive review on recent developments in bentonite-based materials used as adsorbents for wastewater treatment. *J. Mol. Liq.* 241, 1091–1113 (2017).

18. Aliou Guillaume P. L., Chelaru A. M., Visa M. & Lassine O. "Titanium oxide-clay" as adsorbent and photocatalysts for wastewater treatment. *J. Membr. Sci. Technol.* 08, 0–11 (2018).
19. Aslam M. M. A. et al. Functionalized carbon nanotubes (Cnts) for water and wastewater treatment: Preparation to application. *Sustainability.* 13, 1–54 (2021).
20. Ding G. K. C. Wastewater treatment and reuse-The future source of water supply. *Encycl. Sustain. Technol.* 2017, 43–52 (2017).
21. Qasem N. A. A., Mohammed R. H. & Lawal D. U. Removal of heavy metal ions from wastewater: A comprehensive and critical review. *npj Clean Water.* 4, 36 (2021).
22. Ahmad A., Mohd-Setapar S. H., Chuong C. S., Khatoon A., Wani W. A., Kumar R., Rafatullah M. Recent advances in new generation dye removal technologies: Novel search for approaches to reprocess wastewater. *RSC Adv.* 5, 30801–30818 (2015).
23. Rend´on-Villalobos R., Ortíz-S´anchez A., Tovar-S´anchez E., Flores-Huicochea E. The role of biopolymers in obtaining environmentally friendly materials. *Compos. Renew. Sustain. Mater.* 151 (2016).
24. Zia Z., Hartland A. & Mucalo M. R. Use of low-cost biopolymers and biopolymeric composite systems for heavy metal removal from water. *Int. J. Environ. Sci. Technol.* 17, 4389–4406 (2020).
25. Ghaedi M. & Mosallanejad N. Removal of heavy metal ions from polluted waters by using of low cost adsorbents. *J Chem Health Risks.* 3, 7–21 (2013).
26. Dixit R., Malaviya D., Pandiyan K., Singh U. B., Sahu A., Shukla R., Singh B. P., Rai J. P., Sharma P. K., Lade H. Bioremediation of heavy metals from soil and aquatic environment: An overview of principles and criteria of fundamental processes. *Sustainability.* 7, 2189–2212 (2015).
27. Bolan N., Kunhikrishnan A., Thangarajan R., Kumpiene J., Park J., Makino T., Kirkham M.B. & Scheckel K. Remediation of heavy metal(loid)s contaminated soils—To mobilize or to immobilize? *J. Hazard. Mater.* 266, 141–166 (2014).
28. Agarwal M. & Singh K. Heavy metal removal from wastewater using various adsorbents: A review. *J. Water Reuse Desal.* 7(4), 387–419 (2017).
29. Nagar A. & Pradeep T. Clean water through nanotechnology: Needs, gaps, and fulfillment. *ACS Nano.* 14, 6420–6435 (2020).
30. Zak S. Treatment of the processing wastewaters containing heavy metals with the method based on flotation. *Ecol. Chem. Eng. S.* 19, 433–438 (2012).
31. Joshi N. C. & Bahuguna V. Biosorption of copper (II) on to the waste leaves of kafal (myrica esculenta). *Rasayan J. Chem.* 11, 142–150 (2018).
32. Alluri H. K. et al. Biosorption: An eco-friendly alternative for heavy metal removal. *African J. Biotechnol.* 6, 2924–2931 (2007).
33. Dongre R. S. Lead: Toxicological profile, pollution aspects and remedial solutions. *Lead Chem.* (2020).
34. Joshi N. C. Utilization of polypyrrole/ZnO nanocomposite in the adsorptive removal of Cu^{2+}, Pb^{2+} and Cd^{2+} ions from wastewater. *Lett. Appl. NanoBioSci.* 10, 2339–2351 (2020).
35. Hao D. & Liang Y. Adsorption of Cu^{2+}, Cd^{2+} and Pb^{2+} in wastewater by modified chitosan hydrogel. *Environ. Technol. (United Kingdom).* 0, 1–18 (2020).
36. Liu M. et al. Simultaneous removal of Pb^{2+}, Cu^{2+} and Cd^{2+} ions from wastewater using hierarchical porous polyacrylic acid grafted with lignin. *J. Hazard. Mater.* 392, 122208 (2020).
37. Wang M., Wang Z., Zhou X. & Li S. Efficient removal of heavy metal ions in wastewater by using a novel alginate-EDTA hybrid aerogel. *Appl. Sci.* 9, 1–14 (2019).
38. Bhaumik M., Leswifi T. Y., Maity A., Srinivasu V. V. & Onyango M. S. Removal of fluoride from aqueous solution by polypyrrole/F3O4 magnetic nanocomposite. *J. Hazard. Mater.* 186, 150–159 (2011).

6 Polymer–Carbon Nanotubes-Based Composite for Removal of Pollutants in Wastewater

Arunesh K. Mishra, Pratibha Mishra, Anil K. Bahe, Atish Roy, Megha Das, and Ratnesh Das
Dr. Harisingh Gour Central University

CONTENTS

6.1 INTRODUCTION

The currently developing field of nanotechnology will lead to a number of advancements in the current technological revolution. Nanotechnology contributes significantly to the environment and energy, in addition to participating in all fields. According to scientists, current technologies have shown to be very helpful in detecting, measuring, and manipulating materials where atoms and molecules are measured at the nanoscale level, with lengths ranging from 1 to 100 billionths of a meter. Due to our ability to manipulate atoms and create molecules on exceedingly small dimensions, there is a lot of opportunity for tackling both recurring and new poisons in the water [1]. A unique class of materials, nanocomposites have a variety of uses, and special physical and chemical characteristics. Nanocomposites can exhibit unique properties when parent constituent traits are successfully combined into a single substance. The aquatic environment and recreational activities are in danger because of this type of ability to manage dye and heavy metal contamination of wastewater. In order to do this, low-cost, environmentally friendly monitoring and eradication techniques must be developed. One of the most crucial requirements for growth,

DOI: 10.1201/9781003328094-6

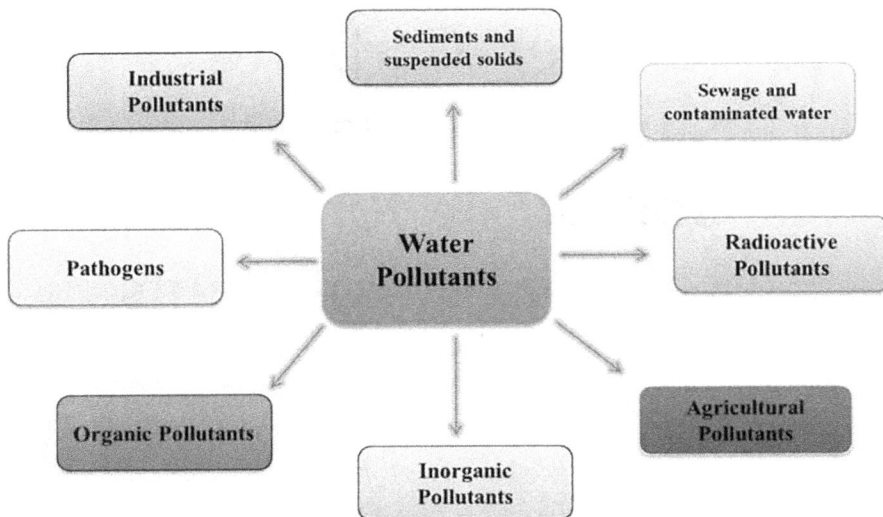

FIGURE 6.1   Different types of water pollutants.

boosting the economy and maintaining health, is wastewater and drinking water quality, and heavy metal adsorption on nanomaterials appears to be a potential method for removing them from contaminated water. This is the revolution of nanoscience and nanotechnology [2–8]. Since the chemical and physical characteristics of materials, such as color, magnetism, and electrical conductivity, may change at this incredibly small scale, completely new mechanisms based on molecule self-assembly can be developed from conventional mechanisms (Figure 6.1) [9,10].

Human survival depends on having access to pure, untainted water [11,12]. The elevated amounts of hazardous chemicals released into the environment as a result of industrial waste [13] provide a major threat to human health [14], the health of other living creatures, and the ecosystem. The water used for delivery was not always clean, so it underwent some sort of treatment to make it smell, taste, and look better or to get rid of bacteria that could cause illness [10]. This water-related need can be resolved by using nanocomposites as prospective adsorbent materials in the nano range in a way that is economical, effective, and flexible [15]. The primary issue nowadays is the radioactive contamination of drinking water, specifically actinides, cesium, strontium, lanthanides, and strontium. The usage of radioisotopes in industry, research labs, and diagnostic medicine are the primary sources of these radionuclides, together with the operation of nuclear power plants. Radionuclide pollution of drinking water is a growing concern to civilian populations [16]. Approximately 70% of the surface of the world is covered in water, with glaciers making up 1.73% of this total. Of this water, 97.5% is found in seas and oceans. On the Earth's surface, there is just 0.77% freshwater that can be used for agriculture and human use [17].

Some of the physicochemical and biological treatment techniques utilized in waste water purification operations include chemical precipitation, coagulation,

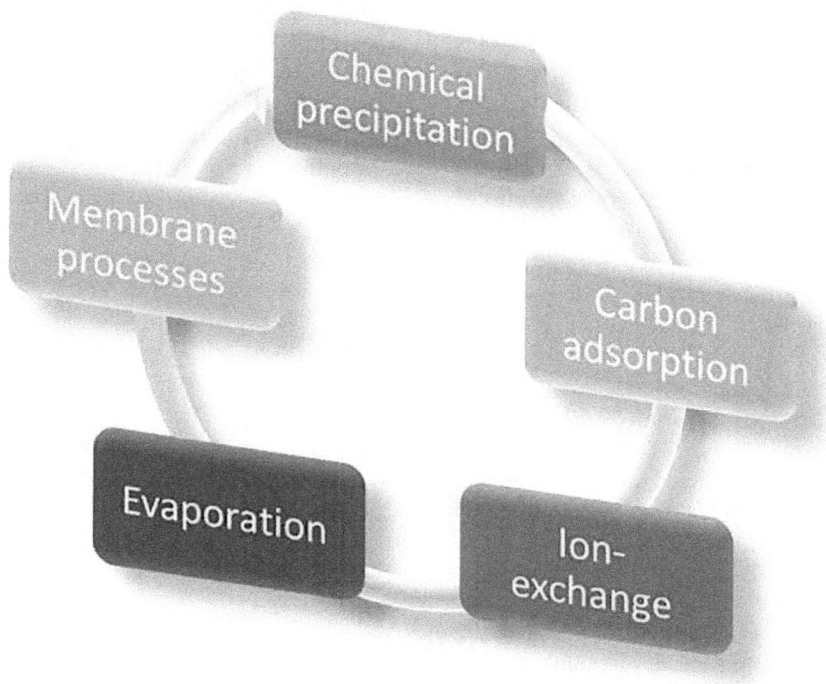

FIGURE 6.2 Conventional methods for wastewater treatments.

ion exchange, chemical oxidation, reduction, flocculation, reverse osmosis, ultrafiltration, electrodialysis membrane separation, and aerobic, anoxic, or anaerobic oxidation methods (Figure 6.2) [18–21].

6.2 POLYMER-CARBON NANOTUBES-BASED COMPOSITES IN SEWAGE AND CONTAMINATED WATER

Carbon nanotubes are made of extremely thin graphite sheets that have been rolled into tubes that can be hundreds of micrometers (microns) long and a few nanometers in diameter. There are many applications for carbon nanotubes in nanocomposite materials, including the filtration and remediation of wastewater [4]. When it comes to waste water applications, carbon nanotubes (CNT) are a unique nanotechnological component that is still in its infancy. Iijima Sumio, a Japanese scientist, established the groundwork for the current, continuous study of CNTs in 1991, continuing this nanotechnological advancement. CNTs can be produced using a variety of methods as either single-walled carbon nanotubes (SWCNTs), which are made up of a single cylindrical layer, or multi-walled carbon nanotubes (MWCNTs), which are composed of overlapping graphene sheets. MWCNTs can be functionalized for effective site-specific targeting. CNTs are appealing options for water filtration due to their versatility in manipulation and covalent or non-covalent functionalization [22]. Water resources can become contaminated by a variety of processes, including metallurgy, mining, tanning, chemical production, fossil fuel refining, battery manufacturing, and plastics production, which uses metal compounds, especially as heat stabilizers, among other things. While traditional pollutants have yet to find a successful way to overcome their issues, there is also a growth in pollutants that is currently forming. Providing everyone with access to clean water and protecting the environment is hampered by the removal of toxic metals [23,24]. The neurological system is harmed, kidney disease, psychological retardation, cancer, and anemia are all brought on by contaminated metal-polluted water, among other problems [25–27]. When carbon nanotubes (CNTs), a unique alloform of the carbon family, were first found in 1991 [22], they displayed extraordinary physical, chemical [28–30], mechanical [29–34], electrical, and mechanical features [34]. This is the consequence of promising future applications in various fields, including applications in nanoelectronics [35], microelectronic devices [36], field emissions [37,38], catalyst support [39,40], chemical sensors [41,42], and strengthening applications for composite materials [43]. Carbon nanotubes are useful candidates for kinetic adsorption due to their large, specific region, superior thermal and chemical stability, and simple, mass manufacture [44]. Activated carbon cannot compare to carbon nanotubes in terms of adsorption capability (AC). This is due to the surface area's expansion, which encourages considerable contact between carbon nanotubes and dioxins [45]. These are potentially heavy-weight carbon nanotubes that are suitable for removing heavier metal ions, including zinc (Zn) [46,47], cadmium (Cd) [48–50], lead (Pb) [51,52], nickel (Ni) [53–55], Cu [56,57], and fluoride [58,59] and radioactive nuclides (Figure 6.3) [60–63].

Carbon nanotubes are also very interested in gas adsorption and have enormous potential. Studies have been detailed in the carbon nanotubes as the adsorption of gases, including ammonia [64], nitrogen and methane [65,66], hydrogen [67–70], ozone [71], and carbon monoxide and carbon dioxide [72], 1,2-dichlorobenzene [73,74],

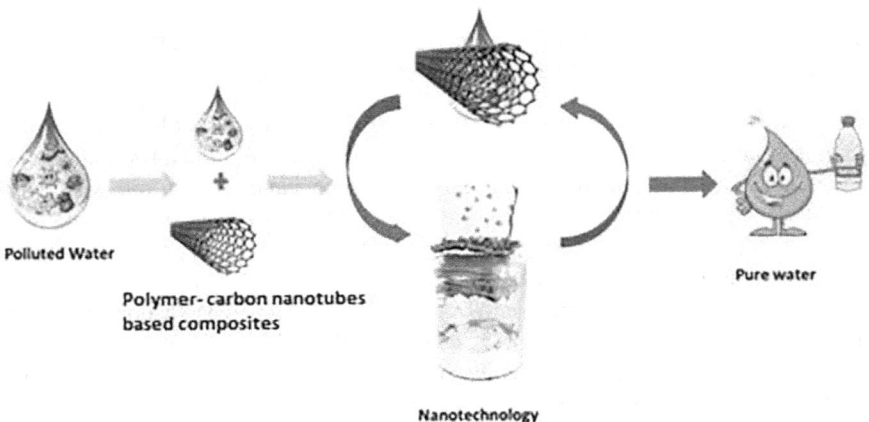

FIGURE 6.3 Advanced purification methods for wastewater treatments.

and dioxin [45,75]. Another important topic for discussion is the use of carbon nanotubes for the adsorption of single or binary/tertiary wastewater heavy metal separation. Therefore, the most promising solutions for heavy metal removal and separation processes from wastewater treatment are carbon nanotubes. The carbon nanotubes have a significant capacity for adsorption because of their porosity, surface area, and wide range of functional surface groups. The attraction, precipitation, and chemical interaction between metal ions and carbon nanotube surface function groups is what leads to the extremely complex integration processes for metal ions with carbon nanotubes [76]. Carbon nanotubes play a vital role in the removal from water of some organic salts, poisonous pigments, and heavy metals. On the other hand, a more extraordinary adsorption ability toward non-polar molecules, such as polycyclic aromatic hydrocarbons, is seen in the unfunctioning of carbon nanotubes [77,78]. Regeneration research reported adsorption and desorption of Ni^{2+} in carbon nanotubes carried out by Lu et al. (2008). Still, the amount of granular activated carbon (GAC) dropped markedly after some cycles [79]. The GAC's porous structure makes it more challenging to desorb Ni^{2+}, since the ions need to travel from the interior area to the exterior surface of the pores. This may be explained. The current technology uses carbon nanotubes as nanofilters to minimize the particles in wastewater, and be a sorbent for organic and inorganic pollutants [80–82]. Similar to sorbents, a unique carbon nanotube filter selectivity may be handled by attaching various functions to the pores [83]. Carbon nanotubes have shown exceptional performance in water transportation, despite their hydrophobic properties. The hydrophobic characteristic of pores of carbon nanotubes indicates weak interactions with water molecules, allowing rapid, almost unripened water circulation molecular dynamic simulations [84].

According to Hummer et al. (2001), the frictionless water flow is due to nanoscale confinement, which causes a narrowing of the interaction energy distribution and decreases the interaction with water [85]. In addition, the recent filtration investigations utilizing carbon nanotubes have shown the capacity of nanofilters' CNT for the removal by treatment in wastewater pathogens from the surface of carbon nanotubes, such as protozoa, bacteria, and viruses, using a deep filtration mechanism pathogen [86]. The efficient technique to eliminate *Escherichia coli* germs using SWCNT filters has been described by Brady-Estévez et al. (2008). The nanotube bundles inside the SWCNT could entirely catch and hold *E. coli* cells as a microporous membrane based on poly vinylidene fluoride (PVDF). A modification including the immobilization of SWCNTs on a microporous ceramic filter was also proposed to improve the filter's durability, reusability, and heat resistance without compromising its performance [87]. In another work, Mostafavi et al. (2009) used a spray pyrolysis technique to create an adjustable nanoscale porosity carbon nanotubes-based filter with the maximum effectiveness in MS2 virus elimination at a pressure of 8–11 bar. Carbon nanotubes use in wastewater treatment is not limited to filtration and sorbent; numerous researches discovered that carbon nanotubes had excellent antibacterial capabilities [88]. Because of this characteristic, carbon nanotubes can replace chemical disinfectants as a novel and effective method of controlling microbial infections. Because carbon nanotubes are not powerful oxidants and are generally inert in water, using them in water disinfection treatment prevents the development of hazardous disinfection by-products (DBPs), such as trihalomethanes, halo acetic acids, and aldehydes [7,89–92]. The recent times have seen the emergence of membrane-based filtering systems as prospective substitutes for waste water purification applications. The best potential polymers include polysulfone, polyamides, cellulose nitrate, and polyethersulfone due to their easy, cost-effective manufacturing, high mechanical strength, and biocompatibility. The major drawbacks of these membranes, nevertheless, are low throughput and fouling because of their extremely small pores, bacterial contamination, and pore clogging caused by the adsorption of inorganic and organic contaminants. Due to their potent antimicrobial properties, variable surface characteristics, and specific strength, carbon nanotubes have become an attractive filler candidate for creating hybrid materials with enhanced corrosion inhibitor features and mechanical characteristics in this context. The multi-walled carbon nanotube/polyamide nanocomposite membrane created by Shawky et al. [93] demonstrated great mechanical strength, exceptionally high salt refusal capability, and good permeability. Excellent tensile stability and high salt refusal capability are primarily because of the ongoing connectivity between the functionally rigid aromatic polyamide matrix and carbon nanotubes, although at the consequence of slightly decreased permeability. A polysulfone/MWCNT hybrid membrane has been created in this way by Choi et al. [94] via a phase inversion process. The MWCNTs that have had their surfaces altered give the membrane conductivity and hydrophilicity. The scientists claimed that increasing the CNT loading up to 1.5 wt% caused the membrane's pores to enlarge; however, as CNTs were loaded further, the mix solution's viscosity increased, and this caused the pores to shrink. The membrane containing 4 wt% of MWCNT contains pores that are just slightly smaller than those of pure polysulfone membranes, and show greater flux and good salt rejection capabilities. However, the CNTs-reinforced polymeric membranes are still a novel concept, and additional work in this area is required to enhance their functionality and solve the detrimental impact on permeability.

6.3 CONCLUSION

The study of nanoparticles is currently receiving a lot of attention, largely because of the wide range of uses they have. From a research standpoint, the use of nanomaterials in environmental monitoring, waste water treatment, and future water remediation is becoming increasingly significant. The features of composites made of polymers and carbon nanotubes used in water purification and remediation equipment are examined in this chapter. Composites made of polymers and carbon nanotubes are a great choice for creating sensor and adsorbent products in the fields of wastewater treatment, industrial emissions monitoring and control, and environmental pollution analysis. The potential development of polymer–carbon nanotube-based composites as a sensing material for environmental monitoring in nanosensors is highlighted in addition to the growth as adsorbents to remove toxins in wastewater treatment. The advantages of composites made of polymers and carbon nanotubes as well as their environmental impact are discussed in the part that concludes the paper. Composites made from polymers and carbon nanotubes have excellent mechanical, electrical, and structural properties that have shown promise for use in environmental applications.

ACKNOWLEDGMENTS

The author is grateful to University Grants Commission, New Delhi, India and Department of Chemistry, Dr. Harisingh Gour Central University, Sagar for financial support.

REFERENCES

1. Brame, J. et al.: Nanotechnology-enabled water treatment and reuse: Emerging opportunities and challenges for developing countries. *Trends Food Sci. Technol.* 22(11), 618–624 (2011).
2. Rickerby, D., Morrison, M.: Nanotechnology and the environment: A European perspective. *Sci. Technol. Adv. Mater.* 8(1), 19–24 (2007).
3. Brumfiel, G.: Nanotechnology: A little knowledge. *Nature* 424(6946), 246–248 (2003).
4. Sadegh, H., Ghoshekandi, R.S., Masjedi, A., Mahmoodi, Z., Kazemi, M.: A review on carbon nanotubes adsorbents for the removal of pollutants from aqueous solutions. *Int. J. Nano Dimens.* 7(2), 109 (2016).
5. Theron, J., Walker, J., Cloete, T.: Nanotechnology and water treatment: Applications and emerging opportunities. *Crit. Rev. Microbiol.* 34(1), 43–69 (2008).
6. Dil, E.A., Ghaedi, M., Asfaram, A.: The performance of nanorods material as adsorbent for removal of azo dyes and heavy metal ions: Application of ultrasound wave, optimization and modeling. *Ultrason. Sonochem.* 34, 792–802 (2017).
7. Savage, N., Diallo, M.S.: Nanomaterials and water purification: Opportunities and challenges. *J. Nanopart. Res.* 7(4–5), 331–342 (2005).
8. Kalpana Sastry, R., Rashmi, H.B., Rao, N.H.: Nanotechnology for enhancing food security in India. *Food Policy* 36, 391–400 (2011).
9. Allhoff, F.: On the autonomy and justification of nanoethics. *Nanoethics* 1, 185–210 (2007).
10. Zinicovscaia, I., Cepoi, L.: Cyanobacteria for bioremediation of wastewaters. *Cyanobacteria for Bioremediation of Wastewaters*, 1–124 (2016). doi:10.1007/978–3–319–26751–7.
11. Ritter, L. et al.: Sources, pathways, and relative risks of contaminants in surface water and groundwater: A perspective prepared for the Walkerton inquiry. *J. Toxicol. Environ. Health – Part A* 65 (2002).
12. Rajasekhar, B., Nambi, I.M., Govindarajan, S.K.: Human health risk assessment of groundwater contaminated with petroleum PAHs using Monte Carlo simulations: A case study of an Indian metropolitan city. *J. Environ. Manage.* 205, 183–191 (2018).
13. Jassby, D., Cath, T.Y., Buisson, H.: The role of nanotechnology in industrial water treatment. *Nat. Nanotechnol.* 13, 670–672 (2018).
14. Rodriguez-Proteau, R., Grant, R.L.: Toxicity evaluation and human health risk assessment of surface and groundwater contaminated by recycled hazardous waste materials. *Handb. Environ. Chem.* 5, 133–189 (2006).
15. Tyagi, P.K., Singh, R., Vats, S., & Kumar, D. Singh Pankaj. : Nanomaterials use in wastewater treatment. International Conference on Nanotechnology and Chemical Engineering (ICNCS'2012), 65–68 (2012).
16. Lytle, D.A., Sorg, T., Wang, L., Chen, A.: The accumulation of radioactive contaminants in drinking water distribution systems. *Water Res.* 50, 396–407 (2014).
17. Ahuja, S.: Overview of global water challenges and solutions. *ACS Symp. Ser.* 1206, 1–25 (2015).
18. Kumar, S. et al.: Nanotechnology-based water treatment strategies. *J. Nanosci. Nanotechnol.* 14, 1838–1858 (2014).

19. Ezugbe, E.O., Rathilal, S.: Membrane technologies in wastewater treatment: A review. *Membranes (Basel)*. 10 (2020).
20. Qu, X., Brame, J., Li, Q., Alvarez, P.J.J.: Nanotechnology for a safe and sustainable water supply: Enabling integrated water treatment and reuse. *Acc. Chem. Res.* 46, 834–843 (2013).
21. Rajasulochana, P., Preethy, V.: Comparison on efficiency of various techniques in treatment of waste and sewage water – A comprehensive review. *Resour. Technol.* 2, 175–184 (2016).
22. Iijima, S.: Helical microtubules of graphitic carbon. *Nature* 354(6348), 56–58 (1991).
23. Muyibi, S.A., Ambali, A.R., Eissa, G.S.: Development-induced water pollution in Malaysia: Policy implication from an econometric analysis. *Water Policy* 10(2), 193–206 (2008).
24. Bansal, R.C., Goyal, M.: *Activated Carbon Adsorption.* Taylor and Francis Group, London, 351–353 (2005).
25. Friberg, L., Nordberg, G.F., Vouk, V.B.: *Handbook on the Toxicology of Metals.* Elsevier, North-Holland, Amsterdam (1979).
26. Calderon, J., Navarro, M.E., Jimenez-Capdeville, M.E., Santos-Diaz, M.A., Golden, A., Rodriguez-Leyva, I., Borja-Aburto, V., Diaz-Barriga, F.: Exposure to arsenic and lead and neuropsychological development in Mexican children. *Environ. Res.* 85(2), 69–76 (2001).
27. Li, Y.H., Wang, S., Wei, J., Zhang, X., Xu, C., Luan, Z., Wu, D., Wei, B.: Lead adsorption on carbon nanotubes. *Chem. Phys. Lett.* 357, 263–266 (2002).
28. Ajayan, P.M.: Nanotubes from carbon. *Chem. Rev.* 99, 1787–1800 (1999).
29. Terrones, M.: Science and technology of the twenty-first century: Synthesis, properties and applications of carbon nanotubes. *Annu. Rev. Mater. Res.* 33, 419–501 (2003).
30. Dai, B.L., Mau, A.W.H.: Controlled synthesis and modification of carbon nanotubes and C60: Carbon nanostructures for advanced polymeric composite materials. *Adv. Mater.* 13, 899–913 (2001).
31. Dresselhaus, M., Dresselhaus, G., Avouris, Ph.: Carbon nanotubes synthesis, structure, properties, and applications. *Topics in Applied Physics*; Dresselhaus M., Dresselhaus G., Avouris PH., eds. (2001); Springer: Berlin, Germany.
32. Rouff, R.S., Lorents, D.C.: Mechanical and thermal properties of carbon nanotubes. *Carbon* 33(7), 925–930 (1995).
33. Treacy, M.M., Ebbesen, T.W., Gibson, J.M.: Exceptionally high Young's modulus observed for individual carbon nanotubes. *Nature* 381(6584), 678–680 (1996).
34. Ebbesen, T.W., Lezee, H.J, Hiura, H., Bennett, J.W., Ghsrmi, H.F., Thio, T.: Electrical conductivity of individual carbon nanotubes. *Nature* 382(6586), 54–56 (1996).
35. Collins, P.G., Zettl, A., Bando, H., Thess, A., Smalley, R.E.: Nanotube nanodevice. *Science* 278, 100–102 (1997).
36. Javey, A., Guo, J., Wang, Q., Lundstrom, M., Dai, H.J.: Ballistic carbon nanotube field-effect transistors. *Nature* 424, 654–657 (2003).
37. Heer, W.A.D., Chatelain, A., Ugarte, D.: A carbon nanotube fieldemission electron source. *Science* 270(5239), 1179–1180 (1995).
38. Wang, Q.H., Setlur, A.A., Lauerhaas, J.M., Dai, J.Y., Seelig, E.W., Chang, R.P.H.: A nanotube-based field-emission flat panel display. *Appl. Phys. Lett.* 72, 2912–2913 (1998).
39. Planeix, J.M., Coustel, N., Coq, B., Brotons, V., Kumbhar, P.S., Dutartre, R., Geneste, P., Bernier, P., Ajayan, P.M.: Application of carbon nanotubes as supports in heterogeneous catalysis. *J. Am. Chem. Soc.* 116, 7935–7936 (1994).
40. Che, G., Lakshmi, B.B., Fisher, E.R., Martin, C.R.: Carbon nanotubule membranes for electrochemical energy storage and production. *Nature* 393, 346–349 (1998).

41. Kong, J., Franklin, N.R., Zhou, C., Chapline, M.G., Peng, S., Cho, K., Dai, H.: Nanotube molecular wires as chemical sensors. *Science* 287(5453), 622–625 (2000).
42. Collins, P.G., Bradley, K., Ishigami, M., Zettl, A.: Extreme oxygen sensitivity of electronic properties of carbon nanotubes. *Science* 287(5459), 1801–1804 (2000).
43. Dalton, A.B., Collins, S., Munoz, E., Razal, J.M., Ebron, V.H., Ferraris, J.P., Coleman, J.N., Kim, B.G., Baughman, R.H.: Super-tough carbonnanotube fibers. *Nature* 423(6941), 703 (2003).
44. Wang, Y., Wei, F., Luo, G., Yu, H., Gu, G.: The large-scale production of carbon nanotubes in a nano-agglomerate fluidized-bed reactor. *Chem. Phys. Lett.* 364(5–6), 568–572 (2002).
45. Long, Q.R., Yang, R.T.: Carbon nanotubes as superior sorbent for dioxin removal. *J. Am. Chem. Soc.* 123, 2058–2059 (2001).
46. Lu, C., Chiu, H.: Adsorption of zinc (II) from water with purified carbon nanotubes. *Chem. Eng. Sci.* 61, 1138–1145 (2006).
47. Lu, C., Chiu, H., Liu, C.: Removal of zinc (II) from aqueous solution by purified carbon nanotubes: Kinetics and equilibrium studies. *Ind. Eng. Chem. Res.* 45(8), 2850–2855 (2006).
48. Li, Y.H., Wang, S.W., Luan, Z.L., Ding, J.D., Xu, C.X., Wu, D.: Adsorption of cadmium (II) from aqueous solution by surface oxidized carbon nanotubes. *Carbon* 41, 1057–1062 (2003).
49. Vukovic, G.D., Marinkovic, A.D., Colic, M., Ristic, M.D., Aleksic, R., Grujic, A.A.P., Uskokovic, P.S.: Removal of cadmium from aqueous solutions by oxidized and ethylenediamine-functionalized multi-walled carbon nanotubes. *Chem. Eng. J.* 57, 238–324 (2010).
50. Gao, Z., Bandosz, T.J., Zhao, Z., Han, M., Liang, C., Qiu, J.: Investigation of the role of surface chemistry and accessibility of cadmium adsorption sites on open-surface carbonaceous materials. *Langmuir* 24(20), 11701–11710 (2008).
51. Li, Y.H., Zhu, Y., Zhao, Y., Wu, D., Luan, Z.: Different morphologies of carbon nanotubes effect on the lead removal from aqueous solution. *Diam. Relat. Mater.* 15, 90–94 (2006).
52. Peng, X., Luan, Z., Di, Z., Zhang, Z., Zhu, C.: Carbon nanotubes–iron oxides magnetic composites as adsorbent for removal of Pb(II) and Cu(II) from water. *Carbon* 43, 880–883 (2005).
53. Kandah, M.I., Meunier, J.L.: Removal of nickel ions from water by multi-walled carbon nanotubes. *J. Hazard. Mater.* 146, 283–288 (2007).
54. Lu, C., Liu, C.: Removal of nickel (II) from aqueous solution by purified carbon nanotubes. *J. Chem. Technol. Biotechnol.* 81, 1932–1940 (2006).
55. Chen, C., Wang, X.: Adsorption of Ni(II) from aqueous solution using oxidized multiwall carbon nanotubes. *Ind. Eng. Chem. Res.* 45(26), 9144–9149 (2006).
56. Li, Y.H., Luan, Z., Xiao, X., Zhou, X., Xu, C., Wu, D., Wei, B.: Removal Cu2+ ions from aqueous solutions by carbon nanotubes. *Adsorp. Sci. Technol.* 21, 475–485 (2003).
57. Li, Y.H., Ding, J., Lun, Z., Di, Z., Zhu, Y., Xu, C., Wu, D., Wei, B.: Competitive adsorption of Pb2+, Cu2+ and Cd2+ ions from aqueous solutions by multiwalled carbon nanotubes. *Carbon* 41, 2787–2792 (2003).
58. Li, Y.H., Wang, S., Cao, A., Zhao, D., Zhang, X., Xu, C., Luan, Z., Ruan, D., Liang, J., Wu, D., Wei, B.: Adsorption of fluoride fromwater by amorphous alumina supported on carbon nanotubes. *Chem. Phys. Lett.* 350, 412–416 (2001).
59. Li, Y.H., Wang, S., Zhang, X., Wei, J., Xu, C., Luan, Z., Wu. D.: Adsorption of fluoride from water by aligned carbon nanotubes. *Mater. Res. Bull.* 38, 469–476 (2003).
60. Wang, X., Chen, C., Hu, W., Ding, A., Xu, D., Zhou, X.: Sorption of 234Am(III) to multiwall carbon nanotubes. *Environ. Sci. Technol.* 39(8), 2856–2860 (2005).

61. Lin, K., Xu, Y., He, G., Wang, X.: The kinetic and thermodynamic analysis of Li ion in multi-walled carbon nanotubes. *Mater. Chem. Phys.* 99, 190–196 (2006).
62. Chen, C., Li, X., Zhao, D., Tan, X., Wang, X.: Adsorption kinetic, thermodynamic and desorption studies of Th(IV) on oxidized multi-wall carbon nanotubes. *Coll. Surf. A: Physicochem. Eng. Asp.* 302, 449–454 (2007).
63. Tuzen, M., Soylak, M.: Multiwalled carbon nanotubes for speciation of chromium in environmental samples. *J. Hazard. Mater.* 147, 219–225 (2007).
64. Bauschlicher, C.W., Ricca, A.: Binding of NH_3 to graphite and to a (9,0) carbon nanotube. *Phys. Rev. B.* 70, 115409–115413 (2004).
65. Bienfait, M., Zeppenfeld, P., Pavlovsky, N.D., Muris, M., Johnson, M.R., Wilson, T., DePies, M., Vilches, O.E.: Thermodynamics and structure of hydrogen, methane, argon, oxygen, and carbon dioxide adsorbed on single-wall carbon nanotube bundle. *Phys. Rev. B.* 70(3), 035410–035419 (2004).
66. Talapatra, S., Migone, A.D.: Adsorption of methane on bundles of closed-ended single-wall carbon nanotubes. *Phys. Rev. B.* 65(4), 045416–045421 (2002).
67. Dillon, A.C., Jones, K.M., Bekkedahl, T.A., Kiang, C.H., Bethune, D.S., Heben, M.J.: Storage of hydrogen in single walled carbon nanotubes. *Nature* 386, 377–379 (1997).
68. Chen, P., Wu, X., Lin, J., Tan, K.L.: High H_2 uptake by alkali-doped carbon nanotubes under ambient pressure and moderate temperature. *Science* 285, 91–93 (1999).
69. Liu, C., Fan, Y.Y., Liu, M., Cong, H.T., Cheng, H.M., Dresselhaus, M.S.: Hydrogen storage in single-walled carbon nanotube at room temperature. *Science* 286, 1127–1129 (1999).
70. Lee, S.M., Lee, Y.H.: Hydrogen storage in single-walled carbon nanotubes. *Appl. Phys. Lett.* 76, 2877–2879 (2000).
71. Yim, W.L., Liu, Z.F.: A reexamination of the chemisorptions and desorption of ozone on the exterior of a (5,5) single-walled carbon nanotube. *Chem. Phys. Lett.* 98, 297–303 (2004).
72. Varghese, O.K., Kichambre, P.D., Gong, D., Ong, K.G., Dickey E.C., Grimes, C.A.: Gas sensing characteristics of multi-wall carbon nanotubes. *Sensors Actuators B* 81(1), 32–41 (2001).
73. Lin, H.F., Ravikrishna, R., Valsaraj, K.T.: Reusable adsorbents for dilute solution separation. 6. Batch and continuous reactors for the adsorption and degradation of 1,2-dichlorobenzene from dilute wastewater streams using titania as a photocatalyst. *Sep. Purif. Technol.* 28(2), 87 (2002).
74. Peng, X., Li, Y., Luan, Z., Di, Z., Wang, H., Tian, B., Jia, Z.: Adsorption of 1,2-dichlorobenzene from water to carbon nanotubes. *Chem. Phys. Lett.* 376(1–2), 154 (2003).
75. Fagan, S.B., Santos, E.J.G., Souza Filho, A.G., Mendes Filho, J., Fazzio, A.: Ab initio study of 2,3,7,8-tetrachlorinated dibenzo-p-dioxin adsorption on single wall carbon nanotubes. *Chem. Phys. Lett.* 437(1–3), 79 (2007).
76. Rao, G.P., Lu, C., Su, F.: Sorption of divalent metal ions from aqueous solution by carbon nanotubes: A review. *Sep. Purif. Technol.* 58, 224–231 (2007).
77. Yang, K., Zhu, L., Xing, B.: Adsorption of polycyclic aromatic hydrocarbons by carbon nanomaterials. *Environ. Sci. Technol.* 40(6), 1855 (2006).
78. Wang, X., Jialong, L.U., Xing, B.: Sorption of organic contaminants by carbon nanotubes: Influence of adsorbed organic matter. *Environ. Sci. Technol.* 42(9), 3207 (2008).
79. Lu, C., Liu, C., Rao, G.P.: Comparisons of sorbent cost for the removal of Ni2+ from aqueous solution by carbon nanotubes and granular activated carbon. *J. Hazard. Mater.* 151(1), 239 (2008).
80. Srivastava, A., Srivastava, O.N., Talapatra, S., Vajtai, R., Ajayan, P.M.: Carbon nanotubes filters. *Nat. Mater.* 3(9), 610 (2004).

81. Jin, S., Fallgren, P.H., Morris, J.M., Chen, Q.: Removal of bacteria and viruses from waters using layered double hydroxide nanocomposites. *Sci. Technol. Adv. Mater.* 8(1–2), 67 (2007).
82. Tahaikt, M., El Habbani, R., Ait Haddou, A., Achary, I., Amor, Z., Taky, M., Alami, A., Boughriba, A., Hafsi, M., Elmidaoui, A.: Fluoride removal from groundwater by nanofiltration. *Desalination* 212(1–3), 46 (2007).
83. Fornasiero, F., Hyung, G.P., Holt, J.K., Stadermann, M., Grigoropoulos, C.P., Noy, A., Bakajin, O.: Ion exclusion by sub-2-nm carbon nanotube pores. *Proc. Natl. Acad. Sci. USA* 105(45), 17250 (2008).
84. Noy, A., Park, H.G., Fornasiero, F., Holt, J.K., Grigoropoulos, C.P., Bakajin, O.: Nanofluidics in carbon nanotubes. *Nano Today* 2(6), 22 (2007).
85. Hummer, G., Rasaiah, J.C., Noworyta, J.P.: Water conduction through the hydrophobic channel of a carbon nanotube. *Nature* 414(6860), 188 (2001).
86. Bohonak, D.M., Zydney, A.L.: Compaction and permeability effects with virus filtration membranes. *J. Membr. Sci.* 254(1–2), 71 (2005).
87. Brady-Estévez, A.S., Kang, S., Elimelech, M.: A single-walled-carbon-nanotube filter for removal of viral and bacterial pathogens. *Small* 4(4), 481 (2008).
88. Mostafavi, S.T., Mehrnia, M.R., Rashidi, A.M.: Preparation of nanofilter from carbon nanotubes for application in virus removal from water. *Desalination* 238(1–3), 271 (2009).
89. Kang, S., Pinault, M., Pfefferle, L.D., Elimelech, M.: Single-walled carbon nanotubes exhibit strong antimicrobial activity. *Langmuir* 23(17), 8670 (2007).
90. Li, Q., Mahendra, S., Lyon, D.Y., Brunet, L., Liga, M.V., Li, D., Alvarez, P.J.J.: Antimicrobial nanomaterials for water disinfection and microbial control: Potential applications and implications. *Water Res.* 42(18), 4591 (2008).
91. Nepal, D., Balasubramanian, S., Simonian, A.L., Davis, V.A.: Strong antimicrobial coatings: Single-walled carbon nanotubes armored with biopolymers. *Nano Lett.* 8(7), 1896 (2008).
92. Cortes, P., Deng, S., Smith, G.B.: The adsorption properties of bacillus atrophaeus spores on singlewall carbon nanotubes. *J. Sensors* 2009 (2009).
93. Shawky, H.A., Chae, S.-R., Lin, S., Wiesner, M.R.: Synthesis and characterization of a carbon nanotube/polymer nanocomposite membrane for water treatment. *Desalination* 272, 46–50 (2011).
94. Choi, J.-H., Jegal, J., Kim, W.-N.: Fabrication and characterization of multi-walled carbon nanotubes/polymer blend membranes. *J. Membr. Sci.* 284, 406–415 (2006).

7 Synthetic Polymer– Graphene/Graphene Derivatives–Based Composites for Wastewater Treatment

Chandra Shekhar Pati Tripathi, Keshav Sharma, Mohd Ali, and Debanjan Guin

Banaras Hindu University

CONTENTS

DOI: 10.1201/9781003328094-7

7.1 INTRODUCTION

Water is one of the most important assets for the existence of life on the planet. With the growth in population and industrialization, along with environmental pollution, universal water issues have aroused (Greve et al. 2018). The worldwide water demand has been evaluated to go on at an equivalent growth rate until 2040, with an estimation of a 20%–30% increment from the existing water utilization. Water pollution is generally caused by the discharge of domestic and industrial effluent wastes, spillage from oil tankers, marine dumping, radioactive waste, and atmospheric deposition. Some examples of pollutants from history include lead, asbestos, dichlorodiphenyltrichloroethane (DDT), polychlorinated biphenyls (PCBs), and chlorofluorocarbons responsible for ozone layer depletion (Landrigan et al. 2018). The major contaminators are industries such as textile, paper and pulp, dye, pharmaceutical, tannery, paint, and kraft bleaching, which are sources of a wide variety of organic pollutants introduced into natural water resources, with the textile industry being the major contributor (54%) followed by dying industry (21%). Figure 7.1a represents the schematic illustration of contamination of water bodies by industrial discharge, and Figure 7.1b represents the contribution of various industries in throwing effluents into the water bodies (Velusamy et al. 2021).

Wastewater-related illnesses include communicable diseases that are mainly water borne, water washed, water based, water related, and vector borne. Chemically polluted water also triggers some non-communicable diseases (Johnson and Paull 2011). According to the Lancet Commission on Pollution and Health (Landrigan 2017), each year a total of around 2.3 million fatalities are caused by water-based diseases mainly because of microbiological contamination, soil pollution, heavy metals, and chemical pollution. The contaminants found in industrial waste are a principal cause of immunological suppression, reproductive problems, and acute poisoning. Infectious illnesses such as cholera and typhoid fever, as well as gastroenteritis, diarrhea, vomiting, and skin and renal problems are spreading due to dirty water (Khan and Ghouri 2011).

Wastewater treatment, therefore, becomes of the utmost importance. Numerous physical and chemical wastewater treatment techniques include gravity separation, air flotation, membrane filtration methods, absorption materials, chemical vapor deposition, coated mesh, techniques based on carbon-based materials, hydrophobic aerogels, sol–gel technique, and sponges. All these methods are beneficial, but they have certain drawbacks, including poor separation rate, fouling, high energy consumption, reusability, and recyclability of the purifying media (Bhol et al. 2021).

The use of adsorption techniques for the removal of various organic pollutants such as dyes, metal ions, etc. from wastewater has gained recent attention among the scientific community. Adsorption is acknowledged to be one of the most effective techniques among the others in terms of its high efficiency, reduced cost, ease of design, simplified adsorbent recovery, and diversity in the choice of adsorbent material (Puri and Sumana 2018). When it comes to adsorption-based wastewater treatment, activated carbon is the material of choice, as carbon-based composites are well-known dye and pollutant adsorbents (Ramesha et al. 2011). Graphene is an atomically thin, two-dimensional (2D) honeycomb sheet of sp^2 carbon atoms.

(a)

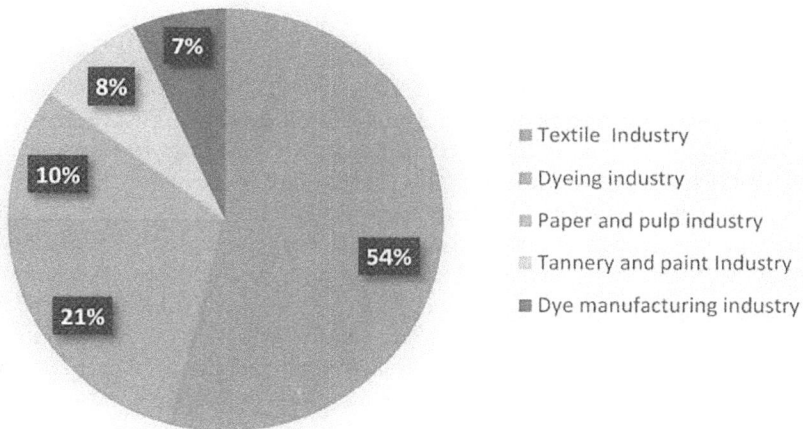

Textile Industry

Dyeing industry

Paper and pulp industry

Tannery and paint Industry

Dye manufacturing industry

(b)

FIGURE 7.1 (a) Schematic illustration of contamination of water bodies by industrial discharge and (b) the contribution of various industries in throwing effluents into the water bodies. [Reproduced from Ref. Velusamy et al. (2021) with permission from The Chemical Society of Japan & Wiley. Copyright 2021 (OPEN ACCESS, NO PERMISSION REQUIRED).]

It enjoys numerous desired features that include strong mechanical strength (Lee et al. 2008), electrical conductivity (Kuilla et al. 2010), and abilities relating to molecular barriers (Cui et al. 2016). Due to these remarkable properties, various research efforts have aimed to integrate graphene into polymers to create polymer-based nanocomposites (Xiao et al. 2002). However, the usage of pristine graphene has proven difficult due to its complex production methods (Zhu et al. 2010), low solubility, and aggregation in solution due to van der Waals interactions. To overcome these problems we use the oxidized form of graphite, which is hydrophilic due to the presence of various oxygen-based functional groups. When graphite is oxidized in acidic

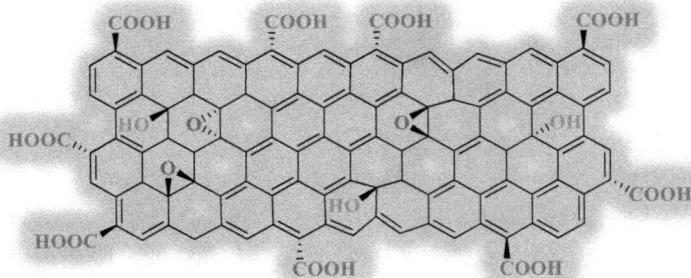

FIGURE 7.2 Structure of a GO sheet displaying the presence of different oxygen-containing functional groups.

solvents, graphite oxide is formed, which is composed of numerous stacked layers of graphene oxide (GO). Figure 7.2 represents the detailed structure of a single GO sheet. It has a hexagonal carbon structure analogous to graphene; besides, it also contains hydroxyl (–OH), alkoxy (C–O–C), carbonyl (C=O), carboxylic acid (–COOH), and other oxygen-based functional groups (Pendolino and Armata 2017).

Apart from the ease of fabrication, these oxygenated groups impart numerous benefits over graphene, such as better adsorption tendency, increased solubility, and the potential for surface functionalization, which has provided many prospects for usage in polymer-based nanocomposite materials. GO is reported to have theoretically a large surface area of around $2630 \, m^2/g$ (Park and Ruoff 2009). As a result of its large specific surface area and presence of various oxygen-containing functional groups, it has a greater capacity for heavy metal ion and dye adsorption. Adsorption can take place by electrostatic attraction, chemical bonding, hydrogen bonding, hydrophobic association, complex formation, and a variety of other mechanisms (Rey and Varsanik 1986). Due to the presence of its ionizable carboxyl groups, GO can function as a weak acid cation exchange resin, allowing ion exchange with metal cations or positively charged organic compounds (Balapanuru et al. 2010). These functional groups can interact with positively charged species such as metal ions, polymers, and biomolecules among others (Yang et al. 2010). Figure 7.3 shows possible interaction between GO and dyes like rhodamine B, methylene blue, Orange G, and methyl violet. It can be concluded from the figure that electrostatic interactions are accountable for the possible interface between GO–methyl violet and GO–methylene blue, whereas solely van der Waals interactions are responsible for the possible interaction between GO and Orange G. GO and rhodamine B share both electrostatic and van der Waals interaction (Ramesha et al. 2011).

The direct use of GO for the adsorption process has several limitations. Solid-phase adsorption using GO is difficult, as it requires sonication each time followed by centrifugation, which is a very time-consuming and inefficient process. GO is hydrophilic, so it easily gets dispersed in the solution, making it very difficult to separate from the solution. As a result of it, the reusability and recyclability of GO

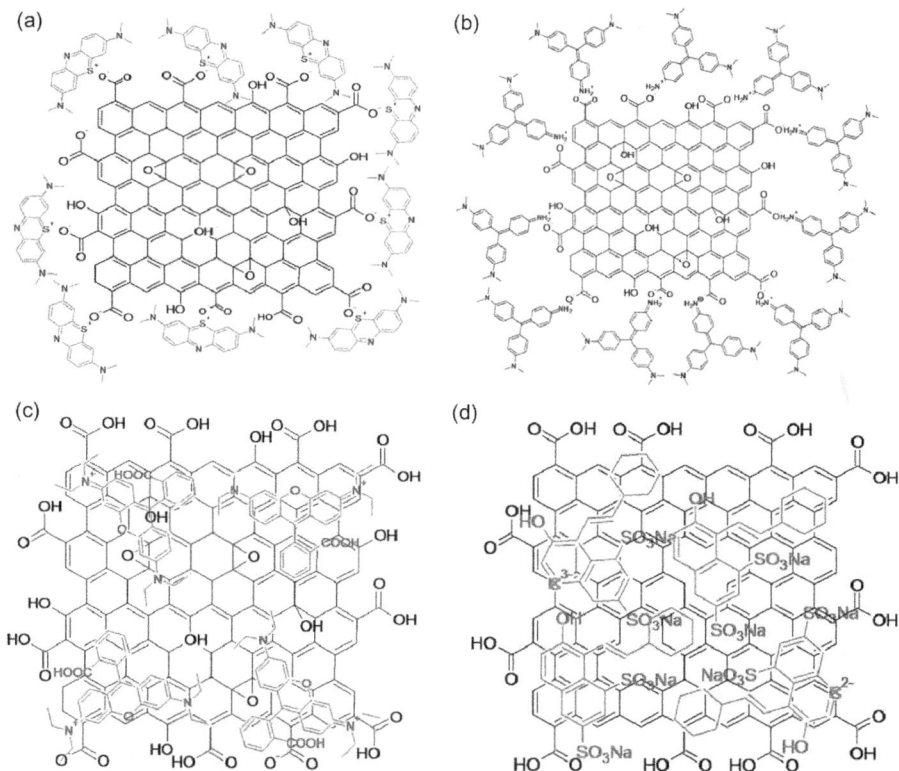

FIGURE 7.3 Possible interaction between functional groups present in GO with (a) methylene blue (b) methyl violet (c) rhodamine B, and (d) rGO and Orange G. [Reproduced from Ref. Ramesha et al. (2011) with permission from Elsevier. Copyright 2011 (PERMISSION OBTAINED).]

are hampered. Even high-speed centrifugation cannot solve the purpose. So, it is of great importance to functionalize GO sheets onto some porous polymeric surfaces to limit the agglomeration and enhance its adsorption properties.

GO has an abundance of oxygen-containing functional groups, such as hydroxyl and epoxide groups present on the basal plane and carboxyl groups present at the marginal planes, which are highly useful in modifying it and mediating its amalgamation with polymers, ceramics, and metal matrices, making it possible for the formation of single-layer GO with homogeneous surface modification. Polymer chains may be easily grafted onto the basal planes of GO via a variety of processes that include carboxylic acid amidation and esterification, as well as the nucleophilic ring-opening reaction of epoxy functional groups (Zhang et al. 2015). In this book chapter, we have summarized various research works associated with the fabrication of GO with various synthetic polymers such as polyvinyl alcohol (PVA), polystyrene (PS), polyvinylpyrrolidone (PVP), polyaniline (PANI), polymethyl methacrylate (PMMA), and polyvinylidene fluoride (PVDF) for applications in wastewater treatment.

7.2   VARIOUS APPROACHES FOR THE SYNTHESIS OF GRAPHENE OXIDE

There are numerous approaches for the synthesis of graphene oxide dating from 1859, when Brodie for the first time reported the synthesis of graphene oxide. Other methods include the Staudenmaier's method, Hofmann's method, and Hummers' method (including their modified and improved forms). Typically, in all these methods, Graphite powder is chemically treated with acids (HCl, H_2SO_4, and HNO_3), followed by the intercalation of alkali metals using alkali metal compounds such as $KClO_3$, $KMnO_4$, $NaNO_3$, and so on into the graphitic layers, which supports in the breakdown of graphitic layers into the minute piece.

7.2.1   BRODIE METHOD (1859)

Brodie (1859) reported on the modifications of graphite when mixed with powerful oxidants, and this work might be considered one of the first preparation of GO. He used the term "graphic acid" for his final product, which we call today "graphene oxide." He used graphite as the source of carbon. $KClO_3$ and HNO_3 were used as oxidizing agents with a reaction time of more than 120 hours.

7.2.2   STAUDENMAIER'S METHOD (1898)

Staudenmaier (1898) reported on the synthesis of graphene oxide using graphite as the starting material. $KClO_3$, HNO_3, and H_2SO_4 were used as oxidizing agents. The reaction was carried out at room temperature, and the time taken for the completion of the reaction was about 96 hours. This method proved to be more efficient than the Brodie method.

7.2.3   HUMMERS' METHOD (1958)

Hummers' method (Hummers Jr and Offeman 1958) is the most widely used method for the synthesis of graphene oxide. Briefly, 100 g of powdered graphite and 50 g of sodium nitrate were mixed with 2.3 L of sulfuric acid. As a precaution, the chemicals were blended in a 15 L battery jar that had been chilled to 0°C in an ice bath. 300 g of potassium permanganate was added to the suspension while vigorously stirring it. The addition of $KMnO_4$ was done in such a way that the reaction temperature did not exceed 20°C. The ice bath was then withdrawn, and the temperature of the suspension was raised to 35°C for 30 minutes. As the reaction proceeded, the reaction mixture became condensed. After 20 minutes, the mixture turned into a brownish-gray paste, with just a little quantity of gas escaping. After 30 minutes, 4.6 L of water was progressively added to the paste, resulting in strong effervescence and a temperature increase to 98°C. The resulting diluted brown solution was kept at this temperature for a further 15 minutes. The solution was then diluted with 14 L of warm water followed by treatment with hydrogen peroxide, turning the reaction mixture to bright yellow color. Filtration of the suspension produced a yellow-brown cake. The resulting yellow-brown cake was washed 3–4 times with around 14 L of warm water.

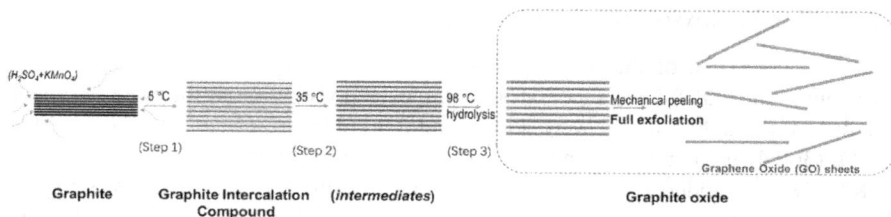

FIGURE 7.4 Schematic illustration of the synthesis of GO using Hummers' method. [Reproduced from Ref. Sun (2019) with permission from Elsevier. Copyright 2019 (PERMISSION OBTAINED).]

The obtained graphitic oxide residue was dispersed in 32 L of water, and treated with resinous anion and cation exchangers. Centrifugation followed by drying at 40°C yielded the final graphitic oxide.

This method has the advantage over the Brodie and Staudenmaier method, because it eliminates the use of dangerous chemicals such as $KClO_3$ and HNO_3. Also, the reaction time is reduced to less than 2 hours. Figure 7.4 shows the schematic representation of the above-discussed Hummers' method of GO synthesis.

7.2.4   GO Synthesis Using Benzoyl Peroxide (Shen et al. 2009)

In 2009, Shen et al. used benzoyl peroxide (BPO) for the synthesis of graphene oxide. Briefly, BPO (10 g) and graphite (0.5 g) were crushed to a fine powder. This powder was then heated for 10 minutes at 110°C. After the reaction was completed, the mixture was cooled to room temperature followed by washing multiple times with water to make the pH of the filtrate neutral. The resulting solid black residue was oven dried. They have also reduced the resulting GO into reduced graphene oxide (rGO) by the use of $NaBH_4$ followed by heating at 125°C for 3 hours. The above method has the advantage of being less time consuming and acid free, as no acids were used, and the reaction time was around 10 minutes only. But BPO has the drawback of being highly explosive when heated in a closed container.

7.2.5   Modified Hummers' Method

This method incorporates the use of additional $KMnO_4$ into the Hummers' reagents. Briefly, 69 mL of concentrated H_2SO_4 was added into a mixture of 3.0 g graphite flakes and 1.5 g $NaNO_3$, and chilled to 0°C in an ice bath. The reaction temperature was maintained below 20°C, and 9.0 g $KMnO_4$ was gently added in parts. The reaction was heated to 35°C and stirred for 7 hours. Further, more $KMnO_4$ (9.0 g) was added, and the reaction was heated for 12 hours under vigorous stirring at 35°C. The reaction mixture was cooled to room temperature and was poured into 400 mL of distilled water ice, and 3 mL of 30% H_2O_2 was added. The mixture was purified using the following techniques of filtration, centrifugation, and decanting after several times washing with distilled water, 30% HCl, and ethanol, respectively. A 4.2 g of solid product was obtained after vacuum drying.

7.2.6   The Improved Method of GO Synthesis

This method is one of the best methods for the synthesis of GO. In this method, Marcano et al. (2010) have successfully synthesized GO without the use of $NaNO_3$. Briefly, 3.0 g graphite flakes were added into a 9:1 mixture of concentrated H_2SO_4/H_3PO_4 (360:40 mL) and kept under vigorous stirring for around 1 hour. Further, 18 g of $KMnO_4$ was added in parts to the reaction mixture and kept under stirring for 1 hour at room temperature. The reaction mixture was then kept at 50°C under stirring for 12 hours. It was then cooled to room temperature and was added to 400 mL of distilled water ice followed by the addition of 3 mL of 30% H_2O_2. For workup, the mixture was sifted through a metal U.S. Standard testing sieve and then filtering it through polyester fiber. The filtrate was centrifuged at 4000 rpm for 4 hour, and the supernatant was decanted away. It was then washed several times with distilled water, 30% HCl and ethanol, respectively. The resulting solid was vacuum-dried overnight at room temperature, yielding around 5.8 g of graphene oxide. Figure 7.5 compares the efficiency of Hummers' method, the modified Hummers' method, and the improved method. It can be concluded from Figure 7.5 that no harmful gases such as NO_x were released in the case of improved method. Also, the unoxidized graphite (recovered as hydrophobic carbon material) is very less in the case of the improved method when compared to Hummers' and modified Hummers' method. Therefore, the improved method turns out to be efficient, less toxic, and yields highly oxidized GO.

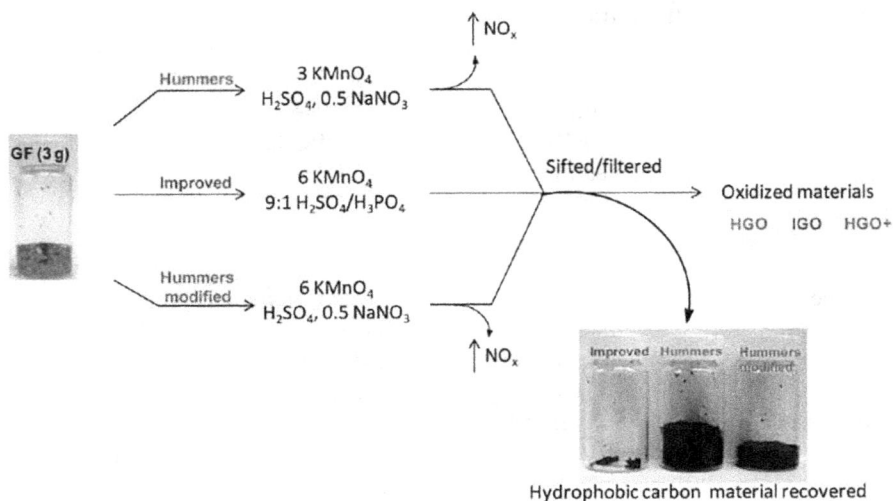

FIGURE 7.5   Illustration of various methods of synthesis of GO using graphite as the starting material. (Reproduced from Ref. Marcano et al. (2010) with permission from the American Chemical Society. Copyright 2010. (PERMISSION OBTAINED).)

7.3 AN INTRODUCTION TO SOME SYNTHETIC POLYMERS

Synthetic polymers, also known as man-made polymers, are polymers that are created artificially in labs. Synthetic polymers are made up of carbon–carbon bonds and are generally generated from petroleum oil in a controlled setting. Here is a brief description of some synthetic polymers that form nanocomposites with GO.

7.3.1 Poly (Vinyl Alcohol) PVA

PVA having the chemical formula $[CH_2CH(OH)]_n$ is a water-soluble, non-toxic, semi-crystalline polymer. It is generally found in powdered form, is white colored and odorless, possessing high dielectric strength and high transparency. It is synthesized through partial or complete hydrolysis of polyvinyl acetate (Razzak and Darwis 2001). The chemical and physical properties of PVA depend on the degree of hydrolysis, which also determines its grade, and molecular weight (DeMerlis and Schoneker 2003). The choice of filler plays a significant role in the surface properties of PVA fillers (Liu et al. 2007). The molecular weight (MW) of PVA products lies in the range of 20,000–400,000 Da. The variance in MW is because of the synthesis parameters such as the length of the initial vinyl acetate polymer, the extent of hydrolysis to reduce the acetate groups, and the hydrolysis condition whether alkaline or acidic (Gaaz et al. 2015; DeMerlis and Schoneker 2003). Owing to its excellent physical, chemical, and optical properties, PVA has been used in a wide range of applications. When it comes to nanocomposites of GO and PVA, they form excellent nanocomposites, as there is intermolecular hydrogen bonding present between the oxygen functional groups of GO and the OH group of the PVA. Figure 7.6 shows the structure of PVA polymer.

7.3.2 Polyvinylpyrrolidone (PVP)

Polyvinylpyrrolidone (PVP), or otherwise polyvidone, is prepared from the monomer *N*-vinylpyrrolidone. It is soluble in water and various organic solvents (Haaf et al. 1985). In dry conditions, it is a white blistering powder and readily absorbs moisture up to 40% of water by its weight. In solution, it has excellent wetting properties and readily forms films, which have been used for coating and binding purposes. PVP is synthesized by free-radical polymerization from its monomer *N*-vinylpyrrolidone in the presence of free-radical initiator like Azobisisobutyronitrile (AIBN) (Haaf et al. 1985). PVP is used as a binder in pharmaceutical industries, moisturizer in various body care products, food additives, adhesives, and so on. Figure 7.7a shows the structure of PVP polymer, and Figure 7.7b shows the possible interaction between GO and PVP molecule (Wu et al. 2017).

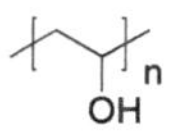

FIGURE 7.6 The structure of PVA polymer.

(a)   (b)

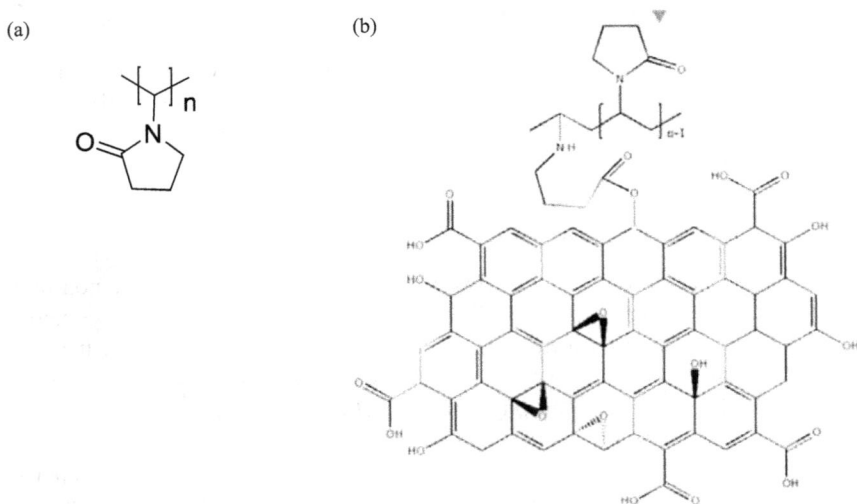

FIGURE 7.7   (a) Structure of PVP, and (b) likely interaction between GO and PVP molecules. (Reproduced from Ref. Wu et al. (2017) with permission from Elsevier. Copyright 2017. (PERMISSION OBTAINED).)

7.3.3   POLYANILINE (PANI)

Polyaniline (PANI) is present as one of the three idealized oxidation states during the polymerization of the aniline monomer, *viz.*, leucoemeraldine (white/clear), emeraldine (salt green /base blue), and pernigraniline (blue/violet). The pernigraniline base is the fully oxidized PANI. Half of the oxidized PANI is reduced to the emerald base, and PANI is completely reduced to the leucomeraldine base. Emeraldine is the most stable and conductive among these. Its conductivity lies in the range of 10^{-9}–10^{0}S/cm. The synthesis method determines the conductivity of PANI. The conductivity is controlled by immersing the emeraldine base in an aqueous acid solution of phosphoric, picric, or sulfonic acid (Blinova et al. 2008). It has an equal amount of amine and imine nitrogen. Protonic acids are used to obtain their salt form through doping. The salt form of emeraldine has a conductivity of 30 s/cm. Depending on the synthetic route and neutralization process, the emeraldine base and its salt assume different types of crystalline arrangements (Figure 7.8) (Beygisangchin et al. 2021).

7.3.4   POLYMETHYL METHACRYLATE (PMMA)

PMMA is a synthetic polymer synthesized from its monomer (methyl methacrylate) using different techniques of polymerization like free-radical and anionic initiations by bulk, solution, suspension, and emulsion techniques (Haaf et al. 1985). It was discovered in the 1930s by British chemists Rowland Hill and John Crawford. PMMA is a transparent, thermoplastic polymer, widely used as a substitute for inorganic glass, because of its high impact strength, lightweight, shatter-resistant, and easy

Basic structure of Polyaniline

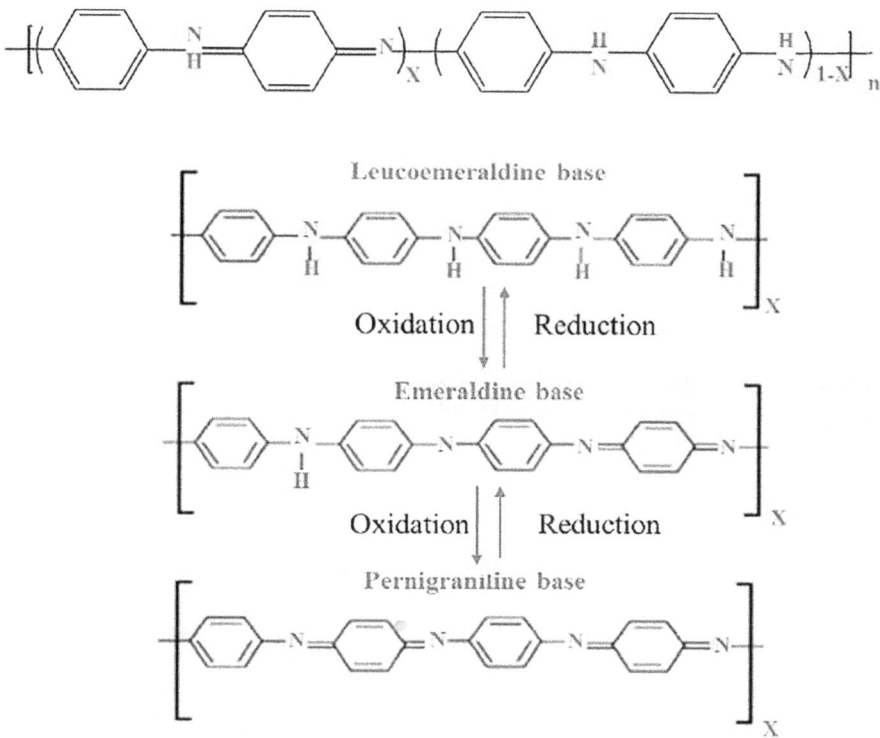

Leucoemeraldine base

Oxidation | Reduction

Emeraldine base

Oxidation | Reduction

Pernigraniline base

FIGURE 7.8 The basic structure of PANI along with three different oxidation forms. [Reproduced from Ref. Beygisangchin et al. (2021) with permission from MDPI. Copyright 2021. (MDPI Open access, no permission required).]

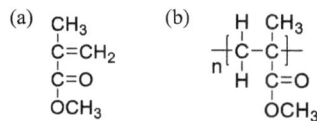

(a) CH_3
$C=CH_2$
$C=O$
OCH_3

(b) H CH_3
$\left[\begin{array}{c} C-C \end{array}\right]_n$
H $C=O$
OCH_3

FIGURE 7.9 (a) Structure of MMA monomer and (b) PMMA polymer.

processing conditions (Demir et al. 2006). It is also weather and scratch resistant. PMMA is an amorphous thermoplastic owing to the presence of an adjacent methyl group in the polymer structure that halts the close packing. PMMA is often used as a lighter, shatter-resistant alternative to glass in everything from windows and aquariums to hockey rinks. The first major application of the polymer took place during World War II, when PMMA was used in aircraft windows, gun turrets for airplanes, submarine periscopes (Figure 7.9).

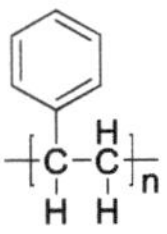

FIGURE 7.10   Structure of PS polymer.

7.3.5   POLYSTYRENE (PS)

PS is a synthetic aromatic hydrocarbon polymer synthesized from monomer styrene (Scheirs and Priddy 2003). It is found in solid or foamed form. Polystyrene consists of a linear polyethylene chain with laterally attached phenyl rings, which account for the enhanced glass transition temperature and high refractive index. Stiffness, brittleness, gloss, and hardness are the main characteristics of this material. PS finds applications in audio/video cassette packs, beakers, transparent food packaging, shower cabinets, lamp covers, and so on. At room temperature, the thermoplastic polymer, polystyrene is present in solid form (glassy); however, upon heating above 100°, it starts flowing. Its rigidity is achieved again by cooling it. Polystyrene is one of the most widely used synthetic polymers with a production scale of several million tonnes per year (Maul et al. 2007). Figure 7.10 represents the structure of polystyrene polymer.

7.3.6   POLYVINYLIDENE FLUORIDE (PVDF)

PVDF is a thermoplastic fluoropolymer with strong chemical resistivity and thermal stability. It has some very unique mechanical qualities, electrical activity, and exceptional aging resistance when compared to other thermoplastics.

Its semi-crystalline structure is responsible for its unique mechanical properties. PVDF consists of repeating CH_2-CF_2 units, as can be seen from its structure in Figure 7.11 (Balli et al. 2019). PVDF polymer possesses high fluoropolymer stability, but interaction groups present gives it a very higher polarity. The polymer's strong polarity has certain advantages such as chemical and oxidative resistance, low hydrophilicity, and substantial swelling in the ionic solutions. Its high polarity groups are also responsible for its interactions with GO.

7.4   PREPARATION OF GO–POLYMER NANOCOMPOSITE

Various approaches such as melt compounding, solution mixing, latex blending, in-situ polymerization, and electropolymerization have been widely employed in the manufacture of graphene-based polymer nanocomposite (Zhang et al. 2015). Nanocomposites of GO with various synthetic polymers such as PVA, PS, PMMA, PANI, PVDF, and PVP were studied. The electrospinning technique is one of the most widely used techniques for the fabrication of GO with synthetic polymers. Recently, L.M.S de Farias et al. (2022) have successfully synthesized polystyrene (PS) film with the help of the electrospinning method followed by coating with graphene oxide (GO-PS). Figure 7.12 represents a typical electrospinning apparatus used

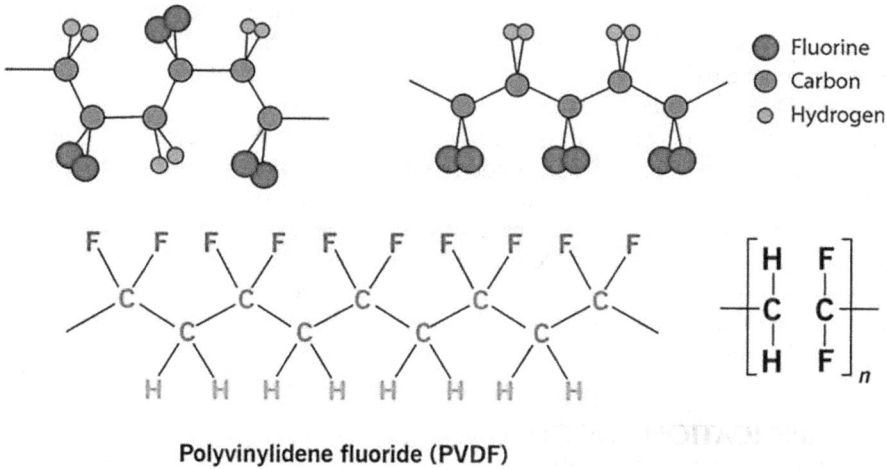

Fluorine
Carbon
Hydrogen

Polyvinylidene fluoride (PVDF)

FIGURE 7.11   Structure of PVDF polymer. [Reproduced from Ref. Balli et al. (2019) with permission from Elsevier. Copyright 2019. (PERMISSION OBTAINED).]

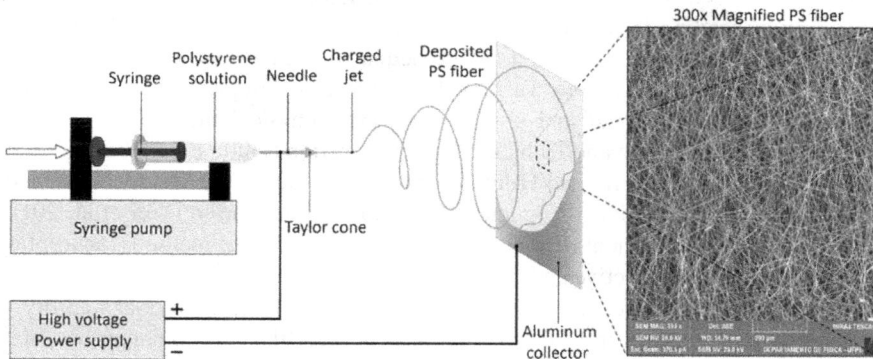

FIGURE 7.12   Typical electrospinning apparatus used for PS fiber synthesis along with TEM image of the as-prepared PS fiber. [Reproduced from Ref. de Farias et al. (2022) with permission from Elsevier. Copyright 2022. (PERMISSION OBTAINED).]

for PS fiber synthesis along with a TEM image of the as-prepared PS fiber (de Farias et al. 2022). In 2018, Ghaffar et al. (2018) synthesized a porous structured polyvinylidene fluoride–graphene oxide (PVDF/GO) nanofibrous membranes by employing an electrospinning approach. Yang et al. (2016) used a chemical fabrication method to fabricate GO–PVA nanocomposite by simply adding GO with PVA in certain compositions followed by mechanical stirring for 2 hours at 90°C. Also, Dai et al. (2016) have synthesized the PVA-supported GO aerogels through the freeze-drying technique, which is an efficient and environment-friendly technique. Li et al. (2014) have reported the preparation of super hydrophobic and superoleophilic graphene/polyvi-nylidene fluoride (G/PVDF) aerogels by making the use of solvothermal reduction of the graphene oxide and PVDF-mixed dispersions using a Teflon-lined autoclave. Solution Blow Spinning (SBS) has recently emerged as a potent fiber-producing process with several benefits over the standard electrospinning technique. Mercante et al. (2017) synthesized rGO/PMMA composite by the use of solution blow spinning (SBS) technique, which turned out to be an efficient adsorbent material. Li et al. (2013) have synthesized the composite of polyaniline and reduced graphene oxide (PANI/RGO) via an in-situ polymerization of aniline with graphene oxide, followed by reduction with hydrate hydrazine. Zhang et al. (2014) also used in-situ polymer-ization technique for the synthesis of polyvinylpyrrolidone-reduced graphene oxide (PVP-rGO) nanocomposite and used it for removal of Cu ions from wastewater.

7.5 APPLICATIONS OF GO–POLYMER NANOCOMPOSITE IN WASTEWATER TREATMENT

7.5.1 DYE REMOVAL

Various dyes in industrial discharge have led to the degradation of water quality and spreading toxicity, thereby, posing a severe environmental problem worldwide. Moreover, these dyes account for low light penetration and insufficient oxygen consumption, leading to marine habitat destruction (Lellis et al. 2019; Moorthy et al. 2021). Therefore, the dyes need to be adequately treated before being expelled into the environment. There are various dye removal techniques. These techniques are classified into chemical and physical methods. Physical methods include ion exchange (Wawrzkiewicz and Hubicki 2015), adsorption (Dutta et al. 2021), and fil-tration/coagulation (Zahrim and Hilal 2013) methods, etc., while chemical methods include Fenton reagent (Poza-Nogueiras et al. 2018), ozonization (Wei et al. 2017), and photocatalytic treatment (Chong et al. 2010). Among these methods, adsorption was found to be very effective and cheap. Adsorption of a dye onto a surface depends on various factors. Surface area and functional groups affect a material's ability to bind substances, and, recently, carbon-based materials like carbon nanotubes, acti-vated carbon, graphene, and graphene oxide (GO) have emerged as prime candidates for the extraction of heavy metal cations, dye adsorption, oil–water separation, etc., due to their high surface areas, lightweight, and potential for high adsorption perfor-mance. Due to its simplicity of synthesis, biocompatibility, and ease of functionaliza-tion, GO has garnered a lot of attention among the various carbon-based materials. Moreover, to stabilize the structure of GO and to provide stability, polymer is used. In this work, we have discussed various dye removal ways of water treatment using the synthetic polymer/graphene nanocomposites. In this direction, (Ghaffar et al. 2018) synthesized a versatile porous structured polyvinylidene fluoride–graphene oxide (PVDF/GO) nanofibrous membranes (NFMs) prepared by electrospinning approach for selective separation and filtration. The GO nanosheets were uniformly distributed throughout the PVDF nanofiber. As a result of the high mechanical strength and surface free energy, the PVDF/GO NFMs results in high permeation and filtration efficiency in comparison to PVDF NFMs. The adsorption process was investigated using pseudo-first order and pseudo-second order kinetic models. The negatively charged PVDF NFMs became more negative with the incorporation of GO. Therefore, they chose positively charged dyes for methylene blue (MB), basic fuchsin (BF), neutral red (NR), rhodamine B (RB), and methyl orange (MO), and a negative one for selective absorption and filtration experiments. These five different organic dyes were chosen based on their variation in electrostatic charge and molecular size as a model representative pollutant to get a detailed study of adsorption kinetics and dynamic molecular filtration. Adsorption kinetics of dyes were studied using an immersion experiment at an initial concentration of 5 mg/L. In short, the sorbates were dissolved in water in which 5 mg NFM samples were immersed in five different kinds of solutions to investigate higher affinity and adsorption capacity toward organic dyes. The experimental adsorption capacity of MB reading 96.48 mg/g was the highest followed by NR (43.74), BF (31.04), and RB (30.39) the lowest, with the maximum capacity for PVDF/GO 5.0% NFM. On the other hand, the NFMs had insignificant adsorption capacity for negatively charged MO molecules, i.e., 1.18, 1.33, and 1.56 mg/g, respectively, for PVDF, PVDF/GO 2.5% and PVDF/GO 5.0% due to electrostatic repulsion. Therefore, it was assumed that the electrostatic interaction was the dominant force between positively charged organic molecules and negatively charged NFMs. The molecular filtration experiments were performed on a Millipore filtration device equipped with N_2 cylinder, a digital balance, and a computer. Four mixtures of organic dyes, i.e., MB/MO, NR/MO, BF/MO, and RB/MO were prepared, MO (negative) being oppositely charged. Three mixtures with the same electrostatic charge (positive/positive), i.e., MB/NR, MB/BF, and MB/RB, were also prepared to further investigate the molecular filtration efficiency among positively charged dye molecules. The mixed dye solutions were allowed to flow through NFMs at a pressure of 1 bar by N_2 gas. UV-Vis spectrophotometer was used to measure the permeate concentration. The separation efficiency (η) of the filtration process was calculated using the following equation (Fu et al. 2014):

$$\eta\% = [OM] \cdot [MO] + [OM]F \times 100$$

where [OM] represents the organic molecules including MB, NR, BF, RB, MO, and [OM]F is the concentrations of MO and the other four organic dyes in the filtrate, respectively. To inspect the regeneration and reusability of as-prepared PVDF/GO 5.0% NFM, the NFMs were recycled three times after the molecular filtration process of mixed dyes and used again for filtration up to three consecutive runs. First, molecular filtration was performed with simple aqueous dye solutions at an average pressure of 1 bar. The PVDF/GO successfully adsorbed the positively charged organic dyes efficiently, and resulted permeate showed 100% purity. On the other hand, there was negligible adsorption for negatively charged MO molecules with 100% permeation. The UV-Vs spectra for all organic dyes were also consistent with the efficient performance of NFMs. It was observed that, in comparison to PVDF/GO NFMs, PVDF NFM was less efficient.

The second molecular filtration performance of as-synthesized NFMs was further inspected with mixtures of various organic dyes. Four mixtures based on opposite charges, i.e., MB/MO, NR/MO, BF/MO, and RB/MO, were used. After permeation, the colored feed solutions turned into the original color of MO molecules, indicating the adsorption of positively charged molecules and allowing only MO (negative) molecules to pass through, highlighting the selective affinity and separation from a mixture only for positively charged molecules. The concentration of MO after filtration was approximately 98.37%, compared to the initial MO feed solution along with simultaneous adsorption of positively charged molecules. After the permeation through PVDF/GO 5.0%, the concentration of positively charged molecules decreased from 40 mg/L to 0.16 (MB) < 0.31 (NR) < 0.37 (BF) < 0.48 (RB) mg/L. They determined the average filtration efficiency for four mixed solutions (MB/MO, NR/MO, BF/MO, and RB/MO) through PVDF/GO 5.0% NFM as 99.20%, 98.47%, 98.17%, and 97.63%, respectively. Moreover, The PVDF/GO 2.5% and PVDF NFMs also adsorbed positively charged molecules, and leftover concentration in permeate was 3.44 (MB) < 3.47 (NR) < 3.61 (BF) < 3.72 (RB) mg/L and 5.52 (MB) < 5.72 (NR) < 5.94 (BF) < 6.21 (RB) mg/L, respectively. The filtrates of PVDF/GO 5.0% showed only the representative peak of MO molecule in UV-Vis spectra for MB/MO, NR/MO, BF/MO, and RB/MO mixtures, whereas obvious peaks were observed in the permeates of PVDF NFMs for both the organic molecules. In conclusion, they reported (99%) selectivity toward positively charged dyes, based on electrostatic attraction, and also a rejection (100%) for negatively charged dye from mixed solutions, thanks to uniform pores and negatively charged PVDF/GO surface. Moreover, they claim recovery of up to three consecutive filtration cycles by regeneration while maintaining efficiency and assuring high stability. This work highlights the importance of PVDF/GO composite for use in practical water treatment and applications, for selective filtration and recycling of dyes.

Quite recently Nawaz et al. (2021) described the synthesis of GO using a modified Hummer's method and its incorporation into pure PVDF and the composite PVDF/PANI/GO membrane using the phase inversion approach. Figure 7.13, shows the schematic diagram for the synthesis of PVDF/PANI/GO composite. Also, the dye filtration has been illustrated. They observed that, on increasing the concentration of GO further, membranes become brittle, and the number of pores in composite membrane decreases. The optimal amount of GO for wastewater treatment membrane was found to be 0.1% w/v. They came to the conclusion that the PVDF-based membranes containing PANI and GO have smaller shrinkage ratios and pore sizes with excessively rough membrane surfaces than pure PVDF membranes, after taking into account the morphological and structural evidence of the membranes. This shows that fillers enhance the permeation capabilities of composite materials.

The inclusion of GO and PANI improves the hydrophilicity of membranes. The presence of oxygenated and hydroxyl functional groups of graphene oxide on the surface of the composite membrane is shown by the contact angle's drop from 90.32° to 56.11°. The composite membrane's pure water flux increases from 112 to 454 L/m²h. Additionally, graphene oxide increased BSA rejection in the pure PVDF membrane from 38.6% to 78.32%. Also, thermal and mechanical stability improves with the addition of GO in the pristine membrane. The addition of GO and PANI to the PVDF membrane raised the degradation temperature from 398°C to 470°C. Composite membranes' Young's modulus and tensile strength increased from 32 to 90 MPa.

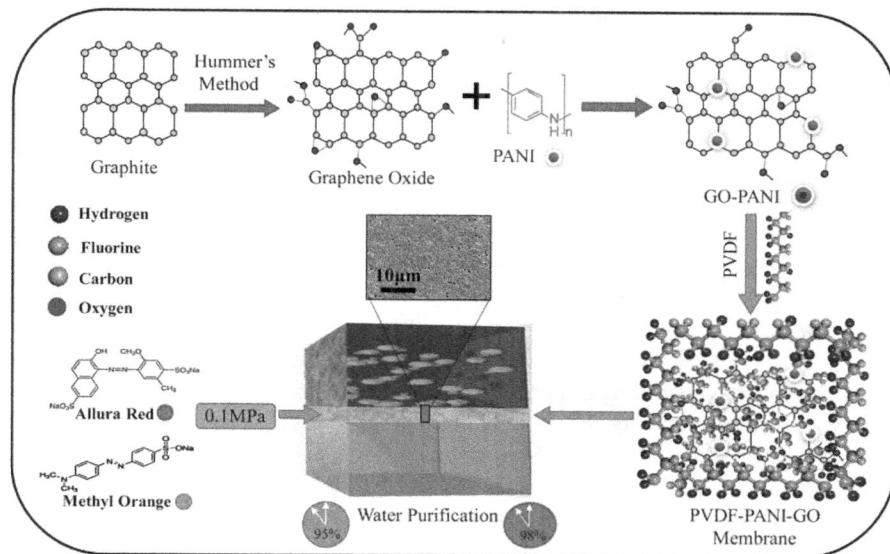

FIGURE 7.13   Shows the schematic representation of PVDF/PANI/GO composite synthesis. Also, The Allura Red and Methyl Orange dye treatment is illustrated. [Reproduced from Ref. Nawaz et al. (2021) with permission from Elsevier. Copyright 2021(PERMISSION OBTAINED).]

The flux recovery ratio of the membrane reached around 94% after several tests, and dyes rejection improved with the increase of GO and PANI content. The composite membrane has a 98% removal efficiency for allura red and a 95% removal efficiency for methyl orange.

Poly(vinyl alcohol)/graphene oxide as effective adsorbents for methylene blue (MB) was first discovered by Yang et al. (2016). The maximum adsorption capacity for the neat PVA used was 196.5 mg/g. The adsorptions of MB depend upon the content of graphene oxide being introduced, pH, temperature, and concentration of adsorbate. At 50% of graphene oxide content, they achieved the maximum adsorption capacity of 476.2 mg/g. The removal efficiencies of 77%–90% were achieved in the pH range of 2.1–9.2. The adsorption processes were best fitted to the Freundlich isotherm model. Kinetic studies suggested that the adsorptions resembled those of the pseudo-second-order kinetic model.

In a separate work, Dai et al. (2016) developed an aerogel based on polyvinyl alcohol (PVA) and graphene oxide (GO), where PVA played a role in stabilizing the 3D porous structure of GO. An environmentally friendly freeze-drying method was used to prepare the PVA-supported GO aerogels. Both GO and PVA exhibit adsorption behaviors for Congo red (CR) and methylene blue (MB). There are primarily two ways that CR molecules can bind to the GO aerogel: through hydrogen bonds with the GO layers of PVA chains and through π–π conjugation interactions with the GO layers. Meanwhile, the key mechanisms by which MB molecules can adsorb on the GO aerogel include hydrogen bonding, π–π conjugation, and charge attraction between MB molecules and GO layers, and also PVA chains can weakly interact with MB molecules. To further investigate the efficiency of PVA-supported GO aerogels in wastewater treatment, other dyes were also employed to measure the selective adsorption of the as-synthesized aerogel. The 0.5GO/0.2PVA aerogel exhibited the highest adsorption efficiency among all the PVA/GO aerogels, hence it was selected for the adsorption studies. Different color changes were observed for different dyes. For the NR, MB, and MG, the colors of solutions become lighter after adsorption. While less obvious color variations were observed after adsorption for ABK and ES. The higher adsorption efficiencies of over 96% were observed for NR, MB, and MG, while relatively lower efficiencies of 81.6%, 8.1%, and 36.7% for ES, ABK, and CR, respectively.

Ali Khan et al. (2021) fabricated GO–PANI nanocomposite using an in-situ chemical polymerization technique. They studied the adsorption of Brilliant green (BG) by the nanocomposite. The abundance and presence of carboxylic and hydroxyl groups over the GO–PANI surface played a significant role during BG adsorption. A pH-dependent BG adsorption on GO–PANI was observed; mechanistically, π–π stacking interaction and electrostatic interaction played a critical role during BG adsorption on GO–PANI. While the maximum of 92% BG removal was achieved at pH 7. They found the GO–PANI nanocomposite to be cost-effective, efficient, and an alternate material for the elimination of BG from water and wastewater. Also, El-Sharkaway et al. used (PANI/GO) and polyaniline/reduced graphene oxide (PANI/rGO) for the effective methylene blue dye removal through the adsorption process (El-Sharkaway et al. 2019). The pH-dependent MB adsorption was observed. The maximum dye adsorption capacities of PANI/GO and PANI/rGO was estimated to be 14.2 and 19.2 mg/g, respectively. From the results, it was concluded that the PANI/rGO is more effective than PANI/GO for the removal of dye pollutants in water. In addition to these, de Farias et al. have successfully synthesized polystyrene (PS) film with the help of the electrospinning method followed by coating with graphene oxide (GO/PS). They investigated the methylene blue dye removal using the composite of polystyrene and graphene oxide (PS/GO). The removal capacity of the composite was approximately 2.3 times greater than that of pure polystyrene. The adsorption followed the pseudo-second-order kinetic model, indicating an adsorption capacity (q_t) of 116.69 mg/g (de Farias et al. 2022).

Mercante et al. (2017) synthesized rGO/PMMA composite using the solution blow spinning (SBS) technique. First, PMMA fibers were synthesized by SBS technique. Then GO were uniformly decorated on the PMMA fibers, resulting in GO/PMMA composite. It was then reduced to rGO/PMMA. They reported having achieved an adsorption capacity of 698.51 mg/g for the MB dye on the rGO/PMMA composite. The dye adsorption kinetics and isotherm were well described by the pseudo-second-order and the Langmuir model, respectively. The π–π stacking interactions were considered the major driving force for the spontaneous adsorption of MB.

7.5.2   HEAVY METAL IONS REMOVAL

Currently, there is a growing interest in graphene derivatives for the treatment of wastewater having metal ions. Graphene oxide (GO)—the oxidized form of graphene that is mainly prepared by the Hummers method and includes a large amount of oxygen-related functional groups, epoxy and hydroxyl groups in the basal planes, and carboxyl acid groups at the edge. Its high mechanical strength and specific surface area make it a suitable material for the efficient removal of heavy metal ions from an aqueous solution. However, GO has a high affinity toward the water, thus making the possibility of leaching and a high cost of synthesis. Moreover, functional groups on GO are very limited. Therefore, the introduction of new functional groups on GO's surface using polymer leads to an efficient nanocomposite with remarkable adsorption properties.

Aniline is a good electron donor, whereas graphene is an excellent electron receiver. Vargas et al. (2017) found that chemical bonding, hydrogen bonding, π–π interactions, and electrostatic interactions are responsible for the GO–PANI interaction in their investigation with GO–PANI. In this context, Zhang et al. (2013) reported the dilute polymerization of polyaniline nanorod-dotted graphene oxide nanosheets below 20°C for the manufacture of polyaniline nanorod-dotted graphene oxide nanosheets. The composite had a high Cr (VI) adsorption capacity of 1149.4 mg/g, and a reduction in Cr (III) was also observed. The nanocomposite was also stable and environmentally friendly, with excellent regeneration and recyclability, and no noticeable capacity loss observed after five cycles. In this work, Fan et al. (2015) studied the adsorption capacity of PANI/GO for Hg(II). SEM analysis revealed that the synthesized PANI/GO is a loose, spongy macroscale porous structure, with sponges having a rough, shaggy structure with numerous pores, which is significantly different from the raw GO. The modified PANI on the GO surface can effectively enhance PANI/GO adsorption capacity for Hg (II). The adsorption isotherms of Hg (II) on PANI/GO is shown in Figure 7.14b. The kinetic adsorption process of Hg (II) on PANI/GO was well described by the pseudo-second-order model under experimental conditions. The ideal PANI-to-GO weight ratio for adsorption was discovered to be 0.64:1.0. At pH 5, the maximum adsorption capacity of 80.7 mg/g toward Hg (II) was achieved. Furthermore, the presence of coexisting ions had only a minor effect on Hg adsorption (II).

In another study, a polyaniline nanocomposite coated with graphene sheets was synthesized by polymerizing aniline on the surface of prepared GO and then reducing it with borohydride (Harijan and Chandra 2016). The nanocomposites had the same coral-like dendritic nanofiber morphology as PANI. The nanocomposite with 10% GO loading was tested for adsorption capacity toward Cr(VI), and it demonstrated the maximum adsorption capacity of 192 mg/g under optimal conditions of pH 2 and 20 minutes of equilibrium time. Ramezanzadeh et al. (2018) fabricated graphene oxide nanosheets coated with highly crystalline PANI nanofibers for Zn(II) removal. PANI nanofibers were successfully linked to the GO and enveloped the graphene oxide surface, according to surface analysis. Microstructural analyses revealed a rough and uneven morphology, implying that fine PANI nanofibers were deposited on the GO surface. Within 20 minutes, the adsorption capacity was 1160 mg/g under optimal conditions of pH 7. The UV-Vis spectral analysis of GO/PANI after adsorption revealed that, despite the high adsorption, the PANI/GO particles' stability was compromised, resulting in agglomeration and sedimentation. Shao et al. (2014) while working with the polyaniline-modified graphene oxide (PANI/GO)

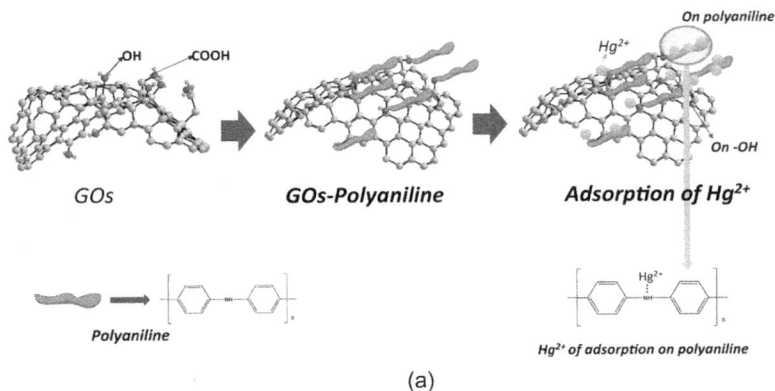

(a)

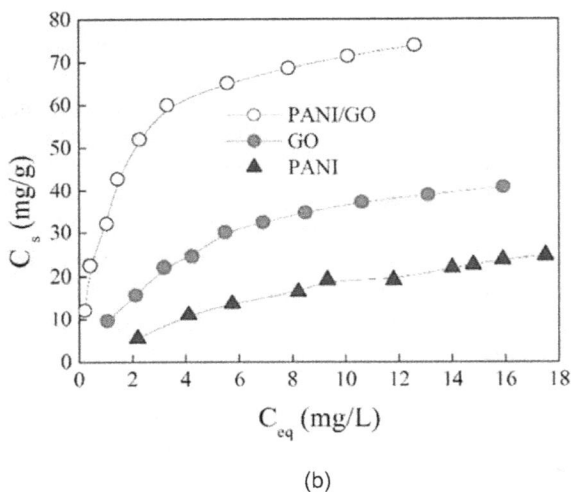

(b)

FIGURE 7.14 (a) Schematic representation of the Hg^{2+} adsorption by PANI/GO. (b) Adsorption isotherms of Hg (II) on PANI/GO. Contact time 24 h, $T = 25 \pm 1°C$, m/V = 0.10 g/L, pH = 5.0 ± 0.1, C[CaCl$_2$]. [Reproduced from Ref. Fan et al. (2015) with permission from Elsevier. Copyright 2015 (PERMISSION OBTAINED).]

composites synthesized by in-situ polymerization technique achieved 1960 mg/g at pH 5.0 and 610 mg/g at pH 3.5 for Uranium (IV). Moreover, they reported that PANI/GO composites show exceptional capability in selective removal of U(VI) from an aqueous solution under a variety of conditions. They claim the composite to be two-fold better than those of traditional adsorbents and today's nanomaterials. Also, the composite exhibited ideal tolerance toward high salt concentrations and excellent regeneration–reuse property.

Zhang et al. observed that the composite of PVP with GO can be used for the treatment of Cu ions. PVP–GO shows an adsorption capability of 1689 mg/g, which is much higher than that of carbon nanotubes, GO, and activated carbon (Zhang et al. 2014).

Density functional theory was employed to study the adsorption mechanism. It reveals that Cu ions are attracted to the surface of reduced graphene oxide by carbon atoms in reduced graphene oxide modified by polyvinylpyrrolidone through physisorption processes responsible for the higher adsorption capacity. Results pointed out that polyvinylpyrrolidone-reduced graphene oxide is an effective adsorbent for treating Cu ions in waste water. Recently, Tang et al. (2021) employed poly(vinyl alcohol)/graphene oxide composite electrodes for efficient electrosorption removal of U (VI) from an aqueous solution. They observed that the electrosorption capacity of U (VI) reaches 333.0 mg/g at 0.9 V. Moreover, excellent regeneration was evident from the electrosorption–desorption cycles. The study suggests the potential application of PVA/GO electrodes for efficient electrosorption removal of U (VI) from wastewater. Li et al. (2013) synthesized PANI/rGO composite through polymerization of aniline in the presence of graphene oxide, and hydrazine hydrate was used as a reducing agent. They observed that, in comparison to PANI, the maximum adsorption capacity of PANI/rGO for Hg (II) was increased from 515.46 to 1000.00 mg/g, and the adsorption process was accelerated at pH 4.0. The superior adsorption was attributed to the presence of reduced graphene oxide, which results in seven times enhancement in the specific surface area and the adsorption sites.

7.5.3   OIL–WATER SEPARATION

Many industries, such as mining, textiles, foods, petrochemicals, and metal/steel industries, produce huge amounts of oily wastewater, which has become a worldwide contaminant and is now a severe global environmental concern. A typical mining operation, for example, produces 140,000 L of oil-contaminated water each day (Guerin 2002). Furthermore, frequent oil spills/leaks during marine transportation or oil production pose a threat to marine habitats and ecology (Wang 2006; Chen and Xu 2013) as well as a significant waste of important natural resources. Thanks to their outstanding physicochemical features, including as high specific surface area, low density, high porosity, and tailorable surface functionality, graphene (G) and graphene oxide (GO) have emerged as key materials in the field of oil/water separation. Because of their increased oil clean-up capability, superior mechanical performance, low cost, and customizable surface chemical composition, the integration of G and GO with polymers to construct functional G/polymer and GO/polymer composites has recently acquired increasing appeal. G/polymer and GO/polymer composite oil clean-up sorbents and filtering membranes in 3D structural forms such as aerogels, foams, sponges, and membranes have benefited from tremendous efforts.

Li et al. reported that (Li et al. 2014) superhydrophobic and superoleophilic graphene/polyvinylidene fluoride (G/PVDF) aerogels were prepared by solvothermal reduction of the graphene oxide and PVDF-mixed dispersions. In a Teflon-lined autoclave, the graphene/PVDF organogels were made by solvothermal reduction of graphene oxide and PVDF-mixed dispersions. The solvent exchange method was used to convert them to hydrogels, which were subsequently freeze-dried to produce the equivalent aerogels. The optimal reaction conditions were as follows: 160 C, GO/PVDF = 2/5, DMF/water = 7/3 or 8/2, The corresponding aerogels were termed AG-1 and AG-2, respectively. FT-IR, XRD, XPS, Raman spectroscopy, and TGA were used to confirm the chemical reduction of the graphene oxide component. The maximum adsorption capacity (Q) of AG-1 and AG-2 was determined by oil-absorption experiments; they showed an absorption capacity of 20–70 for various oils and organic solvents, which was much higher than that of many carbon aerogels. Moreover, even after ten cycles, AG-1 and AG-2's absorption capacity only slightly decreased and continued to function at a high capacity, demonstrating the excellent absorption recyclability of these two compounds. Moreover, the composite efficiency was further tested for different oils and various organic solvents. The maximum adsorption by AG-1 is shown for phenixin (~7500%) and the minimum for ethanol (~2500%). The maximum adsorption for AG-2 is shown for chloroform (~6500%) and the minimum for heptane (~2000%). They concluded that the as-prepared aerogel exhibited a high specific surface area, outstanding oil and organic solvent absorption capacity, exceptional water repellence, great absorption recyclability, and significant mechanical characteristics. As a result, this type of aerogel holds promise for oil–water separation, oil spill clean-up, and organic solvent recovery. Furthermore, using graphene oxide and a hydrophobic polymer, this research paved the way for the easy fabrication of superhydrophobic and superoleophilic graphene-based aerogels.

Dai et al. (2016) investigated the oil–water separation ability of polyvinyl alcohol and graphene oxide (PVA/GO) aerogels. In this work, the microstructure, the structure stability, and the oil/water separation of the aerogels were investigated. The PVA-supported GO aerogels demonstrated great results. It exhibited high structure stability, especially in the non-polar solvent, which was missing in widely reported GO aerogels in the literature, which experienced structure disintegration in the solvent, mainly in water. They reported that the aerogels completely adsorbed Sudan red/cyclohexane solution that floats on the water within 5 seconds, as shown in Figure 7.15. This work indicates the great potential of PVA/GO in wastewater treatment.

In another work, Ghaffar et al. (2019) reported a multifunctional porous structured polyvinyl alcohol (PVA)-based nanofibrous membrane (NFM) with graphene

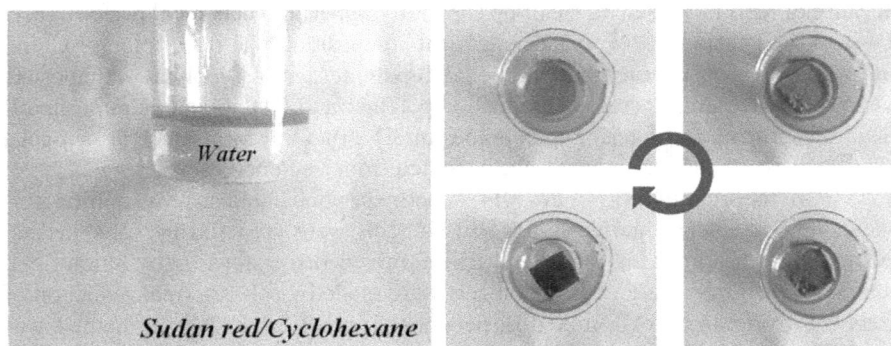

FIGURE 7.15 Photographs showing the process of adsorption of cyclohexane into the 0.1GO/0.2PVA aerogel. [Reproduced from Ref. Dai et al. (2016) with permission from Elsevier. Copyright 2016 (PERMISSION OBTAINED).]

oxide (GO) was prepared by easy and basic electrospinning and used to treat oily water. The PVA–GO NFM was designed with an effective pore shape and selective wettability in mind to promote oily water emulsion separation totally by gravity, with a regulated GO ratio and compatibility to avoid the aggregation and easy spin ability. To understand the mechanism of oily water emulsion separation, Figure 7.6b illustrates the significant difference between the surfactant-free and surfactant-stabilized oil emulsions. Compared to the PVA–GO NFM, the PVA NFM's surface allows for easier oil droplet adhesion. Because the PVA NFM surface has a poor anti-oil property (Figure 7.6b), oil droplets cannot be removed from it easily. This can quickly block nanopores, resulting in a low permeation flux. The PVA–GO NFM, on the other hand, demonstrates superior anti-oil properties and demonstrates good permeation flux, preventing the oil droplets from adhering to the membrane surface easily, and allowing them to move freely and coalesce because of the presence of GO. As a result, the entire oil droplet could be effortlessly separated from the PVA–GO NFM surface, allowing only water to pass through and be recycled for later use. Owing to a homogeneous distribution with high mechanical strength and thermal stability, the shape of the PVA NFM was entirely altered in the presence of GO. As a result of the inclusion of GO, the as-prepared PVA–GO NFM achieves a separation efficiency of above 99% for surfactant-free and surfactant-stabilized oily water emulsions with water fluxes of around 45 and 30 L/m^{2}h, respectively, exclusively under gravitational force. Furthermore, even after recycling, the PVA–GO NFM retains its real selectivity and surface wettability, enabling long-term applications with nearly the same separation efficiency and water flux. The current membrane fabrication methodology is thought to be an effective energy-saving filtration solution with anti-oil properties and reusability, making it attractive for application in water purification and treatment procedures, particularly for oily water treatment.

7.5.4 DESALINATION

Even though the Earth has a lot of water, 97.5% of it is seawater with an average salinity of 35,000 ppm or milligrams per liter (Ibrahim et al. 2017; Shahzad et al. 2017). In other words, the amount of fresh water on Earth is only 2.5%, of which 80% is contained in glaciers, leaving 20% (or 0.5%) to be found in the rivers, lakes, and aquifers. Freshwater is being taken out of many areas of the planet faster than it is being replenished naturally (Richter et al. 2013). A rise in worldwide water use is anticipated due to a population that is both urbanizing and increasing quickly. Water scarcity is spreading and getting worse all across the world, as water demand rises. Around 40% of the world's population is thought to experience severe water shortages, and that percentage is predicted to increase to 60% by 2025 (Ibrahim et al. 2017). This is mostly brought on by the rise in world population, contaminated and overused freshwater resources, and economic activity. Desalination of the ocean's water is an effective way to counter the water crisis. In this direction, carbon-based nanomaterials like carbon nanotubes (CNTs), graphene, and their derivatives (GO and rGO) have acquired a lot of attention from the scientific community because of their unique properties like large surface area, wide porosity range, excellent thermal and electrical conductivity, and high mechanical strength and stiffness (Mishra and

Ramaprabhu 2011; Li et al. 2009; Zhu et al. 2010). The introduction of carbon nano-tubes, graphene, and graphene oxide with these unique features has been suggested as a key potential advance in water desalination. A recent study found that using single-layer non-porous graphene, 100% salt rejection may be obtained for routinely used ions. However, a viable filtration solution can be found in the cost-effective production of graphene oxide membranes with fine control of pore size if one can attain a salt rejection rate of 100% (Zhu et al. 2015). Furthermore, these functional groups serve as a foundation for composite structures, such as graphene oxide–polymer composites, which have been proven to improve membrane mechanical stability and characteristics. Graphene oxide also has hydrophilicity and pH sensitivity because of the oxygenated functional groups (Mi 2014; Ganesh et al. 2013). The functional group density is properly managed during fabrication, since the hydrophobic region governs water flow. Various studies concerning the use of synthetic polymer–graphene oxide nanocomposites for desalination purposes have been discussed here . Recently, there has been a lot of interest in using graphite oxide for membrane distillation-type water desalination. However, due to difficulties in precise pore control, high production costs, and complicated processes, most previous studies on graphite oxide have remained only theoretical.

In this context, Su et al. (2019) had fabricated a highly hydrophobic and permeable block GO-PVP/PVDF membrane for desalination that addresses the above-mentioned challenges utilizing low-cost, readily available raw materials, and processing via a simple vacuum filtration process. The optimal arrangement of GO nanosheets was obtained at a suitable mixing ratio of GO and PVP to achieve the highest vapor permeation. As a result, the fabricated block GO-PVP/PVDF membranes have a high moisture permeability, allowing moisture to pass not only through the overlapping graphene gaps but also through the PVP cracks. They said that the permeable GO-PVP/PVDF membranes exhibit excellent permeability with the vapor diffusivity of $1.82 \times 10^6 \, m^2/s$ (which is almost as permeable as PVDF membranes), and extremely high hydrophobicity with the contact angle of 145.2° without any post-synthesis pore procedure. According to the authors' knowledge, this study is the first to create a highly permeable and extremely hydrophobic GO-assisted desalination membrane by mixing a polymer with the GO suspensions. The current membrane might help increase the effectiveness of desalinating seawater. This new GO-PVP/PVDF membrane may prove to be a practical desalination technology with improvements in permeability and a better comprehension of heat and mass transfer.

In another study, Guo et al. (2020) prepared GO/PVA electrospun nanofibrous membrane (EFMs) using an electrospinning apparatus. The as-synthesized EFMs were successfully employed for solar photothermal desalination of seawater. The uniform nanofibers that are made up of the GO/PVA EFMs had GO immobilized in them as a component for photo absorption. The GO concentration was directly correlated with the photothermal conversion characteristics of GO/PVA EFMs. Under 1 sun (1 kW/m^2) irradiation, the evaporation rate of GO/PVA EFMs with 5% of GO concentration for clean water could reach up to 1.40 kg/m^2h, and the energy conversion efficiency was as high as 90.0%. Their ideal evaporation rate for the simulated seawater was 1.42 kg/m^2h, and the associated energy conversion efficiency was up to 94.2%. The concentrations of five principal ions in the purified water are around

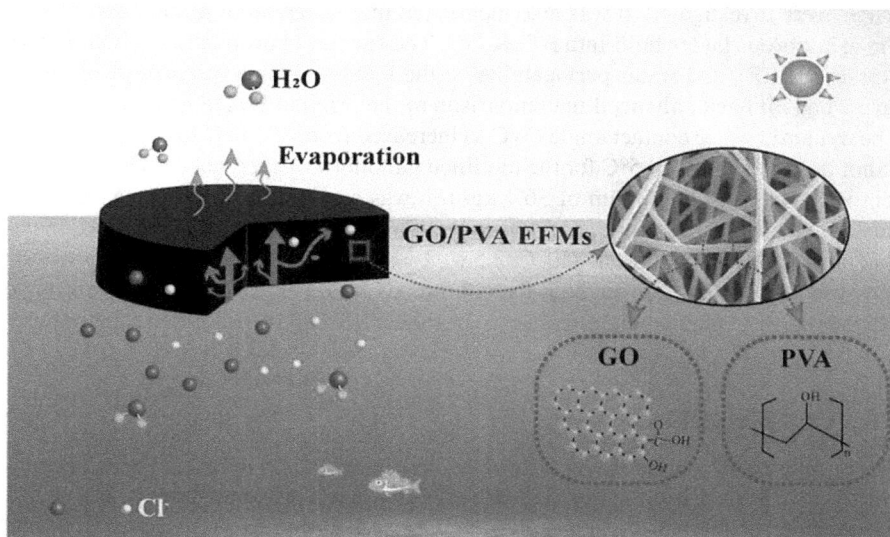

FIGURE 7.16 Schematic representation of desalination of seawater using the GO/PVA. [Reproduced from Ref. Guo et al. (2020) with permission from Elsevier. Copyright 2020 (PERMISSION OBTAINED).]

1 mg/L after utilizing the GO/PVA EFMs-5 percent to desalinate actual seawater from the Yellow Sea, which is significantly below the WHO-recommended limit for drinking water.

The GO/PVA EFMs exhibit excellent photothermal desalination efficiency, as shown in Figure 7.16. The extreme hydrophilicity could guarantee a constant supply of water, the uniformly distributed GO in the fibers could achieve efficient photo absorption and photothermal conversion, the porous non-woven structure could temporarily confine heat energy to help liquid water turn into water vapors, and the numerous mesoporous channels could help water vapor to flow and escape. More importantly, the GO/PVA EFMs have exceptional durability and reusability, as well as great stability in severe environments such as strong acid and alkalis, high-concentration brine, and long-term photothermal evaporation. As far as practical, long-term applications in solar photothermal desalination are concerned, GO/PVA EFMs are the most promising materials. In a separate work, Li et al. (2020) introduced hydrophobic nanofibrous composites for vacuum membrane distillation (VMD), consisting of a hydrophobic polypropylene (PP) non-woven fabric (NWF) substrate and a hydrophobic polyvinylidene fluoride (PVDF) nanofibrous layer. By using the electrospinning technique, a PVDF nanofibrous layer was directly created on the surface of a PP NWF substrate. To increase the hydrophobicity and water flux permeation for VMD, 1H, 1H, 2H, and 2H-perfluorooctyltriethoxysilane (FTES), functionalized GO nanosheets were added to the PVDF nanofiber layer during the electrospinning process. Surface morphology and composition were characterized by SEM and FTIR. The effects of FTES-GO nanosheet content on the desalination capabilities of

VMD were investigated. It was also mentioned that water vapor might pass through the nanofibrous layers built into FTES-GO. The surface hydrophobicity, liquid entry pressure (LEP), and water permeability of the FTES-GO incorporated nanofibrous layers had all been enhanced in comparison to the original PVDF nanofibrous layer. The dynamic water contact angle (WCA) increased from 104.0°C for the neat PVDF nanofibrous layer to 140.5°C for the modified nanofibrous layer when FTES-GO content was 4 wt%. A maximum of 36.4 kg/m²h was reached for the permeation water flux. The salt rejection remained above 99.9% while the water flux was two times that of the original membrane (50°C, 3.5 wt% NaCl aqueous solution and permeation pressure of 31.3 kPa). Throughout the continuous VMD experiment for 60 hours, no obvious wetting phenomenon was seen for the FTES-GO incorporated membrane

7.6   CONCLUSION

Wastewater treatment becomes of utmost importance when it comes to the sustainability of life on Earth. Industrial effluents discarded in water bodies are one of the major contributors to water pollution. Out of the various wastewater treatment techniques, activated carbon-based adsorption techniques show promising results. GO—a derivative of graphene can be synthesized with the help of various methods. The improved method is one of the most efficient methods for the synthesis of GO in bulk. Due to the presence of various oxygenated functional groups, GO is a perfect adsorbing material for organic dyes, metal ions, and desalination of ocean water. GO in its reduced form (rGO), which is hydrophobic, can be a perfect candidate for the separation of oil from water. GO when blended with various synthetic polymers show enhanced adsorption efficiency. Numerous techniques such as electrospinning, melt compounding, in-situ polymerization, solvothermal method, and so on are employed for the synthesis of GO-based polymer nanocomposites. Various research groups have reported the fabrication of GO-based polymer nanocomposites of PVA, PS, PMMA, PANI, and PVDF and have shown that these nanocomposites have a very high affinity toward the adsorption of various pollutants, *viz.* organic dyes, metal ions, oil and hydrocarbons, and desalination. rGO-based nanocomposites of these polymers are great at separating oil from water.

REFERENCES

Ali Khan, M., et al. (2021). "Carbon based polymeric nanocomposites for dye adsorption: Synthesis, characterization, and application." *Polymers* **13**(3): 419.

Balapanuru, J., et al. (2010). "A graphene oxide–organic dye ionic complex with DNA-sensing and optical-limiting properties." *Angewandte Chemie International Edition* **49**(37): 6549–6553.

Balli, B., et al. (2019). Graphene and polymer composites for supercapacitor applications. *Nanocarbon and Its Composites*, Elsevier: 123–151.

Beygisangchin, M., et al. (2021). "Preparations, properties, and applications of polyaniline and polyaniline thin films—A review." *Polymers* **13**(12): 2003.

Bhol, P., et al. (2021). "Graphene-based membranes for water and wastewater treatment: A review." *ACS Applied Nano Materials* **4**(4): 3274–3293.

Blinova, N. V., et al. (2008). "Control of polyaniline conductivity and contact angles by partial protonation." *Polymer International* **57**(1): 66–69.

Brodie, B. C. (1859). "XIII. On the atomic weight of graphite." *Philosophical Transactions of the Royal Society of London* (149): 249–259.

Chen, P.-C. and Z.-K. Xu (2013). "Mineral-coated polymer membranes with superhydrophilicity and underwater superoleophobicity for effective oil/water separation." *Scientific Reports* **3**(1): 1–6.

Chong, M. N., et al. (2010). "Recent developments in photocatalytic water treatment technology: A review." *Water Research* **44**(10): 2997–3027.

Cui, Y., et al. (2016). "Gas barrier performance of graphene/polymer nanocomposites." *Carbon* **98**: 313–333.

Dai, J., et al. (2016). "High structure stability and outstanding adsorption performance of graphene oxide aerogel supported by polyvinyl alcohol for waste water treatment." *Materials & Design* **107**: 187–197.

de Farias, L. M., et al. (2022). "Electrospun polystyrene/graphene oxide fibers applied to the remediation of dye wastewater." *Materials Chemistry and Physics* **276**: 125356.

DeMerlis, C. and D. Schoneker (2003). "Review of the oral toxicity of polyvinyl alcohol (PVA)." *Food and Chemical Toxicology* **41**(3): 319–326.

Demir, M. M., et al. (2006). "PMMA/zinc oxide nanocomposites prepared by in-situ bulk polymerization." *Macromolecular Rapid Communications* **27**(10): 763–770.

Dutta, S., et al. (2021). "Recent advances on the removal of dyes from wastewater using various adsorbents: A critical review." *Materials Advances*.

El-Sharkaway, E., et al. (2019). "Removal of methylene blue from aqueous solutions using polyaniline/graphene oxide or polyaniline/reduced graphene oxide composites." *Environmental Technology*.

Fan, Q., et al. (2015). "Preparation of three-dimensional PANI/GO for the separation of Hg (II) from aqueous solution." *Journal of Molecular Liquids* **212**: 557–562.

Fu, G., et al. (2014). "Photo-crosslinked nanofibers of poly (ether amine)(PEA) for the ultrafast separation of dyes through molecular filtration." *Polymer Chemistry* **5**(6): 2027–2034.

Gaaz, T. S., et al. (2015). "Properties and applications of polyvinyl alcohol, halloysite nanotubes and their nanocomposites." *Molecules* **20**(12): 22833–22847.

Ganesh, B., et al. (2013). "Enhanced hydrophilicity and salt rejection study of graphene oxide-polysulfone mixed matrix membrane." *Desalination* **313**: 199–207.

Ghaffar, A., et al. (2018). "Porous PVdF/GO nanofibrous membranes for selective separation and recycling of charged organic dyes from water." *Environmental Science & Technology* **52**(7): 4265–4274.

Ghaffar, A., et al. (2019). "Underwater superoleophobic PVA–GO nanofibrous membranes for emulsified oily water purification." *Environmental Science: Nano* **6**(12): 3723–3733.

Greve, P., et al. (2018). "Global assessment of water challenges under uncertainty in water scarcity projections." *Nature Sustainability* **1**(9): 486–494.

Guerin, T. F. (2002). "Heavy equipment maintenance wastes and environmental management in the mining industry." *Journal of Environmental Management* **66**(2): 185–199.

Guo, X., et al. (2020). "Scalable, flexible and reusable graphene oxide-functionalized electrospun nanofibrous membrane for solar photothermal desalination." *Desalination* **488**: 114535.

Haaf, F., et al. (1985). "Polymers of N-vinylpyrrolidone: Synthesis, characterization and uses." *Polymer Journal* **17**(1): 143–152.

Harijan, D. K. and V. Chandra (2016). "Polyaniline functionalized graphene sheets for treatment of toxic hexavalent chromium." *Journal of Environmental Chemical Engineering* **4**(3): 3006–3012.

Hummers Jr, W. S. and R. E. Offeman (1958). "Preparation of graphitic oxide." *Journal of the American Chemical Society* **80**(6): 1339–1339.

Ibrahim, A. G., et al. (2017). "Exergoeconomic analysis for cost optimization of a solar distillation system." *Solar Energy* **151**: 22–32.

Johnson, P. T. and S. H. Paull (2011). "The ecology and emergence of diseases in fresh waters." *Freshwater Biology* **56**(4): 638–657.

Khan, M. A. and A. M. Ghouri (2011). "Environmental pollution: Its effects on life and its remedies." *Researcher World: Journal of Arts, Science & Commerce* **2**(2): 276–285.

Kuilla, T., et al. (2010). "Recent advances in graphene based polymer composites." *Progress in Polymer Science* **35**(11): 1350–1375.

Landrigan, P. J. (2017). "Air pollution and health." *The Lancet Public Health* **2**(1): e4–e5.

Landrigan, P. J., et al. (2018). "The Lancet Commission on pollution and health." *The Lancet* **391**(10119): 462–512.

Lee, C., et al. (2008). "Measurement of the elastic properties and intrinsic strength of monolayer graphene." *Science* **321**(5887): 385–388.

Lellis, B., et al. (2019). "Effects of textile dyes on health and the environment and bioremediation potential of living organisms." *Biotechnology Research and Innovation* **3**(2): 275–290.

Li, H., et al. (2020). "Improved desalination properties of hydrophobic GO-incorporated PVDF electrospun nanofibrous composites for vacuum membrane distillation." *Separation and Purification Technology* **230**: 115889.

Li, R., et al. (2013). "Preparation of polyaniline/reduced graphene oxide nanocomposite and its application in adsorption of aqueous Hg (II)." *Chemical Engineering Journal* **229**: 460–468.

Li, R., et al. (2014). "A facile approach to superhydrophobic and superoleophilic graphene/polymer aerogels." *Journal of Materials Chemistry A* **2**(9): 3057–3064.

Li, X., et al. (2009). "Large-area synthesis of high-quality and uniform graphene films on copper foils." *Science* **324**(5932): 1312–1314.

Liu, M., et al. (2007). "Drying induced aggregation of halloysite nanotubes in polyvinyl alcohol/halloysite nanotubes solution and its effect on properties of composite film." *Applied Physics A* **88**(2): 391–395.

Marcano, D. C., et al. (2010). "Improved synthesis of graphene oxide." *ACS Nano* **4**(8): 4806–4814.

Maul, J., et al. (2007). "Polystyrene and styrene copolymers." *Ullmann's Encyclopedia of Industrial Chemistry* **29**: 475–522.

Mercante, L. A., et al. (2017). "Solution blow spun PMMA nanofibers wrapped with reduced graphene oxide as an efficient dye adsorbent." *New Journal of Chemistry* **41**(17): 9087–9094.

Mi, B. (2014). "Graphene oxide membranes for ionic and molecular sieving." *Science* **343**(6172): 740–742.

Mishra, A. K. and S. Ramaprabhu (2011). "Functionalized graphene sheets for arsenic removal and desalination of sea water." *Desalination* **282**: 39–45.

Moorthy, A. K., et al. (2021). "Acute toxicity of textile dye Methylene blue on growth and metabolism of selected freshwater microalgae." *Environmental Toxicology and Pharmacology* **82**: 103552.

Nawaz, H., et al. (2021). "Polyvinylidene fluoride nanocomposite super hydrophilic membrane integrated with Polyaniline-Graphene oxide nano fillers for treatment of textile effluents." *Journal of Hazardous Materials* **403**: 123587.

Park, S. and R. S. Ruoff (2009). "Chemical methods for the production of graphenes." *Nature Nanotechnology* **4**(4): 217–224.

Pendolino, F. and N. Armata (2017). *Graphene Oxide in Environmental Remediation Process*, Springer.

Pouget, J., et al. (1991). "X-ray structure of polyaniline." *Macromolecules* **24**(3): 779–789.

Poza-Nogueiras, V., et al. (2018). "Current advances and trends in electro-Fenton process using heterogeneous catalysts–a review." *Chemosphere* **201**: 399–416.

Puri, C. and G. Sumana (2018). "Highly effective adsorption of crystal violet dye from contaminated water using graphene oxide intercalated montmorillonite nanocomposite." *Applied Clay Science* **166**: 102–112.

Ramesha, G., et al. (2011). "Graphene and graphene oxide as effective adsorbents toward anionic and cationic dyes." *Journal of Colloid and Interface Science* **361**(1): 270–277.

Ramezanzadeh, M., et al. (2018). "Fabrication of an efficient system for Zn ions removal from industrial wastewater based on graphene oxide nanosheets decorated with highly crystalline polyaniline nanofibers (GO-PANI): Experimental and ab initio quantum mechanics approaches." *Chemical Engineering Journal* **337**: 385–397.

Razzak, M. T. and D. Darwis (2001). "Irradiation of polyvinyl alcohol and polyvinyl pyrrolidone blended hydrogel for wound dressing." *Radiation Physics and Chemistry* **62**(1): 107–113.

Rey, P. and R. Varsanik (1986). *Application and Function of Synthetic Polymeric Flocculents in Wastewater Treatment*, ACS Publications.

Richter, B. D., et al. (2013). "Tapped out: How can cities secure their water future?" *Water Policy* **15**(3): 335–363.

Scheirs, J. and D. Priddy (2003). *Modern Styrenic Polymers: Polystyrenes and Styrenic Copolymers*, John Wiley & Sons.

Shahzad, M. W., et al. (2017). "Pushing desalination recovery to the maximum limit: Membrane and thermal processes integration." *Desalination* **416**: 54–64.

Shao, D., et al. (2014). "PANI/GO as a super adsorbent for the selective adsorption of uranium (VI)." *Chemical Engineering Journal* **255**: 604–612.

Shen, J., et al. (2009). "Fast and facile preparation of graphene oxide and reduced graphene oxide nanoplatelets." *Chemistry of Materials* **21**(15): 3514–3520.

Staudenmaier, L. (1898). "Verfahren zur darstellung der graphitsäure." *Berichte der deutschen chemischen Gesellschaft* **31**(2): 1481–1487.

Su, Q.-W., et al. (2019). "Fabrication and analysis of a highly hydrophobic and permeable block GO-PVP/PVDF membrane for membrane humidification-dehumidification desalination." *Journal of Membrane Science* **582**: 367–380.

Sun, L. (2019). "Structure and synthesis of graphene oxide." *Chinese Journal of Chemical Engineering* **27**(10): 2251–2260.

Tang, X., et al. (2021). "Nanoarchitectonics of poly (vinyl alcohol)/graphene oxide composite electrodes for highly efficient electrosorptive removal of U (VI) from aqueous solution." *Separation and Purification Technology* **278**: 119604.

Vargas, L. R., et al. (2017). "Formation of composite polyaniline and graphene oxide by physical mixture method." *Journal of Aerospace Technology and Management* **9**: 29–38.

Velusamy, S., et al. (2021). "A review on heavy metal ions and containing dyes removal through graphene oxide-based adsorption strategies for textile wastewater treatment." *The Chemical Record* **21**(7): 1570–1610.

Wang, L. K. (2006). Waste chlorination and stabilization. *Advanced Physicochemical Treatment Processes*, Springer: 403–440.

Wawrzkiewicz, M. and Z. Hubicki (2015). "Anion exchange resins as effective sorbents for removal of acid, reactive, and direct dyes from textile wastewaters." *Ion Exchange-Studies and Applications*: 37–72.

Wei, C., et al. (2017). "Ozonation in water treatment: The generation, basic properties of ozone and its practical application." *Reviews in Chemical Engineering* **33**(1): 49–89.

Wu, X., et al. (2017). "Polyvinylpyrrolidone modified graphene oxide as a modifier for thin film composite forward osmosis membranes." *Journal of Membrane Science* **540**: 251–260.

Xiao, M., et al. (2002). "Synthesis and properties of polystyrene/graphite nanocomposites." *Polymer* **43**(8): 2245–2248.

Yang, S.-T., et al. (2010). "Folding/aggregation of graphene oxide and its application in Cu2+ removal." *Journal of Colloid and Interface Science* **351**(1): 122–127.

Yang, X., et al. (2016). "Adsorption of methylene blue from aqueous solutions by polyvinyl alcohol/graphene oxide composites." *Journal of Nanoscience and Nanotechnology* **16**(2): 1775–1782.

Zahrim, A. and N. Hilal (2013). "Treatment of highly concentrated dye solution by coagulation/flocculation–sand filtration and nanofiltration." *Water Resources and Industry* **3**: 23–34.

Zhang, M., et al. (2015). "Recent advances in the synthesis and applications of graphene–polymer nanocomposites." *Polymer Chemistry* **6**(34): 6107–6124.

Zhang, S., et al. (2013). "Polyaniline nanorods dotted on graphene oxide nanosheets as a novel super adsorbent for Cr (VI)." *Dalton Transactions* **42**(22): 7854–7858.

Zhang, Y., et al. (2014). "Highly efficient adsorption of copper ions by a PVP-reduced graphene oxide based on a new adsorptions mechanism." *Nano-Micro Letters* **6**(1): 80–87.

Zhu, B., et al. (2015). "Application of robust MFI-type zeolite membrane for desalination of saline wastewater." *Journal of Membrane Science* **475**: 167–174.

Zhu, Y., et al. (2010). "Graphene and graphene oxide: Synthesis, properties, and applications." *Advanced Materials* **22**(35): 3906–3924.

8 Recent Development of Natural Polymers– Graphene Oxide- Based Composites for Wastewater Treatment
Mechanism and Classifications

Ishita Mukherjee
The University of Burdwan

CONTENTS

DOI: 10.1201/9781003328094-8

8.1   INTRODUCTION

From the earliest ages, the most crucial constituents for life construction on earth is water. Even a human adult body is made of 60% of water.[1] According to several scientific experimental reports, human brain and heart are composed of 73%, lungs are of almost 83%, and muscles and kidneys are of 79% water. Moreover, water resources are the main pillars to flourish modern civilisation since ancient era, as most of the industries like paper, plastic, food, carpet, cosmetic, textile and printing, etc. are associated with water components. However, the present century witnessed severe water pollution as a result of excessive and rapid industrial growth, and the situation is becoming worse day by day. This is happening majorly due to absence of consisting process to a spontaneous water cleaning and purification with high recyclability worldwide. According to a scientific report from World Health Organization (WHO, 2012), almost 780 million of people have the lack of accessibility of purified water even in modern world.[2] Additionally, the over usage of limited water resources on the earth in industrial purpose and due to lack of knowledge can lead to a severe shortage of purified drinking water in the near future. This phenomenon desperately challenges the entire existence of human health and civilisations at this time.

Hopefully, awareness is increasing within society in the recent days. Moreover, scientists are putting their best effort to overcome this global challenge. The highest research attention has been paid to recyclability of wastewater in this regard. Generally, industrial waste water is composed of dyes and heavy metals, and a series of materials have been reported for wastewater treatments. Active carbon compound is one of the most investigated substances due to their adsorption capacity of hazardous dyes or heavy metals from water. Several scientific groups have been suggested the graphene, graphene oxide or carbon nanotubes as efficient compounds for wastewater treatment. Similarly natural polymers, specially polysaccharides, proteins, DNA extracted from several plant, animal or microbial sources can be considered by coagulation, flocculation[3] or degradation[4] of hazardous chemical substances to transform wastewater to drinking water. Recent development focused on modified active carbon material, with natural polymers introducing much higher efficacy to adsorption through composite formation. Again, the natural polymers are generally biocompatible and biodegradable in nature; as a result, they introduced the same properties. For example, polysaccharide-based natural polymers can introduce specific properties like water solubility, gelation capacity or important surface properties.

Graphene has recently attracted much interest in the materials field of active carbon compounds due to its unique 2D structure: outstanding physical and mechanical properties. When this graphene is highly oxidised, it transformed to graphene oxide (GO). This compound exhibited an excellent exfoliation ability in water due to presence of several oxygen functionalities. The amorphous carbon plane of GO was decorated with numerous epoxy and hydroxy functional groups with non-stoichiometric atomic composition. This led to an excellent adsorption property compared to normal graphene.[5] Recently, two-dimensional GO surface extensively coupled with natural polymers to generate a three-dimensional composite,[6,7] which can be classified as nanocomposite, nanogels or hydrogels. Moreover, the potential enhancement of GO surface area resulted in after modification with natural polymers. Natural biopolymer backbones are generally composed of multiple functional groups, which can involve intermolecular hydrogen bonding with the functionalities present on the GO sheet surface.[8,9] This effect is much more pronounced for polysaccharide-based natural biomacromolecules, where the epoxy groups on the GO sheets are majorly involved. Epichlorohydrin can be used to activate those epoxy groups at the initial stage.[10,11] Establishment of the H-bonding at the next step resulted efficient nanocomposites possessing each property of excellence coming from original compounds. Moreover, the newly introduced, three-dimensional porous structure can exhibit much greater adsorption efficacy in comparison to traditional two-dimensional GO sheets. Additionally, the synthesised assembly can be easily separated from the aqueous medium by filtration technique. These composite materials can include every useful property of natural biopolymers, such as eco-friendly behaviour, non-cytotoxicity, higher abundance and lower cost-effectiveness, and reported as sustainable material for wastewater purification.[12,13] Thus, composite formation with natural polymers always lead to enhanced mechanical, physical and sometimes electrical properties with excellent adsorption capacity essential for wastewater treatment applications with potential recyclability. Moreover, high pharmacological abilities like antimicrobial,[14,15] anticancer effect[16] with high extent of drug,[17] gene delivery[18] applications are resulted from those composites. Hence, those composites exhibited excellency to perform as an ideal eco-friendly wastewater-treating materials in every aspect because of their oxidation, adsorption, and catalytic properties.[19,20]

Thus, various forms of GO–natural polymer composites such as microspheres, membranes, sponges, foams, gel-like hydrogel, xerogels, aerogel, etc. have been reported so far, along with their efficacy towards removal of contaminants including dyes and metal ions from wastewater, as presented by Figure 8.1. Though, polysaccharide-based natural polymers were reported as the mostly used materials in this regard due to their high cost-effectiveness and high ease of bioavailability, but recently some focus has been shifted to some other natural biopolymers like proteins, enzymes, or deoxy ribonucleic acids (DNA) due to their good potential of selective adsorption towards charged organic dyes or heavy metal ions. In this book chapter, the recent progress of the graphene oxide (GO) incorporated natural polymers have been extensively reviewed along with a clear understanding about their classification and mechanism for wastewater treatment applications.

[Long description] Classification of natural polymer–GO composites used for wastewater treatment into three types; nanocomposites, magnetic nanocomposites, and composite gel. Nanocomposites are classified further into glucose–GO, cyclodextrin–GO, cellulose–GO, chitin–GO, chitosan–GO. Each of them is further categorised into microspheres, membranes, sponges and foams. The magnetic nanocomposites are also categorised into the previously mentioned four types along with gel. Composite gels

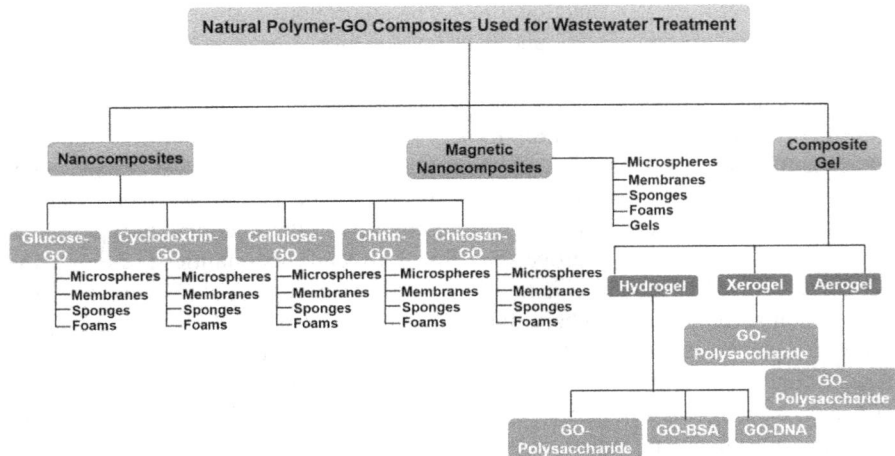

FIGURE 8.1   Various forms of natural polymers–graphene oxide (GO) composites for wastewater treatment.

are classified into hydrogel, xerogel and aerogels, each of them is composed of GO–polysaccharides. Moreover, GO-BSA and GO-DNA are other two types of hydrogels in addition to polysaccharide composed gel.

8.2   MECHANISTIC OUTLINE OF WASTEWATER TREATMENT

Before going to a detail discussion on GO and natural polymer section, a clear and common mechanistic outline of wastewater treatment has to be portrayed. In this context, a wide variety of physicochemical and biological approached are well established. The industrial wastewater mainly composed of several heavy inorganic metal ions or coloured organic conjugated compound with extensive π-electron delocalisation, which are too hazardous for human health. Those wastewaters can be treated using suitable agents by several mechanism like biodegradation or photolytic degradation of organic chemicals[21]; chemical oxidation[22]; complete destruction through electrolysis[23]; separation through membrane[24] or adsorption of hazardous elements[25] through using proper adsorbent materials. Most of these methods have limited applications due to low cost-effectiveness, difficulties to design the whole process and large-scale separation of hazards from wastewater.[26,27] Additionally, lower effectivity towards less amount of pollutant levels resulted in larger extent of chemicals, cost and energy, even a huge amount of toxic byproducts sometimes. This phenomenon not only made those mechanisms industrially and economically non-feasible, but also transform them to environmentally hazardous. Among the above-mentioned mechanism, the most common and effective method is adsorption.[28,29] This mechanism is the most privileged one due to its low cost effectiveness, safe operation process with high ease and eco-friendliness. Another feature of recyclability and reusability of the adsorbent after large number of cyclic regenerations made this mechanism most unique and facile one. The recent researches explored this mechanism in little more details. Adsorption is a well-known physical treatment with very high efficacy of removing pollutants from wastewater with no risk of toxic byproduct generation. Moreover, the scaling up is very easy in the arrangement of fixed bed columns through application as a batch process. The rate of adsorption was accompanied with the treated nano-materials and mostly dependent on size of the particles. Another clear understanding of enhanced effectiveness of adsorbent's physical structure and adsorption capacity was explored through contamination of effluents with pharmacological molecules. For example, graphene oxide, supported to rigid fluorine-containing (TFT or DFB) molecules, exhibited an enhanced adsorption potential for wide range of molecules like sulfamethoxazole, carbamazepine, ibuprofen, sulfadiazine, phenacetin, and paracetamol from solutions.

8.2.1 Contributing Factors on Adsorption Mechanism

The major interactions responsible for adsorption mechanisms are reported as hydro-gen bonding, electrostatic interactions, hydrophilic interactions and π–π interactions. During adsorption, a significant electrostatic force of reaction was already explored between the polar groups of the pollutant molecules and surface charge of adsorbent composite. Moreover, the selective adsorption efficacy towards cationic or anionic pollutants like organic dye molecules in water has been dictated by the net charge present on the adsorbent surface. Several major factors are responsible to decide the total surface charge such as characteristics of the chosen natural polymers, the quan-titative ratio of polymer and graphene-based derivatives, nanocomposite fabrication process, and pH of the wastewater. Variable charge can be measured through zeta potential measurement.[30,31] The way pH of the medium dictated the surface charge of adsorbent can be discussed with some examples as follows. Some polysaccharides such as chitosan consists of amino groups, which are the nitrogen-containing func-tionalities in addition to carboxyl and hydroxyl groups. Generally, the nitrogen and oxygen atoms are responsible for hydrogen bonding with pollutant molecules. Under acidic condition, the carboxyl, hydroxyl and amino groups become protonated, hence lead to selective adsorption to anionic pollutants.[32,33] On the other hand, at higher pH, the deprotonation of those above-mentioned functionalities take place, resulting in a total negative surface charge on the adsorbent site; hence, a selective adsorption of cationic pollutants was observed.[34,35]

Another important factor is π–π interaction, which played an important role in adsorption process due to presence of extensive conjugation of π electrons both in the GO nanosheet adsorbent and structure of organic conjugated pollutants, specifically dye molecules. The whole process has been demonstrated through taking several case studies. For example, in graphene nanosheet (GNS)/WCE composite aero-gels[36] and reduced graphene oxide (GO)–SA gels,[37] the adsorption mechanism was reported as a combination of physical as well as multistep diffusion-based chemical adsorption process. Again, selectivity towards heavy metals like Cr (III) ions as well as organic dye pollutants such as methylene blue (MB) and methyl orange (MO) was demonstrated by a complex polysaccharide, pectin-GO (Pc/GO) nanocomposite.[38] Herein, the ability of the contributing factor to alter the structural pattern, hence

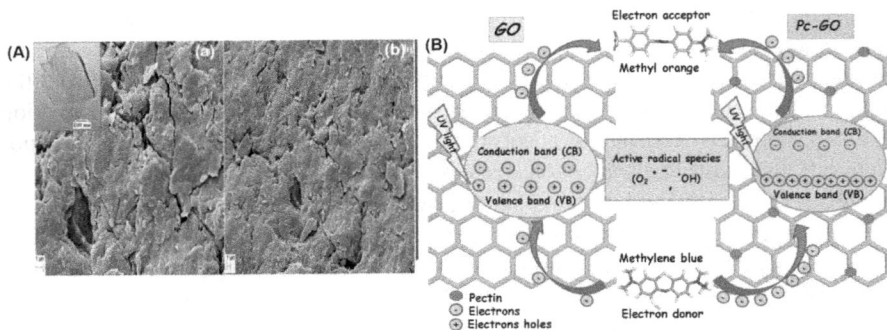

FIGURE 8.2   (a) (i) and (ii) FESEM images of Pc/GO (inset FESEM image of GO); (b) diagrammatic depiction of the degradation process of MB and MO dyes. (Reproduced with permission from Ref. [38]. Copyright 2020 Elsevier.)

adsorption efficacy can be reflected by FESEM image, as presented by Figure 8.2 along with diagrammatic degradation mechanism. Here, the bare GO layer exhibited a single, well-defined flake with three-dimensional multilayer arrangements, and the composite showed a wrinkled flakes along with agglomeration.

(a) (i) and (ii) A wrinkled flakes through agglomeration, and a multilayer and three-dimensional presentation of pc/GO; (b) a schematic representation of photocatalytic decomposition of two dye molecules: electron acceptor methyl orange and electron donor methylene blue on the pc/GO catalyst.

8.3   WHY TO CHOOSE GRAPHENE OXIDE (GO)?

This section demonstrated the specific efficiencies of GO while forming composite with natural polymers for wastewater treatment. The essential requirement to exhibit a potential derivative for treating wastewater is acting as a good adsorbent. Various modified compounds such as carbon nanotubes (CNT),[39,40] activated charcoal,[41,42] metal-organic framework (MOF),[43,44] zeolite,[45] etc. are reported with superadsorbent quality in this regard. But, most of them faced major limitation of usage due to their inability to meet the basic criteria of easy availability and cost-effectiveness. These two points are too important to fulfil the global demand of absorbent materials. These two criteria are well-met by GO. Additionally, the presence of extensive oxidised functionality such as numerous epoxy and hydroxy functional groups introduced high structural flexibility for modification and hence resulted a definite higher adsorption ability than most of others available in market place.

8.3.1   PROPERTIES & SUPERIORITY OF GO OVER GRAPHENE AS ADSORBENT

As per the discussion till now, GO originated from oxidisation of graphene and exhibited several outstanding properties, which are very significant to material science field, such as enhanced specified surface area, free of corrosion, high electrical conductivity with electron transfer capacities along with higher mechanical strength, electrical and thermal stability.[46,47] The superiority of graphene oxide over normal graphene was associated with those mentioned properties. Additionally, the demerits of poor solubility, hydrophobicity, synthetic difficulties of graphene[48] can be overcome through oxidising to GO. Moreover, graphene molecules have high agglomerating tendency in solution as a result of high van der Waals interactions,[49] hence exhibited difficulties in applications. This limitation can be completely eliminated in GO.

When the scientists aimed the wastewater treatment applications of graphene and its derivatives, they explored adsorption physical method as a dominant mechanism. The adsorption quality of GO was very high in this regard due to its structural benefit and several hydrophilic functionalities. The multilayered GO sheet structure was explored in several reports[50,51] with a favourable orientation of carbonyl, carboxyl functionalities at the edges, and hydroxyl and epoxide groups on the base plane. This structural construction led to an efficient hydrogen bonding and electrostatic interaction with adsorbate molecules, hence exhibiting a superior adsorption ability.[52] Moreover, extensive hydrophilicity of GO sheet,[53] originated from above-mentioned oxygen-encompassing functionalities, introduced an easy dispersion nature in water, either by sonication or stirring.[54] This phenomenon introduced easy penetration of water-solvent molecules into the vacancies between the sheet layers and hence resulted in excellent adsorbent efficacy towards pollutants from wastewater.

8.4 SYNTHETIC PROCEDURE OF GO

Established methodologies are available to synthesise GO from graphite induced to a strong and spontaneous oxidation process. Some standard methodologies are named by the scientists Brodie,[55] Staudenmaier[56] and Hummer,[57] which are utilised for development of GO. According to Brodie and Staudenmaier, fuming nitric acid (HNO_3) and Potassium chlorate ($KClO_4$) were the major ingredients. Whereas, in the Hummer's process, concentrated sulfuric acid (H_2SO_4) with potassium permanganate ($KMNO_4$) were used as major ingredients with graphite flakes. In all these three methods the π-conjugation of the graphene sheets was broken to a mixture of graphitic sp^2 along with sp^3 oxidised domains and the carbon vacancies-defects. This phenomenon resulted in the rapid exfoliation forming a single-layer and stable water suspension. In recent days, the most preferentially adopted procedure was modified or improved Hummer's methodology.[58,59] This method resulted in the most enhanced yield and quality of obtained GO. Table 8.1 represents the three major procedures in detail.

8.5 TYPES OF NATURAL POLYMERS FOR COMPOSITE PREPARATION

Till now, I have summarised the common mechanistic outline associated with GO composite. This section mainly demonstrated the types of natural polymers participated in composite formation for wastewater treatment, as demonstrated by Figure 8.3. All the polymers that originated from nature ingredients like flora and fauna are termed as natural polymers. The term includes each biopolymeric

TABLE 8.1

Detail Stepwise Synthetic Process of GO from Graphite

Synthetic Process	Details
Brodie's method[60]	**Step 1**: Fine graphite flakes (10 g) + fuming HNO_3 (200 mL) (taken in a flask placed in an ice bath)
	Step 2: $KClO_4$ (80 g) (gradually added for 1 hour with continuous stirring conditions to avoid any explosion)
	Step 3: Siring the mixture for about 21 hours and 0°C
	Step 4: After complete oxidation, the product, GO was formed. GO was washed well with distilled water, followed by vacuum filteration to bring neutral pH
Staudenmaier's method[61]	**Step 1**: Graphite flakes (1 g) + an ice-cooled mixture (27 mL) of H_2SO_4 and HNO_3 (mixed in 2:1 ratio)
	Step 2: $KClO_3$ (11 g) (gradually added to the mixture, maintaining the reaction temperature below 35°C)
	Step 3: The mixture was kept for 96 hours
	Step 4: Distilled water (800 mL) (added with continuous stirring)
	Step 5: The reaction mixture filtered through nylon membrane
	Step 6: The GO solid was obtained and collected followed by washing repeatedly with 5% HCl solution
Hummer's method[62]	**Step 1**: Graphite (1 g) + sodium nitrate (0.50 g) + concentrated sulfuric acid (23 mL) (kept for continuous stirring at 5°C in an ice bath for 5 minutes)
	Step 2: Slow addition of $KMnO_4$ (3 g)
	Step 3: The reaction mixture kept for another 30 minutes with vigorous stirring
	Step 4: Deionised water (46 mL) added to the suspension, which resulted in the evolution of hydration heat that enhanced the temperature to approximately 98°C.
	Step 5: The bath was kept at this temperature for 30 minutes under continuous stirring condition
	Step 6: Washing with deionised water (140 mL) and hydrogen peroxide (10% v/v, 10 mL) that resulted in a brownish-yellow coloured product finally separated by vacuum filtration.

components to construct plant, animals or microbe bodies like polysaccharides, proteins, enzymes or DNA, RNA along with different natural constituents like rubber, plastic or synthetic polymers like poly(vinyl alcohol) (PVA), poly(vinyl chloride) (PVC) etc. Among them, the most feasible natural polymers for synthesising the composites mainly come from polysaccharide fields. Starch, cellulose, chitin, chitosan cyclodextrins, gums are commonly used polysaccharides to form nanocomposite or bulk gel, with GO exhibiting wide wastewater treatment applications.

Three types of natural polymer–GO composites are presented as polysaccharide–GO composite, protein–GO composite, nucleic acid–GO composites. Glucose, cyclodextrin, cellulose, chitin/chitosan, starch, sodium alginate, xanthan gum, agar as the common polysaccharides; BSA as the protein and DNA as nucleic acid are the common components for GO composite formation.

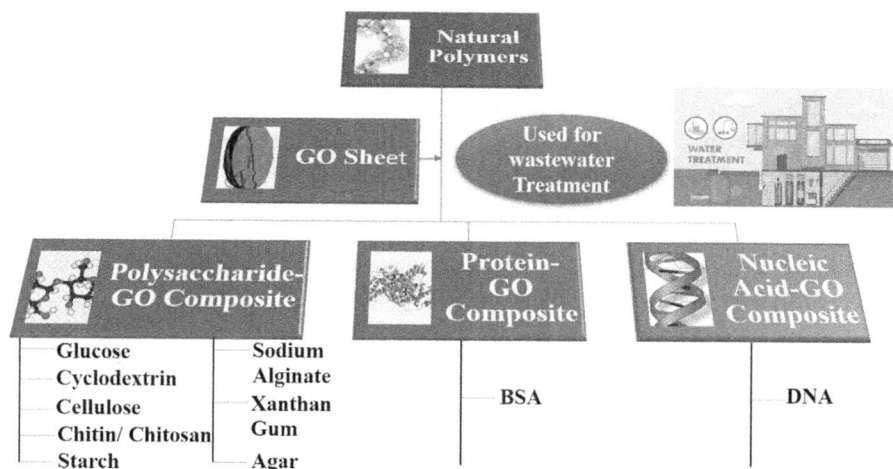

FIGURE 8.3   Types of natural polymers for GO composite preparation.

Moreover, some proteins like BSA and DNA can also form composite gel with efficient recyclability. All of those natural polymers showed an excellent inclusion of desired properties like biocompatibility and biodegradability with high ease of availability. Additionally, the presence of numerous functionalities[63,64] like $-OH$, $-COOH$, $-CONH_2$, $-NH_2$ and $-SO_3$ on these polymers' backbone resulted in structural flexibility and excellent blending efficacy with GO, which maximise the research attention these days.

8.6   CLASSIFICATIONS OF GO COMPOSITES WITH SYNTHETIC PROCEDURE

In order to explore the field of GO composites with available natural polymers in wastewater treatment, a detail discussion about the major types of composite materials is much needed. Based on variable range of internal structure and surface state morphology, all the composite materials are majorly classified as nanocomposites and composite gels with different adsorption properties. Additionally, introduction of magnetic components leads to magnetic nanocomposite with greater efficacy, as discussed further. Moreover, the composite gel might be further categorised as nanogel, hydrogel, aerogel and xerogel materials with excellent wastewater treatment properties.

The synthetic procedure is the major optimising factor to dictate structural variation, hence adsorption ability even in the nanoscale level. As a result, the classification cannot be completed without considering the individual synthetic process during the fabrication technique. Recent researches open a novel direction to eco-friendly synthetic method of GO–natural polymer composites with exceptional adsorption capacity towards a wide range of water pollutants. The classification-wise fabrication process of several recently developed natural polymer-based GO composite material with wastewater treatment applications are briefly described in the following subsections.

8.6.1   NANOCOMPOSITES

8.6.1.1   Glucose–GO Nanocomposites

Though glucose is not directly a natural polymer but a well-known monosaccharide and acts as a base of several polysaccharide-based natural polymers. Hence, the nanocomposite formation ability of this moiety is considered in this section for discussion. In this regard, a glucose-based porous carbon nanosheets (GPCNS)[65] were fabricated through integration of restricted nanospace in two-dimensional GO nanosheet. In the whole process, glucose moiety acted as a source of carbon followed by a potassium hydroxide (KOH) activation to generate porous structural framework.[66,67] An excellent usage of the adsorption sites along with an increased rate of diffusion for adsorbate resulted in very high adsorption ability of this porous nanocomposite frames.

8.6.1.2   Cyclodextrin–GO Nanocomposites

Cyclodextrin (CD) moiety is a well-known natural macrocyclic oligomers of saccharide family, generally processed through enzymatic degradation of starch. They can be classified into three types, α-CD, β-CD and γ-CD composed of 6, 7 and 8 glucose monomeric units, respectively, linked through α-1,4 glycosidic bonds. Among them, the β-form, generated through corn-starch extraction, was majorly reported to generate GO-based nanocomposite along with wastewater treatment applications for high absorptivity through host–guest interaction and encapsulation efficacy for selective pollutants like aromatic molecules, which are highly hazardous materials.[68,69] Several β-CD-GO nanocomposites are reported in this regard. After synthesis of GO by Hummer's process and functionalisation of epoxy groups, a simple coprecipitation technique was designed to graft with β-CD to generate a novel composite—GO-β-CD.[70] Another report showed an improved procedure to get graphene/β-cyclodextrin (GNS/β-CD) composite material.[71] Here, the covalent bond formation ability between β-CD and GNS enhanced the system stability along with a collaborative adsorption capacity of hazardous dye molecules from water through both the components. The design of this nanocomposite can be done in such a manner that they can portray an *in-situ* aggregation using hydrazine solutions under treated condition.[72] Additionally, eco-friendly synthesis[73] has been reported to get the composite material through the reaction between two targeted components—β-CD and GO in 1:1 ratio under basic condition. Graphene oxide–isophorone diisocyanate composites (GO-IPDICDs) was designed through incorporation of isophorone diisocyanate (IPDI) to β-CD functionalised graphene oxide.[74] Recent development of a novel carboxymethyl functionalised β-CD modified GO (CMβ-CD-GO) was reported.[75] Another innovative design and facile synthesis of an adsorbent composed of β-CD and poly(poly(L-glutamic acid) grafted onto GO was also presented by Jiang et al. (2017). Sometimes, synthetic polymers can couple to natural polymers and further form nanocomposite with GO.[76] In this regard, a successful synthesis of an innovative β-cyclodextrin (β-CD)/poly(acrylic acid) grafted onto GO (β-CD/PAA/GO) nanocomposites are reported through a chemical bond farming esterification reactions.[77]

8.6.1.3 Cellulose–GO Nanocomposites

Cellulose, composed of D-glucose monomeric units linked through β-(1–4)-glycosidic bond, is reported as the most abundant natural biopolymers in the polysaccharide family with potentially high renewability and rate of production (almost 1012 tons/year). The polymer backbone is composed of extensively large number of hydroxyl groups due to their presence at C-2, C-3 and C-6 position of glucose monomer units. As a result, the overall polymer backbone can effectively form inter and intra-molecular H-bond with GO, which leads to nanocomposite formation. The major nanocomposites observed from cellulose are cellulose nanofibers (CNFs) or micro-fibrillated cellulose (MFC) and nanocrystalline cellulose (NCC) extracted from natural sources like wood pulp.[78,79] CNFs can be synthesised by treatment of the cellulose wood pulp suspended in water solvent with mechanical homogeniser. Several new techniques for CNFs fabrication process are utilisation of cryogenic grinders, ultrasonic homogenisers,[80] high-pressure homogenisers, electrospinning method[81] and steam explosion[82] that have already been explored in this area. In some cases, the cellulose extracted from newspapers[83] or jute waste[84] are recently reported to utilise in preparation of GO-based nanocomposites. Additionally, hazardous dye-removal capacities from water by carboxymethyl cellulose (CMC) microbeads integrated with carboxylated graphene oxide (GOCOOH) are extensively studied.[85] Composite membrane and nanofibers are reported as another two forms of cellulose/GO nanoadsorbent (CGCMs) with extensively high adsorption efficiency.[86] In some synthetic reports, the nanocellulose composites are designed and prepared through alkaline hydrolysis of cotton materials followed by vacuum filtration.[87] According to Aboamera et al. (2018), a novel cellulose-based nanofibers' composite was designed and synthesised from GO and cellulose acetate *via* chemical cross-linking and surface modification process.[88] When cellulose or its derivatives was compiled with different components along with GO, it could produce an excellent membrane-like nanocomposite structure such as, according to a recent report, selective removal of Pb(II) heavy metal by cellulose acetate/vinyl triethoxysilane-modified GO and gum Arabic (GuA) membranes was demonstrated in this regard *via* an effective fabrication method called dissolution casting technique.[89] The hydrogel bonding-based complexation-NF framework was presented by Figure 8.4.

The hydroxyl groups present in cellulose acetate (CA), vinyl triethoxysilane-modified GO (VTES-GO) and gum Arabic (GuA) membrane resulting in inter- and intra-molecular hydrogen bonding and thus form a complexation-nanofiber (NF) framework.

8.6.1.4 Chitin–GO Nanocomposites

Chitin is the second most abundant natural polymer next to cellulose and consists of β-(1,4)-linked *N*-acetyl-D-glucosamine.[90] This moiety is of polysaccharide family with excellent renewable nature, biocompatibility and biodegradability, along with high water solubility.[91] These properties of superiority along with high bioavailability in nature made this biopolymer an excellent ingredient of cost-effective adsorbent materials and has been studied extensively for the hazardous chemical's removal from water solution.[92,93] Several novel synthetic approaches have been recently developed

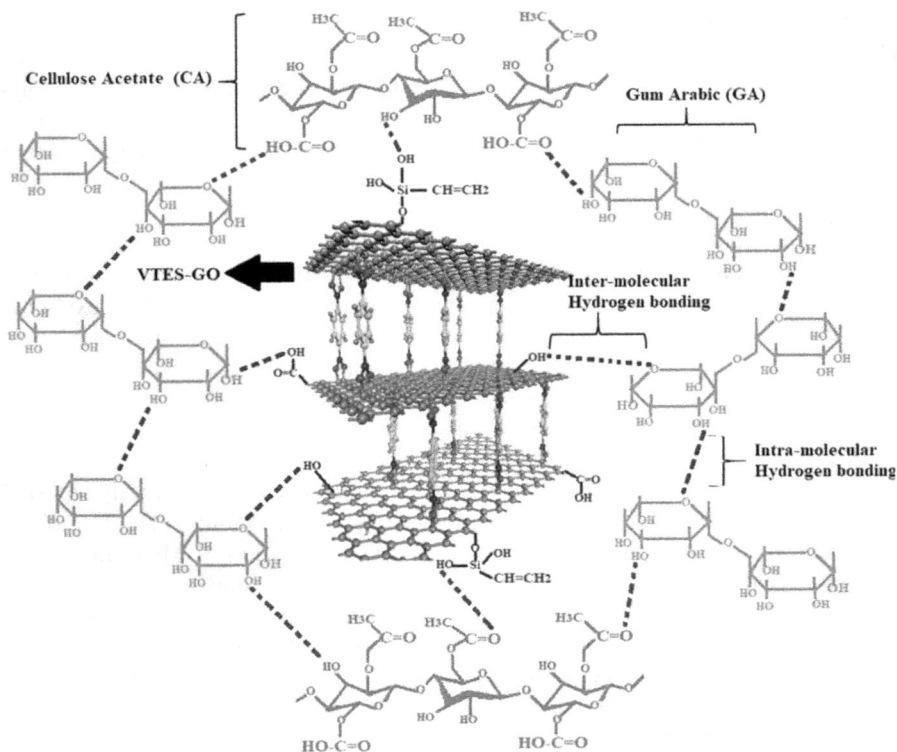

FIGURE 8.4 Complexation-NF network of inter- and intra-molecular hydrogen bonding between CA, GO and GuA molecules. (Reproduced with permission from Ref. [89]. Copyright 2021 Scientific Reports.)

to couple with GO preparing excellent chitin/GO nanocomposites foam of membranes for wastewater treatments. An innovative fabrication of GO/chitin nanofibrils (GO–CNF) composite foam[94] was reported through incorporating CNF into GO as a column adsorbent utilising a homogenising and wet-grinding procedure. A novel composite membrane based on GO, consisting of chitin nanocrystals (ChNCs), modified further to dopamine (D) utilising the vacuum filtration process.[95] Additionally, tannic acid-derivatised reduced graphene oxide (TRGO) strengthened chitin/GO composites. Another innovative process was reported by Liu et al. (2020) for fabrication through facile freezing–thawing cycle and cross-linked *via* epichlorohydrin.[96]

8.6.1.5   Chitosan–GO Nanocomposites

Specificity towards anionic dye adsorption from water can be introduced by choosing chitosan (CS) [97]—a cationic naturally available biopolymers of polysaccharide family as a GO composite formation component. The polymers mainly composed of 4-linked β-2-amino-2-deoxy-glucopyranose; sometimes *N*-acylated derivative has to be considered for composite preparation.[98,99] A series of established and standard procedures are reported such as lyophilising followed by freeze drying,[100] mechanical stirring through ultrasonication,[101] cross-linking *via* tetraethyl orthosilicate (TEOS)[102] and an open exposure under ultrasonic treatment.[103] Another important type of nanocomposites is called the sponge materials with excellent adsorption ability due to several smart characteristic like highly accessible pore size and volume, excellent strength and stability, and toughness even at low density along with enhanced surface area.[104,105] Several chitosan/GO nanocomposites with sponge characteristic are reported with increasing number of practical applications, especially in wastewater treatment. Ultrasonication is a well-established process for GO/chitosan sponge preparation, as reported by Qi et al.[106] Another innovative chitosan–zinc oxide (ZnO)–graphene oxide hybrid composite material was synthesised *via* one-pot chemical method.[107] A recent report demonstrated a β-chitosan/ triethylenetetramine functionalised graphene oxide (CS/TFGO) hybrid extensively used for wastewater treatment.[108] After adsorption, this hybrid exhibited solid–liquid separation problems, and the solution lied on the process as mentioned by Ren et al. (2016).[109] Some reports established further the role of a covalently bonded metal-organic framework (MOF) like $Cu_3(btc)_2$ incorporated to CS/GO composite materials and exhibited enhanced adsorption capacity.[110]

8.6.2   MAGNETIC NANOCOMPOSITES

In the above section, the synthetic concept of different natural polymer–GO-based normal nanocomposites were clearly demonstrated with the help of several recent researches. Sometimes introduction of magnetic properties leads to additional privilege to act as a good and specific wastewater-treating agents. Several reports can be seen in this field. Magnetic properties can be majorly fabricated to natural polymer–GO composite materials through incorporation of specific magnetic nanoparticles such as Fe_3O_4. This integration facilitated an easy adsorbent separation from the dye solution. Magnetic CD/GO (MCGO) nanomaterials utilising a highly feasible chemical process[111] and coprecipitation[112,113] has already been reported. Wang et al. suggested another new procedure to develop Fe_3O_4-based magnetic β-CD/GO nanocomposite, where functionalisation of β-CD to its carboxymethyl derivative (CM-β-CD) through reaction *via* monochloroacetic acid and sodium hydroxide (NaOH) was the first stage.[114,115] Later, the system along with Fe_3O_4 nanoparticle was treated with dispersed GO in water to generate the final magnetic nanocomposite. This system has been used on different hazardous dyes for adsorption from water. Reversible addition-fragmentation chain transfer (RAFT) polymerisation can be used as another facile process to form the resultant magnetic nanocomposite adsorbent through multiconjugation of magnetic nanoparticle (MNPs) and GO with β-CD (MNPs/GO-β-CD).[116] The whole process was demonstrated through the covalent linking of poly(glycidyl methacrylate) onto the surface of MNPs, followed by derivatisation through the mixture of these modified MNPs and the GO with β-CD. Again, adsorption selectivity towards cationic dyes were reported through developing a chitin-based magnetic nanocomposites. Similarly, magnetic properties can be incorporated to cellulose base GO absorbent materials too preparing magnetic cellulose/GO composite (MCGO) through a simple process of coprecipitation.[117] This technique was used at the initial stage of Fe_3O_4 nanoparticle formation through coprecipitation of ferrous and ferric salts in presence of nitrogen gas.[118] The same method was also used in the last stage to form the final magnetic cellulose-based composite materials. Chitin, another common polysaccharide, -based GO composite with magnetic property having selective adsorption towards cationic dye was also reported recently.[119] Similarly, chitosan-based GO magnetic nanocomposite (MCGO) was designed and synthesised through ultrasonic dispersion, as reported by Fan et al.[120] Coprecipitation process can also be introduced here for synthesis of magnetic CS/GO composite (mCs-GO).[121] Sohni et al. reported the detail method based on complete permeability of Cs to Fe_3O_4 nanoparticles followed by cross-linking of the used natural polymer. Another similar report exhibited by Jiang et al. (2016), where the chitosan/GO composite was initially formed followed by insertion of magnetic nanoparticle through a continuous mechanical stirring in presence of nitrogen gas.[122] Additionally, Sheshmani et al. demonstrated a solvothermal process utilising ethylene glycol as reducing agent for preparation of the magnetic GO/CS composite.[123] Another report exhibited the modification with ethylenediaminetetraacetic acid (EDTA) to magnetic GO/CS composite with increased adsorption capacity and higher ease of separation of pollutants from water.[124] Moreover, magnetite–polypyrene/chitosan/graphene oxide (M-PPy/CS/GO)[125] and magnetic chitosan/quaternary ammonium salt GO[126] are also demonstrated in this regard with excellent adsorption ability. A recent report exhibited the facile synthesis of chitosan-based composite system consisting of chitosan, magnetic ferrite (Fe_3O_4) nanoparticles and GO (CFG) resulting in cooperative combination of the chosen properties; the well-known coprecipitation process was used as a basic fabrication method through alkalisation of Fe^{+2} and Fe^{+3} compounds, as presented in Figure 8.5.[127]

(a) Dropwise addition of ammonium hydroxide solution to a mixture of Fe (II) and Fe (III) salts in aqueous media, followed by addition of sodium lauryl sulphate resulting in Fe_3O_4 nanoparticles, which can be separated through magnet and dried through oven. (b) Dispersion of Fe_3O_4 nanoparticles along with GO in chitosan solution in 1 wt% acetic acid resulting in an aqueous suspension containing chitosan, Fe_3O_4 nanoparticles and GO. Vigorous stirring at room temperature and separation through magnet resulted in the chitosan/Fe_3O_4/graphene oxide nanocomposite.

Multiple natural polymers that can couple with GO to generate composite and magnetic properties can be introduced with selective and efficient adsorption ability. In this regard, Fan et al. reported a simple chemical bonding process to prepare cyclodextrin–chitosan/GO nanocomposite (MCGO) with enhanced surface area.[128] Multiple natural polymers like sodium alginate and β-CD can be coupled to magnetic iron-based nanoparticles (Fe_3O_4) and activated charcoal simply through mixing of the polymer matrix with nanofillers showing an excellent wastewater treatment application.[129]

8.6.3   COMPOSITE GEL

8.6.3.1   GO–Natural Polysaccharides Hydrogel

An interesting category of composite materials with GO are hydrogels, where the hydrophilic natural polymer chains are oriented in three-dimensional arrangements *via* cross-linking. These materials are not water soluble but exhibited a

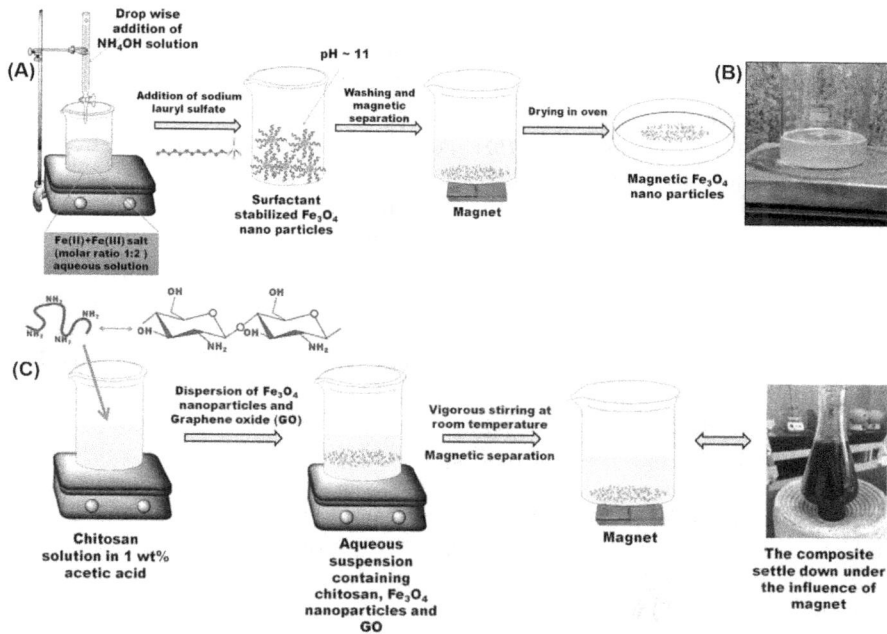

FIGURE 8.5   (a) Schematic representation of synthesis of Fe_3O_4 nanoparticles using copre-cipitation method, (b) aqueous chitosan solution obtained after overnight stirring at room temperature, (c) scheme showing synthesis of chitosan/Fe_3O_4/graphene oxide nanocomposite. (Reproduced with permission from Ref. [127]. Copyright 2020 Journal of Thematic Analysis.)

superior adsorption quality. In this section, I mainly emphasised on the synthesis of natural polysaccharide-based hydrogels, coupled with graphene with improved adsorption quality; hence, they exhibited an excellent wastewater treatment application.

The first natural polymer of polysaccharide family that was considered for discussion was cellulose and its derivative, carboxymethyl cellulose (CMC). This natural polymer-based hydrogel coupled with GO was reported to prepare from tea residue utilising an ionic liquid (1-allyl-3-methylimidazolium chloride) solvent.[130] The initial stage was extraction step as per the reported method of Hu.[131] The resultant composite hydrogel made of graphene oxide/tea cellulose were synthesised *via* a heating–cooling–washing method.[132,133] In some reports, the CMC was considered as starting material to design hydrogel framework with three-dimensional cross-linking with excellent porosity and entrapment capacity of high volume of water solvent, biological fluids along with solute molecules in their structure. Under dried condition, these hydrogels can recover the original form *via* shrinking.[134,135] An interesting report by Varaprasad et al. (2017), CMC–acrylamide–CO (CMC–AM–GO) hydrogel synthesis *via* a free-radical polymerisation technique was demonstrated with variable ratios of CMC and GO.[136] Another report by Dai et al. exhibited the synthetic method of a sustainable and eco-friendly poly(vinyl alcohol)/

(PVA/PCMC/GO/bentonite) hydrogels through extraction from pineapple peel *via* well-known freeze–thaw cyclic process.[137] The adsorption property improvements were observed clearly in eco-friendly PVA/CMC hydrogel, when coupled with GO. The potential usage of this system to selective adsorption of hazardous dyes from wastewater was explored further. Additionally, nanocellulose was reported as another efficient composite hydrogel formation components through derivatisation with GO resulting in a selective adsorption of hazardous organic solvents like cyclohexane, dimethylformamide (DMF) from water.[138] Reduced form of GO (rGO), prepared *via* reduction of GO precursor through hydrazine, was also used as a replacement of normal GO to form nanocomposite hydrogel with cellulose. Hence, these hydrogels can be generated through incorporation of coupled cellulose and rGO together into an additional hydrophilic block poly(ethylene glycol) dimethacrylate *via* photopolymerisation.[139]

Chitin and chitosan are another set of highly abundant natural polymers to generate composite hydrogels with GO. They can be produced to manufacture a column for water purification. Selective adsorption towards anionic dyes was reported through explaining the process of the hydrogel fabrication. In this regard, agitation and incubation were reported as an excellent fabrication process to synthesise a chitin/GO hybrid gel.[140] Another method called semidissolution/acidification/sol–gel transition (SD-A-SGT) was adopted by Chang et al. (2020) to produce a polyacrylate (PAA)-coupled chitosan-based composite gel, chitosan/polyacrylate/GO (CTS/PAA/GO) with enhanced adsorption and mechanical properties.[141]

Starch was reported as another well-known polysaccharide to synthesise porous nanocomposite hydrogel structure with GO. Grafting of 2-acrylamido-2-methylpropanesulfonic acid and acrylamide onto starch in the presence of GO and calcium carbonate ($CaCO_3$) was reported as an excellent synthetic method.[58] Another report exhibited that an efficient removal of a selective dye—malachite green (MG) from water was explored by the starch–graft–poly(acrylamide)/graphene oxide/hydroxyapatite nanocomposite hydrogel adsorbent (NHA).[142] The synthesis was performed *via* a simple free-radical copolymerisation after grafting of acrylamide (AM) monomer to starch with variable compositions in presence of GO nano sheets.

Sodium alginate was also reported as an established natural polysaccharide to generate hydrogel composite with GO. Numerous fabrication methods like solvent exchange, precursor freezing and drying *via* ethanol were reported to produce the three-dimensional porous network based on GO.[143] The consolidating and crosslinking ability of sodium alginate with random orientation of GO structure has been incorporated a great utility while gel formation. This system can be reduced further to prepare reduced GO-based gel *via* glucose reaction.[144] Additionally, an excellent fabrication method involved hydrothermal action of reduced GO and alginate in water, which could undergo an ionic cross-linking of metal ions further.[145] Again, a synthetic approach of simple GO introduced to alginate hydrogel beads (SA/GO) was demonstrated by Gan et al. (2018).[146] Modification of sodium alginate *via* grafting polymerisation method with acrylic acid, followed by encumbered with powder graphite, resulted in a fabrication of sodium alginate cross-linked acrylic acid/graphite hydrogel having increased adsorption capacity.[147]

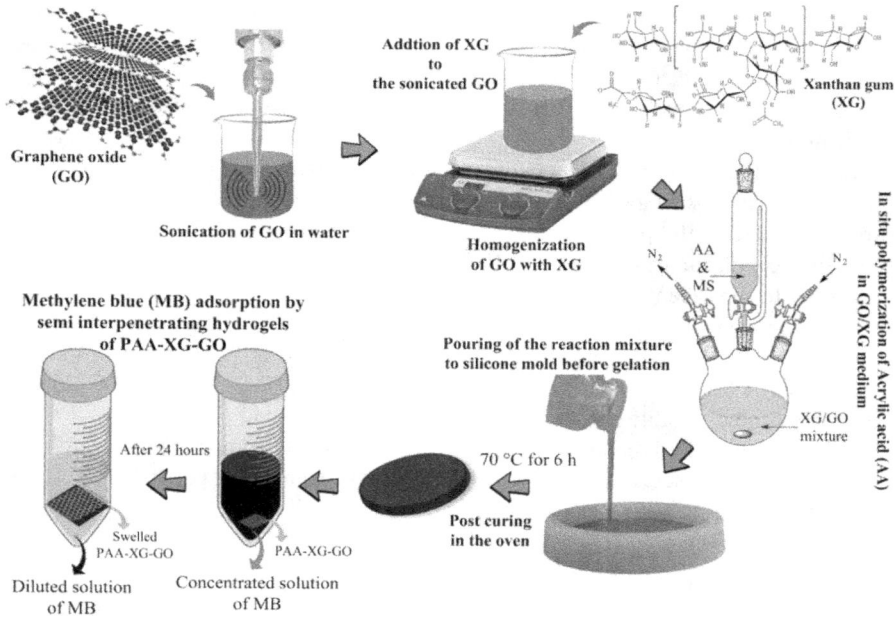

FIGURE 8.6   Different stages of preparation and evaluation of PAA-XG-GO-based superadsorbent hydrogels for adsorption of methylene blue. (Reproduced with permission from Ref. [149]. Copyright 2020 Elsevier.)

Xanthan gum was considered as another efficient natural polysaccharide components to produce GO composite hydrogels with high adsorption capacities. In this regard, microwave-aided copolymerisation method of acrylic acid on xanthan gum chain utilising bis-acrylamide and ammonium persulfate as a cross-linker and initiator, respectively, fabricated a novel gum-based gel matrix.[148] At the next stage, the reduced form of GO was introduced on the matrix through grafting method. Selective adsorption of cation dye materials from wastewater was demonstrated further by an innovative biocompatible xanthan gum/cross-linked poly(acrylic acid)/graphene oxide (PAA-XG-GO) gel system with superabsorbent characteristic (Figure 8.6).[149]

Another gum-based report was also fabricated, where the radical polymerisation approach of 2-Acrylamido-2-methyl-1-propanesulfonic acid)-grafted Tragacanth gum/GO composite hydrogel was demonstrated with high ability to adsorb heavy metal (Pb, Ag and Cd) ions from the wastewater.[150]

Addition of xanthan gum (XG) to the sonicated GO resulted in a XG/GO mixture, which underwent in-situ polymerisation in presence of acrylic acid (AA) and cross-linker (MS), followed by pouring the reaction mixture before gelation to a silicone mould and post curing in the oven at 70°C for 6 hours. The resulting gel exhibited a super adsorption capacity in MB dye aqueous solution.

Agar can also be used as an efficient natural polysaccharide to fabricate GO-based composite hydrogel. In this regard, one report exhibited fabrication of agar-based GO nanocomposite (AGO) gel and GO-filled konjac glucomannan hydrogel materials.[151] Moreover, another agar-based nanocomposite hydrogel could be fabricated *via* decoration of reduced GO derivatised nanoclay, which resulted in polyacrylamide–agar nanocomposite microporous hydrogel (PAAm-Agar/Clay@ rGO) through *in-situ* polymerisation. Activation of epoxy groups of graphene through treated with epichlorohydrin followed by grafting of GO with polysaccharides like xylan, κ-carrageenan resulted in a good composite hydrogel fabrication, according to the recent reports.

Incorporation of multiple polysaccharides-based natural polymers to a single composite gel preparation was demonstrated by Liu et al. (2018) and reported a three-dimensional β-CD/chitosan functionalised GO hydrogel (3D-GO/CS/β-CD) through a simple feasible reduction process in presence of sodium ascorbate.[152] Some other reports exhibited the fabrication of β-CD–CMC–GO composite material[153] and sodium alginate-based adsorbent-immobilised β-CD and GO gel (SCGG)[154] with excellent adsorption ability.

8.6.3.2   GO-BSA Hydrogel

Proteins are also considered as an efficient, naturally available biopolymer to show the composite hydrogel with GO, though very few are reported in terms of wastewater treatment application. Hence, lots more scope is still left over in this field. Bovine serum albumin (BSA) is an efficient protein, which can be utilised to fabricate GO-based gel.[155] The detailed process was demonstrated as follows. Initially, a stock solution of BSA (approximately 4 g dissolved in 20 mL of deionised water through overnight mechanical stirring) was prepared. This was the minimum reported concentration that has to be taken into account for gel fabrication. Next, the solution (0.3 mL) was added to the concentrated suspension of GO (4 g approximately, concentration 5 mg/g). The *in-situ* hydrogel formation occurred *via* violent stirring for only 10 seconds, confirmed by tube inversion process. Under sonication for 3 minutes, a homogeneous gelation can occur followed by freeze-drying process to obtain the final GO/BSA composite hydrogel with a final weight ratio of 20/60 mg and excellent adsorption quality.

8.6.3.3   GO-DNA Hydrogel

DNA (deoxy ribonucleic acid) is another natural biopolymer that can for GO composite hydrogel with excellent adsorption capacity other that common polysaccharide molecules.[155] They are generally extracted from plant, animal or microbe cells composed of nucleic acid as monomer units. A detail synthetic procedure was reported as follows. At the first stage, A DNA stock solution was prepared through dissolution of around 400 mg of DNA in 20 mL of deionised water and left the solution overnight with constant stirring. Next, this solution (around 1 mL) was introduced to a concentrated GO suspension (about 4 g, concentration in weight/weight = 5 mg/g) followed by a violent shaking of the mixture for 10 seconds to generate an instantaneous hydrogel composite. Under sonication for 3 minutes, a homogeneous gelation can occur followed by freeze-drying process to obtain the final GO/DNA composite hydrogel. Several reports exhibited the fabrication of GO sponge through introducing a freeze-drying process with GO (approximately 5 mg/g) colloidal suspension, which are utilised as a control to compare the adsorption ability on various hazards in water.

Selective adsorption capacity of different GO-based natural biopolymers gel on a series of pollutants has been demonstrated by Figure 8.7.[155] From the result, we can get a clear idea about the selective adsorption for cationic dyes and highly comparable capability of GO–biopolymer gels as that of control GO sponge.

(a) The order of adsorption capacity of GO and different GO–natural biopolymers composites against MB and MV are much higher in compared with CR, RhB, BPA and p-NP. The order is (i) against MB: GO > GO-BSA > GO-CS > GO-DNA, (ii) against MV: GO > GO-BSA > GO-CS > GO-DNA, (iii) against CR: GO > GO-DNA > GO-CS > GO-BSA, (iv) against RhB: GO > GO-BSA > GO-CS = GO-DNA, (v) against BPA: GO > GO-DNA > GO-CS > GO-BSA and (vi) against p-NP: GO > GO-DNA > GO-CS > GO-BSA. (b) Coloured solution of MB and MV become completely colourless after the selective adsorption.

8.6.3.4   Natural Polymer-Based Composite Xerogel

Xerogels are another type of composite materials with highly porous structure. The pore size is much smaller in comparison to bulk hydrogels due to some characteristics like low-density solid framework with extensively high surface area. Glucose, the most abundant saccharide molecules, has been reported extensively as a starting component for xerogel formation with GO. In this regard, the mesoporous and monolithic carbon-based xerogels through the process of glucose carbonisation in a hydrothermal way was designed in the presence of GO nanosheet structure.[156] Hence, the gel was termed as hydrothermal carbon (HTC) xerogels. The activation process was carried out with potassium hydroxide (KOH) further.

8.6.3.5   Natural Polymer-Based Composite Aerogel

The concept of aerogels lied on replacement of the liquid components of the wet gel by gas medium or vacuum without disturbing the porous backbone of the original gel structure. Aerogels are other types of solid composite materials with lower density, excellent porosity with enhanced surface area; hence, they resulted in super absorptivity. Many natural biopolymers like cellulose, chitosan, starch or sodium alginate are reported as starting components to generate composite aerogel with GO having excellent wastewater treatment capacity. A highly porous nature with super absorptivity of cellulose nanofibril (CNF) aerogel with magnetic GO has been demonstrated,[157] where CNFs were synthesised from the powder wood of KC (kenaf core) and disintegrated in high speed further.[158] The detailed synthetic process expressed the effective nixing of GO, CNF in suspension and a mixed solution of $FeCl_2$ and $FeCl_3$ through a magnetic mechanical stirring followed by a method of lyophilic freezing. Ultimately the desired composite of CNF/GO in aerogel form was reported. Microcrystalline cellulose (MCC) can be targeted as a superior starting material for designing aerogel composite.[159] For example, an eco-friendly green process to prepare a GO- and MCC-based aerogel hybrids was reported by Wei et al. (2017),

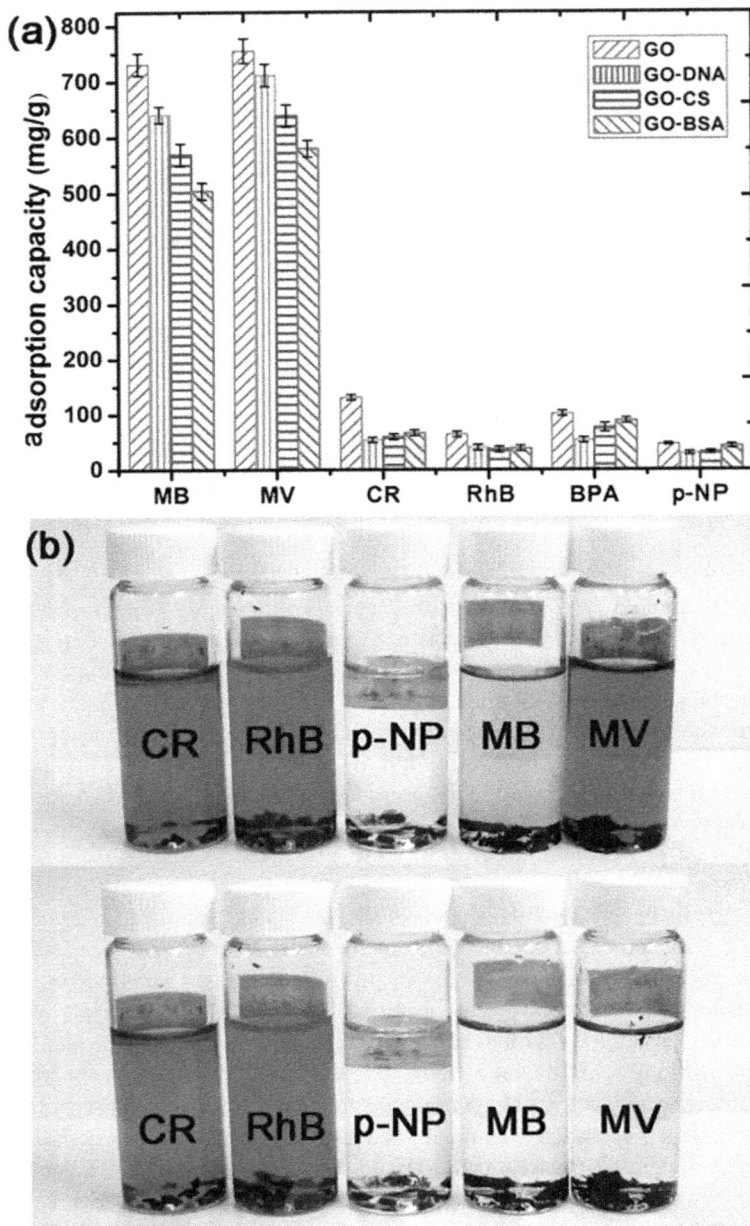

FIGURE 8.7   (a) The adsorption capacity for different pollutants by GO sponge and GO–biopolymer gels after 24 hours. 1 mg of each gel is applied in 20 mL of 100 mol/L pollutant solutions at $25 \pm 1°C$; the pH value of the solution is 7. Data are expressed as the mean $\pm$ S.D. of three independent measurements. (b) Photograph of adsorption of different pollutants at equilibrium but without BPA, since it is colourless and transparent before and after the adsorption. (Reproduced with permission from Ref. [155]. Copyright 2013 Elsevier.)

where lithium bromide (LiBr) solution was used as a solvent for complete dissolution of MCCs. Additionally, a mixture of regenerated cellulose (RCE) and GO can form an aerogel composite with high adsorption ability.[160] Freeze-drying method after the mixing of the solutions is reported as the synthetic procedure. Chitosan was presented as another aerogel composite forming natural polymer, when doped with GO forming CSGO aerogel by freeze-drying and cross-linking procedure, as reported by Wang et al.[161] Another report showed the combination process of dialdehyde starch nanocrystals (SRGO) with GO nanosheet, which led to the high porous and compressible aerogel nanocomposite with excellent mechanical strength, hence high absorptivity and wastewater treatment applications.[162] Sodium alginate, another well-known natural polymers, can generate a porous three-dimensional (3D) graphene oxide-montmorillonite/ sodium alginate (GO-MMT/SA) aerogel beads synthesised through the same common techniques discussed in earlier examples.[163]

8.7 CONCLUSIONS AND FUTURE PERSPECTIVES

This chapter demonstrated a clear understanding of GO, their utility to choose as an excellent adsorbent for wastewater purification and the synthetic procedure. Then the major focus was shifted to the classification of natural polymers to prepare composites after presenting a strong concept on mechanistic outline, among which adsorption is the most feasible one. Highly abundant, renewable and biodegradable natural polymers with potential flexibility to structural modification acquired a special place in the modern world to prepare a sustainable fabrication of natural pollutant adsorbent from wastewater. Furthermore, several types of composite materials with high adsorption efficacy, based on their nature, pour size, structural properties, are demonstrated with facile, eco-friendly and cost-effective synthetic procedure. Those composite materials resulted in highly desirable practical applications *via* industrially feasible process.

   Though in modern era, several researches are ongoing to prepare polysaccharide-based GO composites with potential wastewater treatment applications, more attention needs to be paid in synthesis of GO composites involving polysaccharides in large-scale application and other natural biopolymers except polysaccharides too, as a replacement of other synthetic alternatives. Several biopolymers like proteins, DNA, etc. possess excellent gel-forming ability *in-situ* with GO, possessing high adsorption capacity with a guaranteed role in solving the wastewater purification problems, as discussed in this chapter. But very little attention has been paid while applying to wastewater treatment. Hence, lots of scope are yet to be explored in this field.

ACKNOWLEDGEMENTS

I thank Professor Bimalendu Ray, The University of Burdwan, the Science and Engineering Research Board (SERB), Council of Scientific and Industrial Research (CSIR) and Government of India for my fellowship and research grands during my Ph.D. in Indian Institute of Science Education and Research, Kolkata and Postdoctoral journey in the University of Burdwan.

REFERENCES

1. Jéquier, E., & Constant, F. (2010) Water as an essential nutrient: The physiological basis of hydration. *Eur. J. Clin. Nutr.*, *64*, 115–123. https://doi.org/10.1038/ejcn.2009.111.
2. Cooley, H. et al. (2014). Global water governance in the twenty-first century. In: Gleick, P.H. (eds) *The World's Water*. Island Press, Washington, DC. https://doi.org/10.5822/978-1-61091-483-3_1.
3. Epalza, J., Jaramillo, J., & Guarín, O. Extraction and use of plant biopolymers for water treatment. http://dx.doi.org/10.5772/intechopen.77319Journal.
4. Houghton, J. I., & Quarmbyt, J. (1999) Biopolymers in wastewater treatment. *Curr. Opin. Biotechnol.*, *10*, 259–262. http://biomednet.com/elecref/0958166901000259.
5. Terzopoulou, Z, Kyzas, G. Z. & Bikiaris, D. N. (2015) Recent advances in nanocomposite materials of graphene derivatives with polysaccharides, *Materials*, *8*, 652–683. doi: 10.3390/ma8020652.
6. Shen, Y., Zhu, X., Zhu, L., & Chen, B. (2017) Synergistic effects of 2D graphene oxide nanosheets and 1D carbon nanotubes in the constructed 3D carbon aerogel for high performance pollutant removal. *Chem. Eng. J.*, *314*, 336–346. https://doi.org/10.1016/j.cej.2016.11.132.
7. Wang, M., Cai, L., & Jin, Q. et al. (2017) One-pot composite synthesis of three-dimensional graphene oxide/poly(vinyl alcohol)/TiO$_2$ microspheres for organic dye removal. *Sep. Purif. Technol.*, *172*, 217–226. https://doi.org/10.1016/j.seppur.2016.08.015.
8. Kadokawa, J., Murakami, M., & Kaneko, Y. (2008) A facile preparation of gel materials from a solution of cellulose in ionic liquid. *Carbohydr. Res.*, *343*, 769–772. https://doi.org/10.1016/j.carres.2008.01.017.
9. Tian, S. Y., Guo, J. H., Zhao, C., Peng, Z., Gong, C. H., Yu, L. G., Liu, X. H., & Zhang, J. W. (2019) Preparation of cellulose/graphene oxide composite membranes and their application in removing organic contaminants in wastewater. *J. Nanosci. Nanotechnol.*, *19*, 147–2153. https://doi.org/10.1166/jnn.2019.15808.
10. Qi, Y., Yang, M., & Xu, W. et al. (2017) Natural polysaccharides-modified graphene oxide for adsorption of organic dyes from aqueous solutions. *J. Colloid Interface Sci.*, *486*, 84–96. https://doi.org/10.1016/j.jcis.2016.09.058.
11. Wang, S., Li, Y., & Fan, X. et al. (2015) β-Cyclodextrin functionalized graphene oxide: An efficient and recyclable adsorbent for the removal of dye pollutants. *Front. Chem. Sci. Eng.*, *9*, 77–83. https://doi.org/10.1007/s11705-014-1450-x.
12. Pooresmaeil, M., & Namazi, H. (2018) β-Cyclodextrin grafted magnetic graphene oxide applicable as cancer drug delivery agent: Synthesis and characterization. *Mater. Chem. Phys*, *218*, 62–69. https://doi.org/10.1016/j.matchemphys.2018.07.022.
13. Namazi, H. (2017) Polymers in our daily life. *BioImpacts*, *7*, 73–74. https://dx.doi.org/10.15171%2Fbi.2017.09.
14. Jia, X., Yao, Y., Yu, G., Qu, L., Li, T., Li, Z., & Xu, C. (2020) Synthesis of gold-silver nanoalloys under microwave-assisted irradiation by deposition of silver on gold nanoclusters/triple helix glucan and antifungal activity. *Carbohydr. Polym.*, *238*, 116169. https://doi.org/10.1016/j.carbpol.2020.116169.
15. Amirnejat, S., Nosrati, A., Javanshir, S., & Naimi-Jamal, M. R. (2020) Superparamagnetic alginate-based nanocomposite modified by Larginine: An eco-friendly bifunctional catalyst and an efficient antibacterial agent. *Int. J. Biol. Macromol.*, *152*, 834–845. https://doi.org/10.1016/j.ijbiomac.2020.02.212.
16. Kholiya, F., Chatterjee, S., Bhojani, G., Sen, S., Barkume, M., Kasinathan, N. K., Kode, J., & Meena, R. (2020) Seaweed polysaccharide derived bioaldehyde nanocomposite: Potential application in anticancer therapeutics. *Carbohydr. Polym.*, *240*, 116282. https://doi.org/10.1016/j.carbpol.2020.116282.

17. Collado-Gonzalez, M., Ferreri, M. C., Freitas, A. R., Santos, A.´C., Ferreira, N. R., Carissimi, G., Sequeira, J. A., Díaz Baños, F. G., Villora, G., Veiga, F., & Ribeiro, A. (2020) Complex polysaccharide-based nanocomposites for oral insulin delivery. *Mar. Drugs*, *55*, 1–18. https://doi.org/10.3390/md18010055.

18. Ansari, M. A., Yadav, M. K., Rathore, D., Svedberg, A., & Karim, Z. (2019) Applications of nanostructured polymer composites for gene delivery. In: *Nanostructured Polymer Composites for Biomedical Applications*. *Elsevier*, 211–226. https://doi.org/10.1016/B978-0-12-816771-7.00011-9.

19. Pooresmaeil, M., & Namazi, H. (2020) Application of polysaccharidebased hydrogels for water treatments. In: *Hydrogels Based on Natural Polymers*. *Elsevier*, 411–455. https://doi.org/10.1016/B978-0-12-816421-1.00014-8.

20. Musarurwa, H., & Tavengwa, N. T. (2020) Application of carboxymethyl polysaccharides as bio-sorbents for the sequestration of heavy metals in aquatic environments. *Carbohydr. Polym.*, *237*, 116142. https://doi.org/10.1016/j.carbpol.2020.116142.

21. Jiang, D., Xue, J., Wu, L., Zhou, W., Zhang, Y., & Li, X. (2017) Photocatalytic performance enhancement of CuO/Cu_2O heterostructures for photodegradation of organic dyes: Effects of CuO morphology. *Appl. Catal. B*, 211, 199–204. https://doi.org/10.1016/j.apcatb.2017.04.034.

22. Chakma, S., Das, L., & Moholkar, V. S. (2015) Dye decolorization with hybrid advanced oxidation processes comprising sonolysis/Fentonlike/photo-ferrioxalate systems: A mechanistic investigation. *Sep. Purif. Technol.*, *156*, 596–607. https://doi.org/10.1016/j.seppur.2015.10.055.

23. Han, Y., Li, H., & Liu, M. et al. (2016) Purification treatment of dyes wastewater with a novel microelectrolysis reactor. *Sep. Purif. Technol.*, *170*, 241–247. https://doi.org/10.1016/j.seppur.2016.06.058.

24. Lin, J., Ye, W., & Baltaru, M. et al. (2016) Tight ultrafiltration membranes for enhanced separation of dyes and Na_2SO_4 during textile wastewater treatment. *J. Membr. Sci.*, *514*, 217–228. https://doi.org/10.1016/j.memsci.2016.04.057.

25. Konicki, W., Aleksandrzak, M., Moszynski, D., & Mijowska, E.´ (2017) Adsorption of anionic azo-dyes from aqueous solutions onto graphene oxide: Equilibrium, kinetic and thermodynamic studies. *J. Colloid Interface Sci.*, *496*, 188–200. https://doi.org/10.1016/j.jcis.2017.02.031.

26. Djehaf, K., Bouyakoub, A. Z., Ouhib, R., Benmansour, H., Bentouaf, A., Mahdad, A., Moulay, N., Bensaid, D., & Ameri, M. (2017) Textile wastewater in Tlemcen (Western Algeria): Impact, treatment by combined process. *Chem. Int.*, *3*, 414–419.

27. Oussama, N., Bouabdesselam, H., Ghaffour, N., & Abdelkader, L. (2019) Characterization of seawater reverse osmosis fouled membranes from large scale commercial desalination plant. *Chem. Int.*, *5*, 158–167. https://doi.org/10.5281/zenodo.1568742.

28. Thakur, K., & Kandasubramanian, B. (2019) Graphene and graphene oxide-based composites for removal of organic pollutants: A review. *J. Chem. Eng. Data*, *64*, 833–867. https://doi.org/10.1021/acs.jced.8b01057.

29. Noreen, S., Bhatti, H. N., Iqbal, M., Hussain, F., & Sarim, F. M. (2020) Chitosan, starch, polyaniline and polypyrrolebiocomposite with sugarcane bagasse for the efficient removal of Acid Black dye. *Int. J. Biol. Macromol.*, *147*, 439–452. https://doi.org/10.1016/j.ijbiomac.2019.12.257.

30. Zhao, L., Tang, P., Sun, Q., Zhang, S., Suo, Z., Yang, H., Liao, X., & Li, H. (2020) Fabrication of carboxymethyl functionalized β-cyclodextrinmodified graphene oxide for efficient removal of methylene blue. *Arabian J. Chem.*, *13*, 7020–7031. https://doi.org/10.1016/j.arabjc.2020.07.008.

31. Halouane, F., Oz, Y., Meziane, D., Barras, A., Juraszek, J., Singh, S. K., Kurungot, S., Shaw, P. K., Sanyal, R., Boukherroub, R., Sanyal, A., & Szunerits, S. (2017) Magnetic reduced graphene oxide loaded hydrogels: Highly versatile and efficient adsorbents for dyes and selective Cr(VI) ions removal. *J. Colloid Interface Sci.*, *507*, 360–369. https://doi.org/10.1016/j.jcis.2017.07.075.

32. Banerjee, P., Barman, S. R., Mukhopadhayay, A., & Das, P. (2017) Ultrasound assisted mixed azo dye adsorption by chitosan–graphene oxide nanocomposite. *Chem. Eng. Res. Des.*, *117*, 43–56. https://doi.org/10.1016/j.cherd.2016.10.009.

33. Jiang, Y., Gong, J. L., Zeng, G. M., Ou, X. M., Chang, Y. N., Deng, C. H., Zhang, J., Liu, H. Y., & Huang, S. Y. (2016) Magnetic chitosangraphene oxide composite for antimicrobial and dye removal applications. *Int. J. Biol. Macromol.*, *82*, 702–710. https://doi.org/10.1016/j.ijbiomac.2015.11.021.

34. de Figueiredo Neves, T., Dalarme, N. B., da Silva, P. M. M., Landers, R., Picone, C. S. F., and Prediger, P. (2020) Novel magnetic chitosan/ quaternary ammonium salt graphene oxide composite applied to dye removal. *J. Environ. Chem. Eng.*, *8*, 103820. https://doi.org/10.1016/j.jece.2020.103820.

35. Qi, C., Zhao, L., Lin, Y., & Wu, D. (2018) Graphene oxide/chitosan sponge as a novel filtering material for the removal of dye from water. *J. Colloid Interface Sci.*, *517*, 18–27. https://doi.org/10.1016/j.jcis.2018.01.089.

36. Feng, C., Ren, P., & Li, Z. et al. (2020) Graphene/waste-newspaper cellulose composite aerogels with selective adsorption of organic dyes: Preparation, characterization, and adsorption mechanism. *New J. Chem.*, *44*, 2256–2267. https://doi.org/10.1039/C9NJ05346H.

37. Ma, T., Chang, P. R., Zheng, P., Zhao, F., & Ma, X. (2014) Fabrication of ultra-light graphene-based gels and their adsorptionof methylene blue. *Chem. Eng. J.*, *240*, 595–600. https://doi.org/10.1016/j.cej.2013.10.077.

38. Kaushal, S., Kaur, N., Kaur, M., & Singh, P. P. (2020) Dual-responsive pectin/graphene oxide (Pc/GO) nano-composite as an efficient adsorbent for Cr (III) ions and photocatalyst for degradation of organic dyes in waste water. *J. Photochem. Photobiol. A: Chem.*, *403*, 112841. https://doi.org/10.1016/j.jphotochem.2020.112841.

39. Gu, Y., Yang, M., Wang, W., & Han, R. (2019) Phosphate adsorption from solution by zirconium-loaded carbon nanotubes in batch mode. *J. Chem. Eng. Data*, *64*, 2849–2858. https://doi.org/10.1021/acs.jced.9b00214.

40. Ortega, P. F. R., Trigueiro, J. P. C., Santos, M. R., Denadai, A. M. L., Oliveira, L. C. A., Teixeira, A. P. C., Silva, G. G., & Lavall, R. L. (2017) Thermodynamic study of methylene blue adsorption on carbon nanotubes using isothermal titration calorimetry: A simple and rigorous approach. *J. Chem. Eng. Data*, *62*, 729–737. https://doi.org/10.1021/acs.jced.6b00804.

41. Hasanzadeh, M., Simchi, A., & Shahriyari Far, H. (2020) Nanoporous composites of activated carbon-metal organic frameworks for organic dye adsorption: Synthesis, adsorption mechanism and kinetics studies. *J. Ind. Eng. Chem.*, *81*, 405–414. https://doi.org/10.1016/j.jiec.2019.09.031.

42. Ho, S. (2020) Removal of dyes from wastewater by adsorption onto activated carbon: Mini review. *J. Geosci. Environ. Prot.*, *8*, 120–131. http://www.scirp.org/journal/Paperabs.aspx?PaperID=100280.

43. Li, C., Xiong, Z., Zhang, J., & Wu, C. (2015) The strengthening role of the amino group in metal–organic framework MIL-53 (Al) for methylene blue and malachite green dye adsorption. *J. Chem. Eng. Data*, *60*, 3414–3422. https://doi.org/10.1021/acs.jced.5b00692.

44. Wo, R., Li, Q.-L., Zhu, C., Zhang, Y., Qiao, G., Lei, K., Du, P., & Jiang, W. (2019) Preparation and characterization of functionalized metalorganic frameworks with core/ shell magnetic particles Fe3O4@SiO2@ MOFs) for removal of congo red and methylene blue from water solution. *J. Chem. Eng. Data*, *64*, 2455–2463. https://doi.org/10.1021/acs.jced.8b01251.

45. Pahlavanzadeh, H., & Motamedi, M. (2020) Adsorption of nickel, Ni (II), in aqueous solution by modified zeolite as a cation-exchange adsorbent. *J. Chem. Eng. Data, 65,* 185–197. https://doi.org/10.1021/acs.jced.9b00868.
46. Stoller, M. D., Park, S., Zhu, Y., An, J., & Ruoff, R. S. (2008) Graphene-based ultracapacitors. *Nano Lett., 8,* 3498–3502. https://doi.org/10.1021/nl802558y.
47. Novoselov, K. S., Geim, A. K., Morozov, S. V., Jiang, D., Katsnelson, M. I., Grigorieva, I. V., Dubonos, S. V., & Firsov, A. A. (2005) Twodimensional gas of massless Dirac fermions in graphene. *Nature, 438,* 197–200. https://doi.org/10.1038/nature04233.
48. Niyogi, S., Bekyarova, E., Itkis, M. E., McWilliams, J. L., Hamon, M. A., & Haddon, R. C. (2006) Solution properties of graphite and graphene. *J. Am. Chem. Soc., 128,* 7720–7721. https://doi.org/10.1021/ja060680r.
49. Kuilla, T., Bhadra, S., Yao, D., Kim, N. H., Bose, S., & Lee, J. H. (2010) Recent advances in graphene-based polymer composites. *Prog. Polym. Sci., 35,* 1350–1375. https://doi.org/10.1016/j.progpolymsci.2010.07.005.
50. Szabo, T., Berkesi, O., & De´kány, I. (2005) Drift study of deuterium-´exchanged graphite oxide. *Carbon, 43,* 3186–3189. http://dx.doi.org/10.1016%2Fj.carbon.2005.07.013Journal.
51. He, H., Klinowski, J., Forster, M., & Lerf, A. (1998) A new structural model for graphite oxide. *Chem. Phys. Lett., 287,* 53–56. https://doi.org/10.1016/S0009–2614(98)00144–4.
52. Bai, H., Chen, J., Wang, Z., Wang, L., & Lamy, E. (2020) Simultaneous removal of organic dyes from aqueous solutions by renewable alginate hybridized with graphene oxide. *J. Chem. Eng. Data, 65,* 4443–4451. https://doi.org/10.1021/acs.jced.0c00277.
53. Stankovich, S., Dikin, D. A., Piner, R. D., Kohlhaas, K. A., Kleinhammes, A., Jia, Y., Wu, Y., Nguyen, S. T., & Ruoff, R. S. (2007) Synthesis of graphene-based nanosheets *via* chemical reduction of exfoliated graphite oxide. *Carbon, 45,* 1558–1565. https://doi.org/10.1016/j.carbon.2007.02.034.
54. Khan, Z. U., Kausar, A., Ullah, H., Badshah, A., & Khan, W. U. (2016) A review of graphene oxide, graphene bucky paper, and polymer/ graphene composites: Properties and fabrication techniques. *J. Plast. Film Sheeting, 32,* 336–379. https://doi.org/10.1177%2F8756087915614612.
55. Brodie, B. C. (1859) On the atomic weight of graphite. *Philos. Trans. R. Soc. London, 149,* 249–259. https://doi.org/10.1098/rstl.1859.0013.
56. Staudenmaier, L. (1898) Verfahren zur darstellung der graphitsaure.¨ *Ber. Dtsch. Chem. Ges., 31,* 1481–1487.
57. Hummers, W. S., & Jr Offeman, R. E. (1958) Preparation of graphitic oxide. *J. Am. Chem. Soc., 80,* 1339–1339. https://doi.org/10.1021/ja01539a017.
58. Pourjavadi, A., Nazari, M., Kabiri, B., Hosseini, S. H., & Bennett, C. (2016) Preparation of porous graphene oxide/hydrogel nanocomposites and their ability for efficient adsorption of methylene blue. *RSC Adv., 6,* 10430–10437. https://doi.org/10.1039/C5RA21629J.
59. Sarkar, N., Sahoo, G., & Swain, S. K. (2020) Nanoclay sandwiched reduced graphene oxide filled macroporous polyacrylamide-agar hybrid hydrogel as an adsorbent for dye decontamination. *Nano-Struct. Nano-Obj., 23,* 100507. https://doi.org/10.1016/j.nanoso.2020.100507.
60. Botas, C., Alvarez, P., Blanco, P., Granda, M., Blanco, C., Santamaría, R., Romasanta, L. J., Verdejo, R., Lopez-Manchado, M. A., & Menendez, R. (2013) Graphene materials with different structures prepared from the same graphite by the Hummers and Brodie methods. *Carbon, 65,* 156–164. https://doi.org/10.1016/j.carbon.2013.08.009.
61. Sali, S., Mackey, H., & Abdala, A. (2019) Effect of graphene oxide synthesis method on properties and performance of polysulfonegraphene oxide mixed matrix membranes. *Nanomaterials, 9,* 769. https://doi.org/10.3390/nano9050769.
62. Drashya, Lal, & Hooda, S. (2017) Magnetic graphene oxide for adsorption of organic dyes from aqueous solution. *AIP Conf. Proc.,* 1–4. https://doi.org/10.1063/1.5032617.

63. Esquerdo, V. M., Cadaval, T. R. S., Jr, Dotto, G. L., & Pinto, L. A. A. (2014) Chitosan scaffold as an alternative adsorbent for the removal of hazardous food dyes from aqueous solutions. *J. Colloid Interface Sci.*, *424*, 7–15. https://doi.org/10.1016/j. jcis.2014.02.028.

64. Zhou, Y., Zhang, M., Hu, X., Wang, X., Niu, J., & Ma, T. (2013) Adsorption of cationic dyes on a cellulose-based multicarboxyl adsorbent. *J. Chem. Eng. Data*, *58*, 413–421. https://doi.org/10.1021/je301140c.

65. Xie, A., Dai, J., Cui, J., Lang, J., Wei, M., Dai, X. H., Li, C., & Yan, Y. (2017) Novel graphene oxide–confined nanospace directed synthesis of glucose-based porous carbon nanosheets with enhanced adsorption performance. *ACS Sustain. Chem. Eng.*, *5*, 11566–11576. https://doi.org/10.1021/acssuschemeng.7b02917.

66. Yuan, K., Zhuang, X., Fu, H., Brunklaus, G., Forster, M., Chen, Y., Feng, X., & Scherf, U. (2016) Two-dimensional core-shelled porous hybrids as highly efficient catalysts for the oxygen reduction reaction. *Angew. Chem., Int. Ed.*, *55*, 6858–6863. https://doi. org/10.1002/ange.201600850.

67. Yuan, K., Guo-Wang, P., Hu, T., Shi, L., Zeng, R., Forster, M., Pichler, T., Chen, Y., & Scherf, U. (2015) Nanofibrous and graphene templated conjugated microporous polymer materials for flexible chemo sensors and supercapacitors. *Chem. Mater.*, *27*, 7403–7411. https://doi.org/10.1021/acs.chemmater.5b03290.

68. Alzate-Sanchez, D. M., Smith, B. J., Alsbaiee, A., Hinestroza, J. P., & Dichtel, W. R. (2016) Cotton fabric functionalized with a β-cyclodextrin polymer captures organic pollutants from contaminated air and water. *Chem. Mater.*, *28*, 8340–8346. https://doi. org/10.1021/acs.chemmater.6b03624.

69. Alsbaiee, A., Smith, B. J., Xiao, L., Ling, Y., Helbling, D. E., & Dichtel, W. R. (2016) Rapid removal of organic micropollutants from water by a porous β-cyclodextrin polymer. *Nature*, *529*, 190–194. https://doi.org/10.1038/nature16185.

70. Wang, S., Li, Y., Fan, X., Zhang, F., & Zhang, G. (2015) β-Cyclodextrin functionalized graphene oxide: An efficient and recyclable adsorbent for the removal of dye pollutants. *Front. Chem. Sci. Eng.*, *9*, 77–83. https://doi.org/10.1007/s11705-014-1450-x.

71. Tan, P., & Hu, Y. (2017) Improved synthesis of graphene/βcyclodextrin composite for highly efficient dye adsorption and removal. *J. Mol. Liq.*, *242*, 181–189. https://doi. org/10.1016/j.molliq.2017.07.010.

72. Zheng, H., Gao, Y., Zhu, K., Wang, Q., Wakeel, M., Wahid, A., Alharbi, N. S., & Chen, C. (2018) Investigation of the adsorption mechanisms of Pb (II) and 1-naphthol by b-cyclodextrin modified graphene oxide nanosheets from aqueous solution. *J. Colloid Interface Sci.*, *530*, 154–162. https://doi.org/10.1016/j.jcis.2018.06.083.

73. Rathour, R. K. S., Bhattacharya, J., & Mukherjee, A. M. (2019) β-Cyclodextrin conjugated graphene oxide: A regenerative adsorbent for cadmium and methylene blue. *J. Mol. Liq.*, *282*, 606–616. https://doi.org/10.1016/j.molliq.2019.03.020.

74. Yan, J., Zhu, Y., Qiu, F., Zhao, H., Yang, D., Wang, J., & Wen, W. (2016) Kinetic, isotherm and thermodynamic studies for removal of methyl orange using a novel β-cyclodextrin functionalized graphene oxide-isophorone diisocyanate composites. *Chem. Eng. Res. Des.*, *106*, 168–177. https://doi.org/10.1016/j.cherd.2015.12.023.

75. Zhao, L., Tang, P., & Sun, Q. et al. (2020) Fabrication of carboxymethyl functionalized β-cyclodextrinmodified graphene oxide for efficient removal of methylene blue. *Arabian J. Chem.*, *13*, 7020–7031. https://doi.org/10.1016/j.arabjc.2020.07.008.

76. Jiang, L., Liu, Y., Liu, S., Hu, X., Zeng, G., Hu, X., Liu, S., Liu, S., Huang, B., & Li, M. (2017) Fabrication of b-cyclodextrin/poly (L-glutamic acid) supported magnetic graphene oxide and its adsorption behaviour for 17b-estradiol. *Chem. Eng. J.*, *308*, 597–605. https://doi.org/10.1016/j.cej.2016.09.067.

77. Liu, J., Liu, G., & Liu, W. (2014) Preparation of water-soluble bcyclodextrin/poly (acrylic acid)/graphene oxide nanocomposites as new adsorbents to remove cationic dyes from aqueous solutions. *Chem. Eng. J.*, *257*, 299–308. https://doi.org/10.1016/j.cej.2014.07.021.

78. Turbak, A. F., Snyder, F. W., & Sandberg, K. R. (1983) Microfibrillated cellulose, a new cellulose product: Properties, uses, and commercial potential. *J. Appl. Polym. Sci.*, *37*, 815–827.

79. Herrick, F. W., Casebier, R. L., Hamilton, J. K., & Sandberg, K. R. (1983) Microfibrillated cellulose: Morphology and accessibility. *J. Appl. Polym. Sci.*, *37*, 797–813.

80. Yano, H., Seki, N., & Ishida, T. (2008) Manufacture of nanofibers and nanofibers manufactured thereby. *JP*, *17*, 2229.

81. Fung, W. Y., Yuen, K. H., & Liong, M. T. (2011) Agrowaste-based nanofibers as a probiotic encapsulant: Fabrication and characterization. *J. Agric. Food Chem.*, *59*, 8140–8147. https://doi.org/10.1021/jf2009342.

82. Cherian, B. M., Leao, A. L., de Souza, S. F., de Costa, L. M.˜ M., Olyveira, G. M., Kottaisamy, M., Nagarajan, E. R., & Thomas, S. (2011) Cellulose nanocomposites with nanofibers isolated from pineapple leaf fibers for medical applications. *Carbohydr. Polym.*, *86*, 1790–1798. https://doi.org/10.1016/j.carbpol.2011.07.009.

83. Feng, C., Ren, P., & Li, Z. et al. (2020) Graphene/waste-newspaper cellulose composite aerogels with selective adsorption of organic dyes: Preparation, characterization, and adsorption mechanism. *New J. Chem.*, *44*, 2256–2267. https://doi.org/10.1039/C9NJ05346H.

84. Zaman, A., Orasugh, J. T., Banerjee, P., Dutta, S., Ali, M. S., Das, D., Bhattacharya, A., & Chattopadhyay, D. (2020) Facile one-pot *in-situ* synthesis of novel graphene oxide-cellulose nanocomposite for enhanced azo dye adsorption at optimized conditions. *Carbohydr. Polym.*, *246*, 116661. https://doi.org/10.1016/j.carbpol.2020.116661.

85. Eltaweil, A. S., Elgarhy, G. S., El-Subruiti, G. M., & Omer, A. M. (2020) Novel carboxymethyl cellulose/carboxylated graphene oxide composite microbeads for efficient adsorption of cationic methylene blue dye. *Int. J. Biol. Macromol.*, *154*, 307–318. https://doi.org/10.1016/j.ijbiomac.2020.03.122.

86. Tian, S. Y., Guo, J. H., Zhao, C., Peng, Z., Gong, C. H., Yu, L. G., Liu, X. H., & Zhang, J. W. (2019) Preparation of cellulose/graphene oxide composite membranes and their application in removing organic contaminants in wastewater. *J. Nanosci. Nanotechnol.*, *19*, 147–2153. https://doi.org/10.1166/jnn.2019.15808.

87. Cai, J., & Zhang, L. (2005) Rapid dissolution of cellulose in LiOH/urea and NaOH/urea aqueous solutions. *Macromol. Biosci.*, *5*, 539. https://doi.org/10.1002/mabi.200400222.

88. Aboamera, N. M., Mohamed, A., Salama, A., Osman, T. A., & Khattab, A. (2018) An effective removal of organic dyes using surface functionalized cellulose acetate/graphene oxide composite nanofibers. *Cellulose*, *25*, 4155–4166. https://doi.org/10.1007/s10570-018-1870-8.

89. Idress, H., Zaidi, S. Z. J., Sabir, A., Shafq, M., Khan, R. U., Harito, C., Hassan S., & Walsh, F. C. (2021) Cellulose acetate based ComplexationNF membranes for the removal of Pb(II) from waste water. *Sci. Rep.*, *11*, 1806. https://doi.org/10.1038/s41598-020-80384-0.

90. Li, G. X., Du, Y. M., Tao, Y. Z., Deng, H. B., Luo, X. G., & Yang, J. H. (2010) Iron(II) cross-linked chitin-based gel beads: Preparation, magnetic property and adsorption of methyl orange. *Carbohydr. Polym.*, *82*, 706–713. https://doi.org/10.1016/j.carbpol.2010.05.040.

91. Kittle, J. D., Wang, C., Qian, C., Zhang, Y.; Zhang, M., Roman, M., Morris, J. R., Moore, R. B., & Esker, A. R. (2012) Ultrathin chitin films for nanocomposites and biosensors. *Biomacromolecules*, *13*, 714–718. https://doi.org/10.1021/bm201631r.

92. Dolphen, R., Sakkayawong, N., Thiravetyan, P., & Nakbanpote, W. (2007) Adsorption of Reactive Red 141 from wastewater onto modified chitin. *J. Hazard. Mater.*, *145*, 250–255. https://doi.org/10.1016/j.jhazmat.2006.11.026.

93. Kyzas, G. Z., Kostoglou, M., Vassiliou, A. A., & Lazaridis, N. K. (2011) Treatment of real effluents from dyeing reactor: Experimental and modeling approach by adsorption onto chitosan. *Chem. Eng. J.*, *168*, 577–585. https://doi.org/10.1016/j.cej.2011.01.026.

94. Ma, Z., Liu, D., Zhu, Y., Li, Z., Tian, H., & Liu, H. (2016) Graphene oxide/chitin nanofibril composite foams as column adsorbents for aqueous pollutants. *Carbohydr. Polym.*, *144*, 230–237. https://doi.org/10.1016/j.carbpol.2016.02.057.

95. Ou, X., Yang, X., Zheng, J., & Liu, M. (2019) Free-standing graphene oxide–chitin nanocrystal composite membrane for dye adsorption and oil/water separation. *ACS Sustain. Chem. Eng.*, *7*, 13379–13390. https://doi.org/10.1021/acssuschemeng.9b02619.

96. Liu, C., Liu, H., Tang, K., Zhang, K., Zou, Z., & Gao, X. (2020) High strength chitin based hydrogels reinforced by tannic acid functionalized graphene for congo red adsorption. *J. Polym. Environ.*, *28*, 984–994. https://doi.org/10.1007/s10924-020-01663-5.

97. Crini, G. (2006) Non-conventional low-cost adsorbents for dye removal: A review. *Bioresour. Technol.*, *97*, 1061–1085. https://doi.org/10.1016/j.biortech.2005.05.001.

98. Cesaro, R., Fabbricino, M., Lanzetta, R., Mancino, A., Naviglio, B., Parrilli, M., Sartorio, R., Tomaselli, M., & Tortora, G. (2008) Use of chitosan for chromium removal from exhausted tanning baths. *Water Sci. Technol.*, *58*, 735–739. https://doi.org/10.2166/wst.2008.692.

99. Gad, Y. H. (2008) Preparation and characterization of poly(2- acrylamido-2-methyl-propane-sulfonic acid)/chitosan hydrogel using gamma irradiation and its application in wastewater treatment. *Radiat. Phys. Chem.*, *77*, 1101–1107. https://doi.org/10.1016/j.radphyschem.2008.05.002.

100. Huyen, N. T. M., Trang, P. T. T., Dat, N. M., & Hieu, N. H. (2017) Synthesis of chitosan/graphene oxide nanocomposite for methylene blue adsorption. *AIP Conf. Proc.*, *1*, 020013. https://doi.org/10.1063/1.5000181.

101. Guo, X., Qu, L., Tian, M., Zhu, S., Zhang, X., Tang, X., & Sun, X. (2016) Chitosan/graphene oxide composite as an effective adsorbent for reactive red dye removal. *Water Environ. Res.*, *88*, 579. https://doi.org/10.2175/106143016X14609975746325.

102. Kamal, M. A., Bibi, S., Bokhari, S. W., Siddique, A. H., & Yasin, T. (2017) Synthesis and adsorptive characteristics of novel chitosan/graphene oxide nanocomposite for dye uptake. *React. Funct. Polym.*, *110*, 21–29. https://doi.org/10.1016/j.reactfunctpolym.2016.11.002.

103. Banerjee, P., Barman, S. R., Mukhopadhayay, A., & Das, P. (2017) Ultrasound assisted mixed azo dye adsorption by chitosan–graphene oxide nanocomposite. *Chem. Eng. Res. Des.*, *117*, 43–56. https://doi.org/10.1016/j.cherd.2016.10.009.

104. Sun, Y., Wu, Q., & Shi, G. (2011) Graphene based new energy materials. *Energy Environ. Sci.*, *4*, 1113–1132. https://doi.org/10.1039/C0EE00683A.

105. Chabot, V., Higgins, D., Yu, A., Xiao, X., Chen, Z., & Zhang, J. (2014) A review of graphene and graphene oxide sponge: Material synthesis and applications to energy and the environment. *Energy Environ. Sci.*, *7*, 564–1596. https://doi.org/10.1039/C3EE43385D.

106. Qi, C., Zhao, L., Lin, Y., & Wu, D. (2018) Graphene oxide/chitosan sponge as a novel filtering material for the removal of dye from water. *J. Colloid Interface Sci.*, *517*, 18–27. https://doi.org/10.1016/j.jcis.2018.01.089.

107. Sanmugam, A., Vikraman, D., Park, H. J., & Kim, H. S. (2017) One-pot facile methodology to synthesize chitosan-ZnO-graphene oxide hybrid composites for better dye adsorption and antibacterial activity. *Nanomaterials*, *7*, 363. https://doi.org/10.3390/nano7110363.

108. Chiu, C.-W., Wu, M.-T., Lin, C.-L., Li, J.-W., Huang, C.-Y., Soong, Y.-C., Lee, J. C.-M., Lee Sanchez, W. A., & Lin, H.-Y. (2020) Adsorption performance for reactive blue 221 dye of β-chitosan/polyamine functionalized graphene oxide hybrid adsorbent with high acid– alkali resistance stability in different acid–alkaline environments. *Nanomaterials*, *10*, 748. https://doi.org/10.3390/nano10040748.

109. Ren, Q., Feng, L., Fan, R., Ge, X., & Sun, Y. (2016) Water-dispersible triethylenetetramine-functionalized graphene: Preparation, characterization and application as an amperometric glucose sensor. *Mater. Sci. Eng. C*, *68*, 308–316. https://doi.org/10.1016/j.msec.2016.05.124.

110. Samuel, M. S., Suman, S., Venkateshkannan, Selvarajan, E., Mathimani, T., & Pugazhendhi, A. (2020) Immobilization of $Cu_3(btc)_2$ on graphene oxide-chitosan hybrid composite for the adsorption and photocatalytic degradation of methylene blue. *J. Photochem. Photobiol.*, *204*, 111809. https://doi.org/10.1016/j.jphotobiol.2020.111809.

111. Li, L., Fan, L., Duan, H., Wang, X., & Luo, C. (2014) Magnetically separable functionalized graphene oxide decorated with magnetic cyclodextrin as an excellent adsorbent for dye removal. *RSC Adv.*, *4*, 37114. https://doi.org/10.1039/C4RA06292B.

112. Ma, Y. X., Shao, W. J., & Sun, W. et al. (2018) One-step fabrication of β-cyclodextrin modified magnetic graphene oxide nanohybrids for adsorption of Pb(II), Cu(II) and methylene blue in aqueous solutions. *Appl. Surf. Sci.*, *459*, 544–553. https://doi.org/10.1016/j.apsusc.2018.08.025.

113. Liu, X., Yan, L., Yin, W., Zhou, L., Tian, G., Shi, J., Yang, Z., Xiao, D., Gu, Z., & Zhao, Y. (2014) Magnetic graphene hybrid functionalized by beta-cyclodextrins for fast and efficient removal of organic dyes. *J. Mater. Chem. A*, *2*, 12296–12303. https://doi.org/10.1039/C4TA00753K.

114. Wang, D., Liu, L., Jiang, X., Yu, J., Chen, X., & Chen, X. (2015) Adsorbent for p-phenylenediamine adsorption and removal based ongraphene oxide functionalized with magnetic cyclodextrin. *Appl. Surf. Sci.*, *329*, 197–205. https://doi.org/10.1016/j.apsusc.2014.12.161.

115. Wang, D., Liu, L., Jiang, X., Yu, J., & Chen, X. (2015) Adsorption and removal of malachite green from aqueous solution using magnetic beta-cyclodextrin-graphene oxide nanocomposites as adsorbents. *Colloids Surf., A*, *466*, 166–173. https://doi.org/10.1016/j.colsurfa.2014.11.021.

116. Cao, X. T., Showkat, A. M., Kang, I., Gal, Y. S., & Lim, K. (2016) β –cyclodextrin multiconjugated magnetic graphene oxide as a nano adsorbent for methylene blue removal. *J. Nanosci. Nanotechnol.*, *16*, 1521–1525. https://doi.org/10.1166/jnn.2016.11987.

117. Shi, H., Li, W., Zhong, L., & Xu, C. (2014) Methylene blue adsorption from aqueous solution by magnetic cellulose/graphene oxide composite: Equilibrium, kinetics, and thermodynamics. *Ind. Eng. Chem. Res.*, *53*, 1108–1118. https://doi.org/10.1021/ie4027154.

118. Luo, X., & Zhang, L. (2009) High effective adsorption of organic dyes on magnetic cellulose beads entrapping activated carbon. *J. Hazard. Mater.*, *171*, 340–347. https://doi.org/10.1016/j.jhazmat.2009.06.009.

119. Gautam, D., & Hooda, S. (2020) Magnetic graphene oxide/chitin nanocomposites for efficient 2 adsorption of methylene blue and crystal violet from aqueous solutions. *J. Chem. Eng. Data*, *65*, 4052–4062. https://doi.org/10.1021/acs.jced.0c00350.

120. Fan, L., Luo, C., Li, X., Lu, F., Qiu, H., & Sun, M. (2012) Fabrication of novel magnetic chitosan grafted with graphene oxide to enhance adsorption properties for methyl blue. *J. Hazard. Mater.*, *215–216*, 272–279. https://doi.org/10.1016/j.jhazmat.2012.02.068.

121. Sohni, S., Gul, K., Ahmad, F., Ahmad, I., Khan, A., Khan, N., & Khan, S. B. (2018) Highly efficient removal of acid red-17 and bromophenol blue dyes from industrial wastewater using graphene oxide functionalized magnetic chitosan composite. *Polym. Compos.*, *39*, 3317–3328. https://doi.org/10.1002/pc.24349.

122. Jiang, Y., Gong, J. L., Zeng, G. M., Ou, X. M., Chang, Y. N., Deng, C. H., Zhang, J., Liu, H. Y., & Huang, S. Y. (2016) Magnetic chitosangraphene oxide composite for antimicrobial and dye removal applications. *Int. J. Biol. Macromol.*, *82*, 702–710. https://doi.org/10.1016/j.ijbiomac.2015.11.021.

123. Sheshmani, S., Ashori, A., & Hasanzadeh, S. (2014) Removal of acid orange 7 from aqueous solution using magnetic graphene/chitosan: A promising nano-adsorbent. *Int. J. Biol. Macromol.*, *68*, 218–224. https://doi.org/10.1016/j.ijbiomac.2014.04.057.

124. Marnani, N. N., & Shahbazi, A. (2019) A novel environmental-friendly nanobiocomposite synthesis by EDTA and chitosan functionalized magnetic graphene oxide for high removal of Rhodamine B: Adsorption mechanism and separation property. *Chemosphere*, *218*, 715–725. https://doi.org/10.1016/j.chemosphere.2018.11.109.

125. Salahuddin, N. A., EL-Daly, H. A., Sharkawy, R. G. E., & Nasr, B. T. (2020) Nanohybrid based on polypyrrole/chitosan/graphene oxide magnetite decoration for dual function in water remediation and its application to form fashionable colored product. *Adv. Powder Technol.*, *31*, 1587–1596. https://doi.org/10.1016/j.apt.2020.01.030.

126. de Figueiredo Neves, T., Dalarme, N. B., da Silva, P. M. M., Landers, R., Picone, C. S. F., & Prediger, P. (2020) Novel magnetic chitosan/ quaternary ammonium salt graphene oxide composite applied to dye removal. *J. Environ. Chem. Eng.*, *8*, 103820. https://doi.org/10.1016/j.jece.2020.103820.

127. Bhandari, H., Ruhi, G., Gaba, R., Chaudhary, A., Johar, R., Singh, T., Rawat, A., Kapoor, S., Sharma, V., & Chadha, Y. (2020) Eco-friendly magnetic biopolymer nanocomposites for removal of organic dye/heavy metals from waste water. *Vantage: J. Themat. Anal.*, *1*, 17–31.

128. Fan, L., Luo, C., Sun, M., Qiu, H., & Li, X. (2013) Synthesis of magnetic β-cyclodextrin–chitosan/graphene oxide as nanoadsorbent and its application in dye adsorption and removal. *Colloids Surf., B*, *103*, 601–607. https://doi.org/10.1016/j.colsurfb.2012.11.023.

129. Yadav, S., Asthana, A., Chakraborty, R., Jain, B., Singh, A. K., Carabineiro, S. A. C., & Susan, M. A. B. H. (2020) Cationic dye removal using novel magnetic/activated charcoal/β-cyclodextrin/alginate polymer nanocomposite. *Nanomaterials*, *10*, 170. https://doi.org/10.3390/nano10010170.

130. Liu, Z., Li, D., Dai, H., Huang, H. (2017) Enhanced properties of tea residue cellulose hydrogels by addition of graphene oxide. *J. Mol. Liq.*, *244*, 110–116. https://doi.org/10.1016/j.molliq.2017.08.106.

131. Hu, X., Hu, K., Zeng, L., Zhao, M., & Huang, H. (2010) Hydrogels prepared from pineapple peel cellulose using ionic liquid and their characterization and primary sodium salicylate release study. *Carbohydr. Polym.*, *82*, 62–68. https://doi.org/10.1016/j.carbpol.2010.04.023.

132. Kadokawa, J., Murakami, M., & Kaneko, Y. (2008) A facile preparation of gel materials from a solution of cellulose in ionic liquid. *Carbohydr. Res.*, *343*, 769–772. https://doi.org/10.1016/j.carres.2008.01.017.

133. Kadokawa, J., Murakami, M., Takegawa, A., & Kaneko, Y. (2009) Preparation of cellulose–starch composite gel and fibrous material from a mixture of the polysaccharides in ionic liquid. *Carbohydr. Polym.*, *75*, 180–183. https://doi.org/10.1016/j.carbpol.2008.07.021.

134. Namazi, H., Hasani, M., & Yadollahi, M. (2019) Antibacterial oxidized starch/ZnO nanocomposite hydrogel: Synthesis and evaluation of its swelling behaviours in various pHs and salt solutions. *Int. J. Biol. Macromol.*, *126*, 578–584. https://doi.org/10.1016/j.ijbiomac.2018.12.242.

135. Ahmadian, A., Bakravi, A., Hashemi, H., & Namazi, H. (1967) Synthesis of polyvinyl alcohol/CuO nanocomposite hydrogel and its application as drug delivery agent. *Polym. Bull.*, *76*, 1983. https://doi.org/10.1007/s00289-018-2477-9.

136. Varaprasad, K., Jayaramudu, T., & Sadiku, E. R. (2017) Removal of dye by carboxymethyl cellulose, acrylamide and graphene oxide *via* a free radical polymerization process. *Carbohydr. Polym.*, *164*, 186–194. https://doi.org/10.1016/j.carbpol.2017.01.094.

137. Dai, H., Huang, Y., & Huang, H. (2018) Eco-friendly polyvinyl alcohol/ carboxy-methyl cellulose hydrogels reinforced with graphene oxide and bentonite for enhanced adsorption of methylene blue. *Carbohydr. Polym.*, *185*, 1–11. https://doi.org/10.1016/j.carbpol.2017.12.073.

138. Wang, Y., Zhang, X., He, X., Zhang, W., Zhang, X., & Lu, C. (2014) *In situ* synthesis of MnO_2 coated cellulose nanofibers hybrid for effective removal of methylene blue. *Carbohydr. Polym.*, *110*, 302–308. https://doi.org/10.1016/j.carbpol.2014.04.008.

139. Halouane, F., Oz, Y., Meziane, D., Barras, A., Juraszek, J., Singh, S. K., Kurungot, S., Shaw, P. K., Sanyal, R., Boukherroub, R., Sanyal, A., & Szunerits, S. (2017) Magnetic reduced graphene oxide loaded hydrogels: Highly versatile and efficient adsorbents for dyes and selective Cr(VI) ions removal. *J. Colloid Interface Sci.*, *507*, 360–369. https://doi.org/10.1016/j.jcis.2017.07.075.

140. Gonzalez, J. A., Villanueva, M. E., Piehl, L. L., & Copello, G. J.' (2015) Development of chitin/grapheme oxide hybrid composite for the removal of pollutant dyes: Adsorption and desorption study. *Chem. Eng. J.*, *280*, 41–48. https://doi.org/10.1016/j.cej.2015.05.112.

141. Chang, Z., Chen, Y., Tang, S., Yang, J., Chen, Y., Chen, S., Li, P., & Yang, Z. (2020) Construction of chitosan/polyacrylate/graphene oxide composite physical hydrogel by semi-dissolution/acidification/solgel transition method and its simultaneous cationic and anionic dye adsorption properties. *Carbohydr. Polym.*, *229*, 115431. https://doi.org/10.1016/j.carbpol.2019.115431.

142. Hosseinzadeh, H., & Ramin, S. (2018) Fabrication of starch-graftpoly(acrylamide)/gra-phene oxide/hydroxyapatite nanocomposite hydrogel adsorbent for removal of mala-chite green dye from aqueous solution. *Int. J. Biol. Macromol.*, *106*, 101–115. https://doi.org/10.1016/j.ijbiomac.2017.07.182.

143. Ma, T., Chang, P. R., Zheng, P., Zhao, F., & Ma, X. (2014) Fabrication of ultra-light gra-phene-based gels and their adsorptionof methylene blue. *Chem. Eng. J.*, *240*, 595–600. https://doi.org/10.1016/j.cej.2013.10.077.

144. Ma, T. T., Chang, P. R., Zheng, P. W., & Ma, X. F. (2013) The composites based on plasticized starch and graphene oxide/reduced graphene oxide, *Carbohydr. Polym.*, *93*, 63–70. https://doi.org/10.1016/j.carbpol.2013.01.007.

145. Xiao, D., He, M., Liu, Y., Xiong, L., Zhang, Q., Wei, L., Li, L., & Yu, X. (2020) Strong alginate/reduced graphene oxide composite hydrogels with enhanced dye adsorption performance. *Polym. Bull.*, *77*, 1–15. https://doi.org/10.1007/s00289-020-03105-7.

146. Gan, L., Li, H., Chen, L., Xu, L., Liu, J., Geng, A., Mei, C., & Shang, S. (2018) Graphene oxide incorporated alginate hydrogel beads for the removal of various organic dyes and bisphenol A in water. *Colloid Polym. Sci.*, *296*, 607–615. https://doi.org/10.1007/s00396-018-4281-3.

147. Verma, A., Thakur, S., Mamba, G., Prateek, Gupta, R. K., Thakur, P., & Thakur, V. K. (2020) Graphite modified sodium alginate hydrogel composite for efficient removal of malachite green dye. *Int. J. Biol. Macromol.*, *148*, 1130–1139. https://doi.org/10.1016/j.ijbiomac.2020.01.142.

148. Makhado, E., Pandey, S., Nomngongo, P., & Ramontja, J. (2017) Xanthan gum-cl-poly (acrylic acid)/reduced graphene oxide hydrogel nanocomposite as adsorbent for dye removal. In: *9th International Conference on Advances in Science, Engineering, Technology & Waste Management (ASETWM-17)*. Parys, South Africa, Nov. 27–28, 159–164. https://doi.org/10.17758/EARES.EAP1117058.

149. Hosseini, S. M., Shahrousvand, M., Shojaei, S., Khonakdar, H. A., Asefnejad, A., & Goodarzi, V. (2020) Preparation of superabsorbent ecofriendly semi-interpenetrating network based on cross-linked poly acrylic acid/ xanthan gum/graphene oxide (PAA/XG/GO): Characterization and dye removal ability. *Int. J. Biol. Macromol.*, *152*, 884–893. https://doi.org/10.1016/j.ijbiomac.2020.02.082.

150. Sahraei, R., & Ghaemy, M. (2017) Synthesis of modified gum tragacanth/graphene oxide composite hydrogel for heavy metal ions removal and preparation of silver nanocomposite for antibacterial activity. *Carbohydr Polym, 157*, 823–833. https://doi.org/10.1016/j.carbpol.2016.10.059.

151. Gan, L., Shang, S., Hu, E., Yuen, C. W. M., & Jiang, S. (2015) Konjac glucomannan/graphene oxide hydrogel with enhanced dyes adsorption capability for methyl blue and methyl orange. *Appl. Surf. Sci., 357*, 866–872. https://doi.org/10.1016/j.apsusc.2015.09.106.

152. Liu, Y., Huang, S., Zhao, X., & Zhang, Y. (2018) Fabrication of three dimensional porous β-cyclodextrin/chitosan functionalized graphene oxide hydrogel for methylene blue removal from aqueous solution. *Colloids Surf., A, 539*, 1–10. https://doi.org/10.1016/j.colsurfa.2017.11.066.

153. Yuan, J., Qiu, F., & Li, P. (2017) Synthesis and characterization of βcyclodextrin–carboxymethyl cellulose–graphene oxide composite materials and its application for removal of basic fuchsin. *J. Iran. Chem. Soc., 14*, 1827–1837. https://doi.org/10.1007/s13738-017-1122–0.

154. Wu, Y., Qi, H., Shi, C., Ma, R., Liu, S., & Huang, Z. (2017) Preparation and adsorption behaviors of sodium alginate-based adsorbent- immobilized bcyclodextrin and graphene oxide. *RSC Adv., 7*, 31549–31557. https://doi.org/10.1039/C7RA02313H.

155. Chenga, C., Denga, J., Lei, B., Hea, A., Zhanga, X., Maa, L., Li, S., & Zhao, C. (2013) Toward 3D graphene oxide gels based adsorbents for high-efficient water treatment *via* the promotion of biopolymers. *J. Hazard. Mater., 263*, 467–478. https://doi.org/10.1016/j.jhazmat.2013.09.065.

156. Martin-Jimeno, F. J., Suarez-Garcia, F., Paredes, J. I., MartinezAlonso, A., & Tascon, J. M. D. (2015) Activated carbon xerogels with a cellular morphology derived from hydrothermally carbonized glucosegraphene oxide hybrids and their performance towards $CO2$ and dye adsorption. *Carbon, 81*, 137–147. https://doi.org/10.1016/j.carbon.2014.09.042.

157. Sajab, M. S., C, H., Chan, C. H., Zakaria, S., Kaco, H., Chook, S. W., Chin, X., & Noor, A. A. M. (2016) Bifunctional graphene oxide-cellulose nanofibril aerogel loaded with Fe (III) for removal of cationic dye *via* simultaneous adsorption and Fenton oxidation. *RSC Adv., 24*, 19638–20452. https://doi.org/10.1039/C5RA26193G.

158. Chan, C. H., Chia, C. H., Zakaria, S., Sajab, M. S., & Chin, S. X. (2015) Cellulose nanofibrils: A rapid adsorbent for the removal of methylene blue. *RSC Adv., 5*, 18204–18212. https://doi.org/10.1039/C4RA15754K.

159. Wei, X., Huang, T., Yang, J. H., Zhang, N., Wang, Y., & Zhou, Z. W. (2017) Green synthesis of hybrid graphene oxide/microcrystalline cellulose aerogels and their use as super absorbents. *J. Hazard. Mater., 335*, 28–38. https://doi.org/10.1016/j.jhazmat.2017.04.030.

160. Ren, F., Li, Z., Tan, W. Z., Liu, X. H., Sun, Z. F., Ren, P. G., & Yan, D. X. (2018) Facile preparation of 3D regenerated cellulose/graphene oxide composite aerogel with high-efficiency adsorption towards methylene blue. *J. Colloid Interface Sci., 532*, 58–67. https://doi.org/10.1016/j.jcis.2018.07.101.

161. Wang, Y., Xia, G., Wu, C., Sun, J., Song, R., & Huang, W. (2015) Porous chitosan doped with graphene oxide as highly effective adsorbent for methyl orange and amido black 10B. *Carbohydr. Polym., 115*, 686–693. https://doi.org/10.1016/j.carbpol.2014.09.041.

162. Chen, Y., Dai, G., & Gao, Q. (2019) Starch nanoparticles–graphene aerogels with high supercapacitor performance and efficient adsorption. *ACS Sustain. Chem. Eng., 7*, 14064–14073. https://doi.org/10.1021/acssuschemeng.9b02594.

163. E, T., Ma, D., Yang, S., Hao, X. (2020) Graphene oxidemontmorillonite/sodium alginate aerogel beads for selective adsorption of methylene blue in wastewater. *J. Alloys Compd., 832*, 154833. https://doi.org/10.1016/j.jallcom.2020.154833.

9 Polymer–Graphitic Nitride Composites for Wastewater Treatment

Chandrashekhar S. Patil and Akhilesh N. Bendre
JAIN University

Anil H. Gore
UkaTarsadia University

Tukaram D. Dongale
Shivaji University

Mahaveer D. Kurkuri
JAIN University

CONTENTS

DOI: 10.1201/9781003328094-9

9.1   INTRODUCTION

Water pollution is a global issue, and the pollutant's effects on the environment have become the world's most serious concern.[1] Various inorganic and organic contaminants such as hazardous pharmaceuticals, PCPs, metal ions, dyes, discharge of household sewage, etc. are the main sources of water pollution.[2] The geological processes and anthropogenic activities are the primary sources of such pollutants.[3] As a result, it's extremely important to remove such hazardous contaminants in a safe, practical, and environmentally sustainable manner. Adsorption, membrane separation, photocatalysis, and sensor systems, which are primarily designed for capturing, isolating, degrading, and detecting environmental pollutants/toxins, are among the prominent remediation technologies.[4]

So far many materials have been considered in water purification, such as biomass-based materials,[5] carbon and its derivatives,[2,6] MOFs,[7] magnetic nanoparticles,[8] COFs,[9] polymers-based materials,[10] clay-minerals,[11] etc., in the form of adsorbents,[12] photocatalyst,[13] sensors,[14] membrane,[15] etc.

In this regard, graphitic carbon nitride (g-C$_3$N$_4$) has been receiving wide attention due to its flexible physiochemical properties. It primarily consists of carbon (C) and nitrogen (N) with hydrogen (H), where they are covalently held together by a tri-s-triazine construction and have been proven to be highly capable compared to the majority of carbon materials due to the strong electron-rich properties. These are also observed to be the most stable allotrope under all ambient conditions and as such are an exceptional replacement for traditional carbon material.[16]

Berzelius and their group invented the first carbon nitride polymer termed "melon" in 1834.[17,18] A significant number of interesting findings on advanced frameworks of g-C$_3$N$_4$ and its composites have been reported, particularly over the last few years. Although several excellent reviews on the effective modification of pristine g-C$_3$N$_4$ for wastewater treatment have been reported, a systematic and comprehensive review on heterojunction construction and nanostructure design has not yet been reported and is urgently needed to promote further developments for its targeted wastewater applications. This chapter provides an updated outline of the progress on g-C$_3$N$_4$ and its hybrid materials in the field of water purification applications, including preparation methodology.[19–21]

9.2   PROPERTIES OF g-C$_3$N$_4$

Graphitic carbon nitride is a unique two-dimensional polymeric material. A lot of work has been done to increase their performance and productivity in terms of various methodologies, variations in precursor materials, and synthesis parameters as well. The detailed configuration of g-C$_3$N$_4$ consists of two major units, viz., tri-s-triazine (C$_6$N$_7$) and the s-triazine (C$_3$N$_3$), as shown in Figure 9.1. According to the literature, it can be seen that the tri-s-triazine is more stable under ambient conditions as compared to the s-triazine (C$_3$N$_3$).

The g-C$_3$N$_4$ has similar functionality to graphene, which is a well-reported material in the field of research.[22] g-C$_3$N$_4$ is an sp^2-hybridized carbonaceous material that comprises π-conjugated graphitic sheets created due to the hybridization of

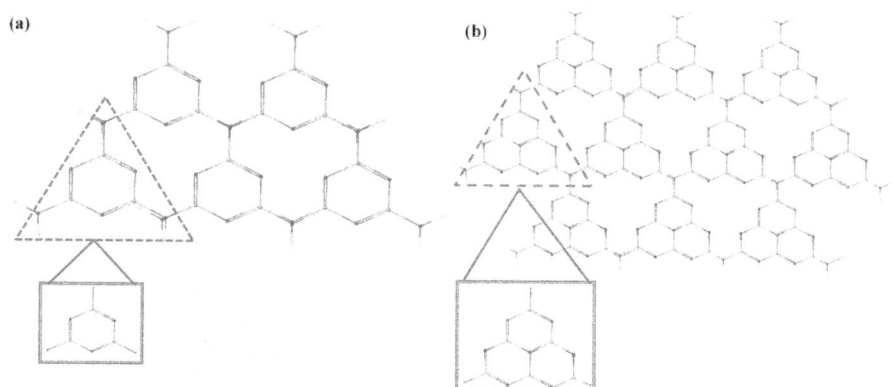

FIGURE 9.1 Basic structures of (a) s-triazine type of g-C_3N_4 and (b) tri-s-triazine type of g-C_3N_4.

carbon and nitrogen atoms. Recently, Fina and coworkers have launched a three-dimensional (3D) structure of g-C_3N_4, which could be a turning point in upcoming research. The results from the X-ray diffraction (PXRD) and neutron diffraction confirm the successful fabrication of the 3D g-C_3N_4. Additionally, it was noticed that the tri-s-triazine-based layers were misaligned, which circumvents the repulsive forces of electrons in adjacent layers.[19,23]

In the present scenario, it has shown a high potential for a variety of applications, such as energy storage and conversion, biomedical, electronics, energy production, sensors, catalysis, environmental remediation, etc. as shown in Figure 9.2.[17,20,23,24]

9.3 SYNTHESIS OF POLYMER–GRAPHITIC NITRIDE COMPOSITES

Due to the extraordinary properties of polymers, they were extensively investigated and, thus, play a very important role in the industrial sector as well as in daily life.[25] They have functional properties such as self-healing, stimuli response, biodegradability, and electrical conductivity, which are of particular interest today.[26] Consequently, g-C_3N_4 is one of the trending 2D polymeric materials that has gotten wide attention in recent days.[27] However, combining conventional polymers with novel, metal-free semiconductors and trending g-C_3N_4 materials would be extremely beneficial. Although g-C_3N_4 is being studied as a metal-free semiconductor, poor processing prevents it from being scaled up, which limits its real-world applicability. Recent studies reveal that the combination of g-C_3N_4 and polymers appears to be a useful solution, as the properties of both material classes can be coupled to have a synergetic effect. As a result, the g-C_3N_4 component improves the physical and mechanical properties of the composite material, while the composite material helps to improve the processability and productivity of the material.[28–30]

The g-C_3N_4 and its hybrid materials have been synthesized via different routes, such as thermal polymerization/decomposition, solvothermal synthesis, microwave heating, solid-state reactions, sol–gel synthesis, chemical vapor deposition, and

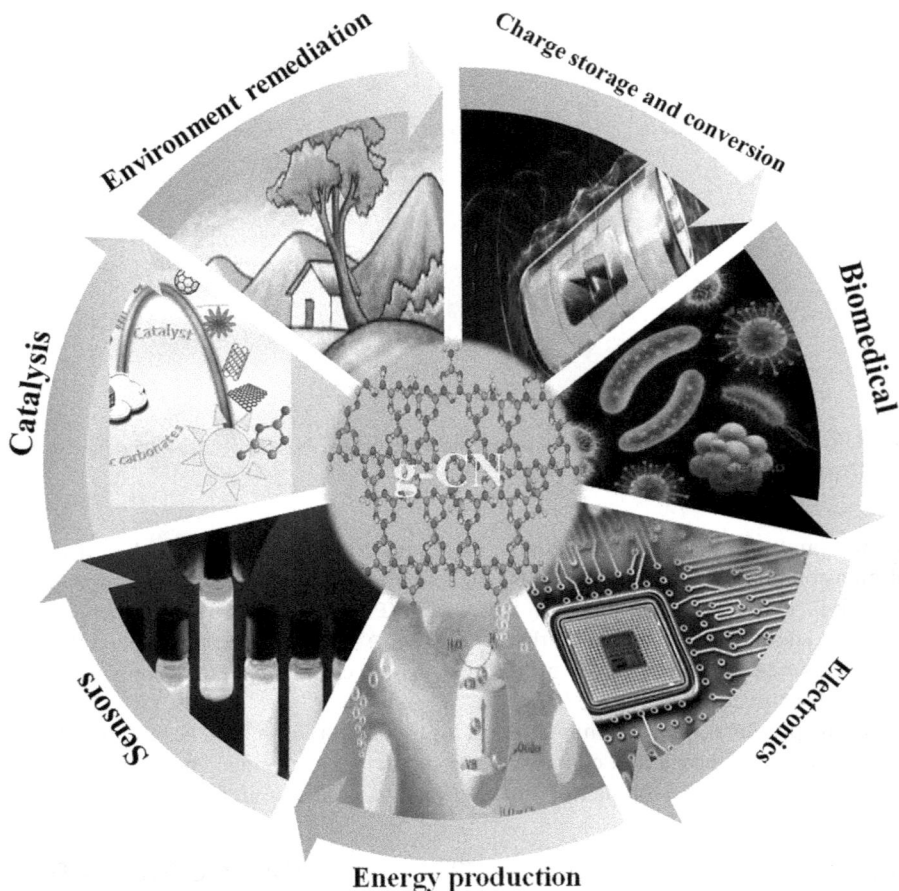

FIGURE 9.2   Graphic representation of different applications of g-C₃N₄.

electro-deposition, as shown in Figure 9.3.[31] Among these, thermal polymerization has been frequently utilized due to its simplified processing and high productivity. Moreover, researchers have used the high nitrogen and oxygen-containing precursors materials such as melamine, cyanamide, thiourea, and so on, because of their huge thermal and chemical stability, and high specific surface area, which makes them a forerunner in the wastewater purification technology.[32] In brief, it can be concluded that both the outer and inner structures, as well as the functional properties of g-C₃N₄ nanostructures, are closely related to the different synthetic strategies employed. Therefore, the study of synthesis methodologies is a very important aspect, concerning the synthesized products along with the other operational parameters.

The design of g-C₃N₄ is achieved via two primary ways, *viz.*, templating (hard or soft) and non-templating methods.[33] The main advantages of using a suitable template during the production are precise control over the size with an outer framework

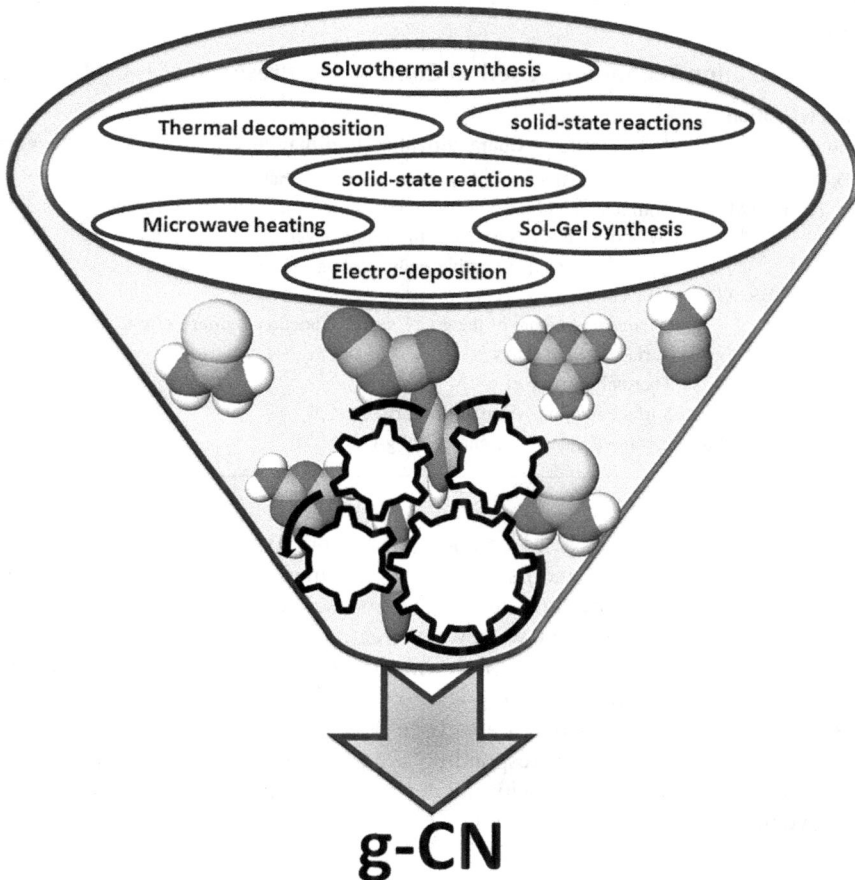

FIGURE 9.3   Schematic illustration of different synthesis routes for g-C_3N_4-based materials (g-CNs).

as well as other desired properties. But it can be dangerous to use templates, as they might contain hazardous and environmentally harmful fluorine-containing agents. In contrast to this, in non-templating techniques, nanostructures are made without fluorine-containing agents. As a result, non-templating methods are preferred for designing g-C_3N_4 nanostructures.[34]

9.4   CHARACTERIZATION OF POLYMER–GRAPHITIC NITRIDE COMPOSITES FOR WASTEWATER TREATMENT

Due to their wide application in various disciplines, g-CNs are being used in large quantities. Many researchers have tried to alter their structural and functional properties as per their expectations.[25] As we already discussed, the general

TABLE 9.1

Characterization Techniques Used to Confirm the Presence of g-C₃N₄

Characterization Techniques	Characterization Information	References
XRD, Raman	Crystalline structures, d-spacing, and phase analysis of g-CN	35
SEM, TEM, STM	Both outer and inner surface morphology along with cross-section structural analysis of CN at the atomic level	36,37
AAS, XPS, CDSICP-OES, XRF	Qualitative and quantitative analysis of various trace elements present in the g-CN	38–40
FTIR	Chemical structure of the g-CN, surface functional groups of g-CN	37
BET	Surface area of g-CN	39
TGA	Thermal stability of g-CN	39
AFM	Surface roughness and uniformity of g-CN	38
Surface zeta potential	Identified the surface charges of g-CN	37

nitrogen-containing and oxygen-free compounds were used as a precursor material for the production of g-CN and its composites. As these materials are already pre-bonded with C–N core structures of triazine and heptazine derivatives, g-CN has become a potential material for wastewater treatment.[30] Various surface modifications and functionalities have been performed to obtain desired outcomes using the above-mentioned methods (see Figure 9.3).

Additionally, enormous work is also been done on the multi-functional applications, properties, and their corresponding synthesis methodologies. Therefore, to confirm the expected materials with the desired properties, the analysis techniques that have been used are shown in Table 9.1.

9.5   POLYMER–GRAPHITIC NITRIDE AND THEIR COMPOSITES FOR WASTEWATER TREATMENT

9.5.1   POLYMER–GRAPHITIC NITRIDE COMPOSITES IN ADSORPTION OF POLLUTANTS

The g-C₃N₄ exhibits graphene-like properties; therefore, it is a leading material in water purification technology. It also has a high porosity with a huge specific surface area like graphene; consequently, it is also widely considered an absorbent material for the removal of miscellaneous contaminants such as metal ions, dyes, microorganisms, and industrial waste, as shown in Figure 9.4. In addition, the availability of large amounts of nitrogen and carbon contents, with excellent porous and textured surfaces, makes g-C₃N₄ a powerful absorbent material in water treatment technology.[29,41]

For instance, Haque et al. demonstrated the first carbon nitride-based material, *viz.*, mesoporous carbon nitride (MCN) for wastewater treatment applications for the selective removal of phenol. It was utilized for the selective removal of phenol from aqueous media and showed an excellent adsorption capacity compared to the other previously reported activated carbons (AC) and CMK-3–150. It was found to be 2.7

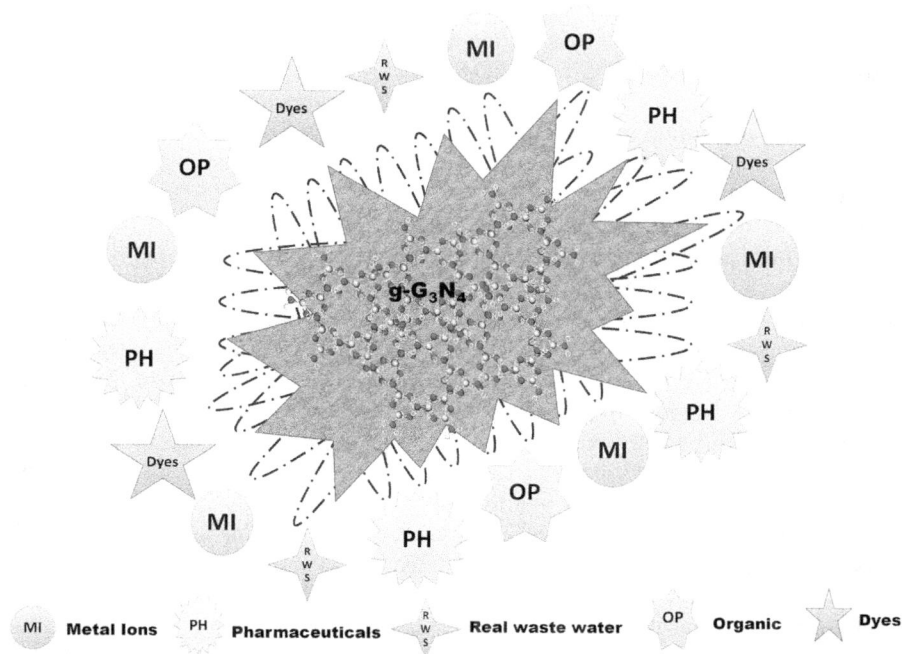

FIGURE 9.4 g-C$_3$N$_4$ and its hybrid materials as adsorbent materials for wastewater treatment.

and 1.29 times greater than that of AC and CMK-3–150, respectively. This may be due to a large number of available amides groups.[42]

Subsequently, similar derivatives of MCNs with slight modifications were also developed using silica nanoparticles as a precursor material, for the removal of perfluorooctanesulfonate (PFOS) and Ni(II) ions, respectively. From the results, it was observed that the surface area of composites increased due to the precursor silica material, which subsequently increased the composite's adsorption efficiency for PFOS and Ni(II). Furthermore, Huang et al. synthesized a highly protonated MCN functionalized with cyanamide that can be employed to remove carcinogenic microcystins (MCs), *viz.*, MC-LR, and MC-RR. The resulting protonation not only increased MCN's uptake capacity for MCs but also sped up the MCs sorption capacity.[43]

The separation of the exhausted adsorbent from aqueous media after the adsorption process is as important as the extraction of pollutants, because exhausting adsorbents may also act as a kind of water pollutant if they stay in aqueous media for a longer period. It is particularly more dangerous, as it is well known that g-C$_3$N$_4$/g-C$_3$N$_4$-based materials are available in nano size. However, even after the adsorption process is complete, the absolute separation of g-C$_3$N$_4$ material from aqueous media is still a challenging task. Keeping this in mind, in the year 2014, Yan et al. and their team proposed magnetic MCN for PFOS and perfluorooctanoic acid (PFOA) removal. The key part of the proposed work is that we can easily separate the adsorbent MMCN from adsorbate by applying an external magnetic field. The maximum adsorption capacity for PFOS and PFOA was found to be 454.6 and 370.4 mg/g, respectively. It was also observed that increasing the temperature from 278 to 318 K increased the adsorption of PFOS and PFOA on MMCN. In conclusion, the adsorption is facilitated due to electrostatic and hydrophobic interactions between PFOS and PFOA, and the MMCN.[44]

Following this, the research was boosted and several researchers have tried to subject numerous surface functionalization to enhance the affinity of g-C_3N_4 toward a variety of pollutants with selective adsorption properties, which are in demand for environmental application. However, Kumar et al. and the team were successful in this, they developed next-level oxygen functionalized ternary composite referred to as Ox-g-C_3N_4/PANI-NF for selective Cr (VI) removal. The modification of the surface with the positive charge of oxygen leads the more adsorption of Cr (VI). Furthermore, the adsorption of Cr (VI) was highly dependent on solution pH, temperature, and Cr concentration (VI). Following that, Liu et al. created a PANI/oxidation composite using g-C_3N_4 exfoliation, which was also used to remove radioactive uranium (VI) from uranium-containing wastewater. As a result, it was found that the oxidized g-C_3N_4 showed better adsorption performance as compared to the unoxidized g-C_3N_4.[45]

Further, in recent days, Shen and coworkers made unique and ecofriendly polymeric graphitic carbon nitride, in-situ grafted, alginate-based hydrogel beads (g-C_3N_4/SA). It is synthesized via a facile one-step crosslinking method in which polymerization occurs between g-C_3N_4/SA and the calcium chloride. The as-developed g-C_3N_4/SA composite effectively removes Pb (II), Ni (II), and Cu (II) metal ions from aqueous media and shows tremendous adsorption capacity toward every metal ion, *viz.*, 383.4, 306.3, and 168.2 mg/g for Pb(II), Ni(II), and Cu(II), respectively. More importantly, the investigated g-C_3N_4/SA composite showed high regeneration potential up to five successive cycles with appropriate adsorption capacity, which may guide the upcoming water technology.[46]

Meanwhile, Talukdar and the team developed a surface-modified mesoporous g-C_3N_4@FeNi$_3$ as a prompt and effective magnetic adsorbent for crude oil–water separation from aqueous media. Herein, the authors combined a porous g-C_3N_4 with magnetic FeNi$_3$ and fatty acid, and it is found that the as-developed adsorbent g-C_3N_4@FeNi$_3$ separates crude oil from water effectively. The surface functionalization causes the enhancement of the surface area and the high mesopore nature of the adsorbent, which contributes to more adsorption of oil. In addition to this, the developed system recovers crude oil rapidly under an external magnetic field, which may be helpful in upcoming water technology.[47]

9.5.2   POLYMER–GRAPHITIC NITRIDE COMPOSITES IN PHOTOCATALYTIC DEGRADATION OF POLLUTANTS

There are several technologies to treat wastewater; photodegradation (photocatalysis) is one among them and is getting wide attention along with the adsorption process in recent years.[48] Nowadays, there is a huge list of catalysts that have been created for effective cleaning-up of wastewater. Among them, g-C_3N_4 and its hybrid material-based materials have received a lot of attention as multi-functional catalysts for the remediation of multiple environmental contaminants. This is mainly because of its unique and versatile properties, which include low cost, environmental friendliness, metal-free catalysis, visible light response, and, most importantly, tunable electronic bandgaps.[49] Thus, in this section, we have discussed the artistic progress of g-C_3N_4 and its hybrid materials as a photocatalyst for wastewater treatment.

Yan and coworkers developed a boron-doped g-C_3N_4 composite for rhodamine B (RhB) and methyl orange (MO) degradation. It was fabricated through a simple, straightforward, and one-step heating technique by a mixture of melamine and boron oxide. Due to boron doping, it showed an excellent photocatalytic activity toward RhB as compared to the inherent g-C_3N_4. The RhB degradation follows a direct hole oxidation reaction mechanism, while MO degradation follows the overall reaction mechanism.[50]

Ge et al. made a novel visible-light-induced g-C_3N_4/Bi_2WO_6 composite for methyl orange (MO)-contaminated wastewater. It is the first report on a heterojunction photocatalyst. The as-prepared catalyst showed excellent photocatalytic activity toward MO degradation. This could be due to the synergistic effect of polymeric g-C_3N_4 and Bi_2WO_6, as well as the impactful separation of photogenerated electron-hole pairs.[51] Furthermore, the same group has used the same catalyst with a little modification for MO degradation. They were decorated with Ag nanoparticles and used for the same. It was found that the Ag/g-C_3N_4 photocatalysts exhibited significantly enhanced photocatalytic performance for the degradation of MO when compared with pure g-C_3N_4 (Figure 9.5).[52]

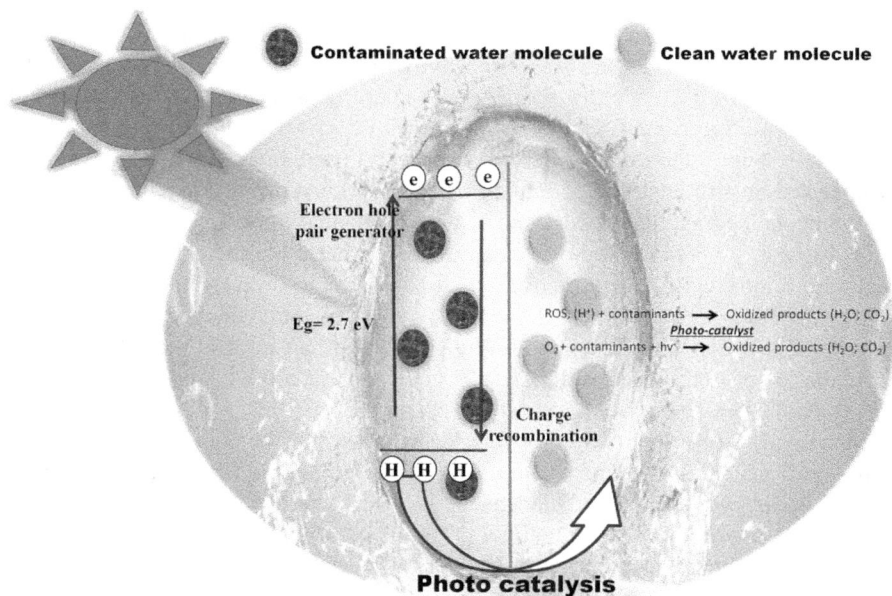

FIGURE 9.5 Mechanistic view of g-C_3N_4 as a photocatalyst for the degradation of pollutants from aqueous solution.

Later on, in 2013, Ye et al. developed an inorganic–organic BiOBr–g–C_3N_4 composite at a very low temperature by an uncomplicated one-step chemical bath method. For rhodamine B (RhB) degradation, the BiOBr–g–C_3N_4 composite demonstrated significantly higher visible-light-driven (VLD) photocatalytic activity as compared to the pristine g-C_3N_4 and BiOBr. According to the photocatalytic mechanism analysis, the interaction between BiOBr and g-C_3N_4 is a type of facet coupling between Bi-O-Brand and g-C_3N_4. The photo-induced charge transfer between these facets resulted in efficient charge separation, according to the active species isolation and quantification experiments.[53]

Right away, Zhou and their team developed a novel, highly porous, and magnetic Fe_3O_4/g-C_3N_4 nanospheres for the very first time by hydrothermal method. Herein, it is found that the as-developed catalyst shows excellent MO degradation performance under visible light, which is the key feature of this work. In addition to this, the Fe_3O_4/g-C_3N_4 nanospheres show outstanding reusability, strong magnetic properties, stagnant MO degradation, and efficiency for up to five successive cycles, which may give valuable inputs in the upcoming water purification technology.[54]

Rong and the group prepared a ternary ZnO–Ag_2O/porous g-C_3N_4 (ZnO–Ag_2O/pg-C_3N_4) composite for the degradation of ciprofloxacin (CIP) under visible light irradiation. It is observed that the obtained composite shows favorable photocatalytic activity and a higher separation charge carriers system under visible light irradiation due to its relatively narrow bandgap. It shows the highest degradation efficiency (97.4%) as compared with other previously reported catalysts within only 48 minutes due to the incorporation of ZnO–Ag_2O.[55]

It was reported that Bi_2O_3 and WO_3 have excellent photodegradation performance in visible light. However, their response is only limited to visible light with a shorter wavelength (460 nm), which is close to solar energy conversion. As a result, it is a difficult task to build a g-C_3N_4-based system with a narrow bandgap semiconductor photocatalyst for efficient use. Herein, in the proposed work Hong and the group synthesized in-situ inserted Z-schemeV_2O_5/g-C_3N_4 heterojunction composites for simultaneous degradation of MO and MB under visible light. The incorporation of V_2O_5 enhances the efficiency of the catalyst. The basic photocatalytic mechanism was determined with help of active species trapping and electron spin resonance (ESR) experiment. According to the mechanism, it was found that the catalyst exhibits strong oxidation and reduction ability, which will be useful in the future for the development of other direct solid-state Z-scheme photocatalytic systems for use in energy conversion and environmental remediation.[29]

Meanwhile, according to a study, it was confirmed that the photocatalytic activity of the catalysts certainly depends on the preparation conditions as well as on exposing g-C_3N_4 to various chemical modifications.[56] For example, Lamkhao and coworkers developed a low-cost charcoal/g-C_3N_4 composite under different atmospheric conditions, viz., first under air atmosphere and another one under oxygen atmosphere. Their photoactivity toward the degradation of MB and Cr (VI) was tested. Finally, it was concluded that the composite prepared in an air atmosphere had higher photocatalytic activity for MB, whereas composites prepared in an oxygen atmosphere had good photoreduction of Cr (VI).[57]

Later on, Hu and their coworkers studied the effect of different co-dopants on the doped performance. In this study, they have prepared several g-C_3N_4-based catalysts co-doped with the following materials such as iron, phosphorus, and sulfur as well as checked their photocatalysis performance on various dyes and phenols.[58] Herein, we have discussed the photocatalytic performance of the sulfur self-doped g-C_3N_4 catalyst. They have developed several sulfur-doped catalysts in the presence of different temperatures ranging from 450°C to 575°C, likewise, they have made a sulfur self-doped g-C_3N_4 at 550°C. However, according to the results, it was observed that the surface of the composite had increased dramatically, and the band gap had also increased, suggesting that sulfur doping may play a vital role in increasing the surface area of any material.[59]

However, recently, researchers have been working to develop novel, high-performance bio-inspired materials known as Z-scheme heterojunction photocatalyst that has an exceptionally well performance of visible-light harnessing as well as optimum band edge potentials. Studies have attempted to hybridize metal-free polymeric conjugated graphitic carbon nitride (g-C_3N_4) with several other aspects to tailor an effective noticeably oriented photocatalyst that prompted development in the field of photocatalysis.

For example, Zhang et al. have investigated a 2D/2D/BiOBr/CDs/g-C_3N_4 Z-scheme nanohybrid combined with carbon dots (CDs) for the removal of hazardous antibiotic drug ciprofloxacin (CIP) and tetracycline (TC). The incorporated CDs play an important role in light-harvesting and promote the separation of the photogenerated charge carriers while also increasing the specific surface area. The developed catalyst shows not only excellent photocatalytic performance under visible light but also remarkable interfacial charge transfer abilities and a large interface contact area. Furthermore, it was found that the photodegradation productivity is also much higher than the pristine BiOBrnanosheets and the g-C_3N_4 ultrathin nanosheets. Also even after four cycles of CIP removal, the composites still exhibit excellent photostability and reusability, which is one of the key features of the as-developed material 2D/2D BiOBr/CDs/g-C_3N_4.[19]

Jindal and their team summarized the current progress in the use of polymeric graphitic carbon nitride-based photocatalysts for dye degradation. Herein the authors have mentioned the progress of polymeric graphitic carbon nitride-based photocatalysts in terms of the synthesis, characterization, and achievements for all dyes. They have discussed in detail the modifications of catalyst through doping, composite formation, morphology control, and their corresponding results, etc. They also discussed the current challenges and future perspectives as well.[60]

Yu and coworkers manufactured a novel hybrid $TiO_2/SiO_2/g$-C_3N_4 superstructure composite via the ternary-soft-template method for seawater purification. Due to its exceptional superstructural properties, it successfully removed hazardous berberine hydrochloride (BH) via both adsorption and degradation, which may strongly recommend that the as-developed $TiO_2/SiO_2/g$-C_3N_4 composite may be a "breakthrough" in upcoming water technology. It may help in the large-scale conversion of seawater into drinking water.[61]

We hope that these findings will provide invaluable guidance to tackle and overcome the challenges related to water and wastewater pollution so that the problem of drinking water does not arise in the future. Also, this will help to facilitate their use in the rural areas and remote area villages.

9.5.3   Polymer–Graphitic Nitride-Based (g-C$_3$N$_4$) Sensors for Pollutant Detection

Wastewater and its negative effects on the health of living beings have emerged as major global issues. However, there is an urgent global need to measure, remove, and control such harmful toxins before they have a significant impact on the ecosystem. In this scenario, we have used a highly selective and long-lasting sensing/detection technique as well as materials to identify such hazardous contaminants before affecting the ecosystem while the concentration was at an extremely low level. It was a groundbreaking invention by researchers to monitor or prevent it either directly on-field or at the laboratory as well. Nanotechnology has so far offered a list of array of materials for the detection of applications qualitatively and quantitatively. However, because of its versatile physicochemical properties, graphitic carbon nitride (g-C$_3$N$_4$) and its composites have attracted widespread interdisciplinary interest, nowadays, as shown in Figure 9.6.

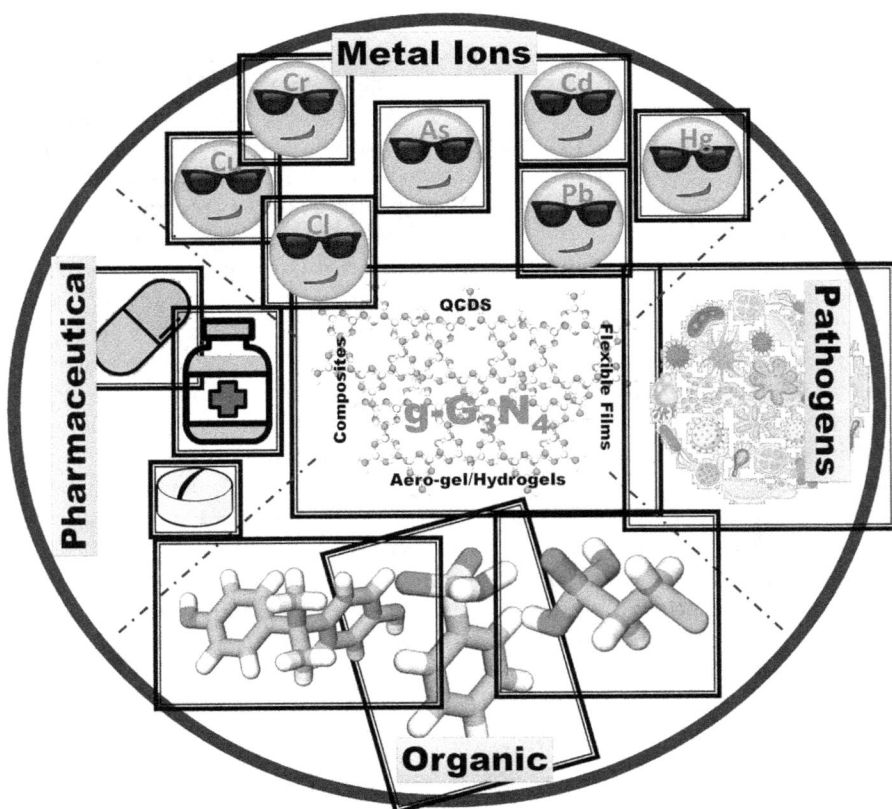

FIGURE 9.6   g-C$_3$N$_4$ and its composites materials as a sensor for the detection of pollutants present in wastewater.

For instance, Lee and their coworkers developed a mesoporous graphitic carbonitride-based Cu^{2+}-c-mpg-C_3N_4 turn-on fluorescence sensor for the selective detection of cyanide (CN^-). Herein, it was fabricated through a simple mixing method. The cubic mesoporous graphitic carbon nitride (c-mpg-C_3N_4) was mixed into an aqueous $Cu(NO_3)_2$ solution. The detection limit of CN^- was found to be 80 nM. Noticeably, it not only works in aqueous solution but also in human blood serum in the same way. The foremost feature of the proposed work is that it responds swiftly, i.e., within only 10 minutes. Though, the as-developed composite material Cu^{2+}-c-mpg-C_3N_4 can be used as a sensor in both aqueous and physiological solutions. As a result, it may have a wide range of demands in the future in the field of sensors, due to its elegant biocompatibility and magnificent photostability.[62]

Analogously, similar kinds of results were also found. The polymer nanodots of graphitic carbon nitride have been developed as effective fluorescent probes for the detection of Hg^{2+}, Cu^{2+}, and Ag^+. They were fabricated through simple microwave-assisted solvothermal and thermal polymerization methods. The overall detection performance of as-developed composites probes might make them a potential candidate for highly sensitive detection of contaminants in wastewater applications in the coming days.[63]

Whereas, Sun and colleagues were reported that Fe-doped g-C_3N_4 nanosheets for glucose detection fabricated through a simple, one-step hydrothermal method. It shows high sensitive optical detection having a 0.5 μM of the limit of detection. In this report, the authors have introduced for the very first time that the ultrathin nanosheets of graphitic carbon nitride (g-C_3N_4) consist of peroxidase activity. However, it was observed that Fe doping leads to increase in the catalytic performance of the as-invented composite Fe–g-C_3N_4.[64] Along with this, the authors also demonstrate that the same composite with a few modifications can be used as an electrochemical sensor for glucose sensing from both aqueous media and human blood with a detection limit of 11 and 45 μM, respectively.[65]

Further, in 2015, Xu and the team developed a g-CNQDs-based chemosensor for quick, sensitive, and selective detection of glutathione (GSH) from food samples. The limit of detection for GSH was found to be 37 nM at the optimal laboratory conditions. More importantly, the as-developed sensor shows excellent regeneration performance up to three successive cycles for GSH in both standard and food samples. Furthermore, with excellent recoveries, this can be used to successfully determine GSH in various types of food samples. However, this development may boost the new era of pollutant detection in real food samples.[66]

Wang and the group designed a novel graphitic carbon nitride-based PEC sensor for the selective detection of copper ions (Cu^{2+}). It was easily fabricated through carbon nitride quantum dots (g-CNQDs) in-situ combined with Bi_2MoO_6 nanohybrid nanoparticles via a simple, easy solvothermal method. The as-developed g-CNQDs were explored for their prospective use as a sensitizer to enhance the photoelectrochemical properties of Bi_2MoO_6. It possesses excellent stability, water solubility, and exceptional electronic properties. The subsequent sensor performed well under optimized conditions, with a wide linear range of 3–40 nM and good selectivity, indicating that g-CNQDs/Bi_2MoO_6 nanohybrids could be a promising photoactive material in wastewater preservation as a pollutant sensor in the forthcoming days.[67]

Further away, Liu and coworkers launched a novel NiO/Co_3O_4 ternary co-doped g-C_3N_4 nanocomposite in sensing applications for the very first time and have succeeded in that. The method has been utilized for the detection of Tetrabromobisphenol-A (TBBP-A) and has been fabricated by a simple, straightforward, low cost, and effective pyrolysis method. Herein, the limit of detection was found to be 0.1 µmol L for sensitive detection of Tetrabromobisphenol-A (TBBP-A). More importantly, the authors have tried its applicability on real wastewater samples to test its functionality in real-world analysis and got satisfactory results.[68]

Nowadays, apta-sensors have gotten wide attention due to their selectivity, which could provide a promising approach for accurate pollutant monitoring. For instance, Feng et al. developed a novel label-free PEC-based apta-sensor for the detection of tetracycline (TC) from aqueous media.[69] It is designed with a coupling plasmon Au with MOF-derived In_2O_3@g-C_3N_4 nano-architectures (AuInCN). The as-designed apta-sensor shows an outstanding sensing activity toward the TC molecules with a 3.3 pM limit of detection, which indicates a potential sensor for the detection of TC from aqueous media in the future.[69] In parallel, concerning the same sensing mechanisms, Deiminiat and Rounaghi developed a label-free PEC apta-sensor for the detection of bisphenol A (BPA),[70] in which authors made a composite between g-C_3N_4 and the gold nanoparticles (AuNPs).

Although high-cost equipment is used, as well as functional materials are being used and are available nowadays, still sometimes they are unable to handle such critical situations as the current COVID-19 pandemic situation. Thereby, there is an extreme urgency to perform more research on this, as it is needed to deal with difficult situations that help to preserve the earth's ecosystem.

9.5.4   POLYMER–GRAPHITIC NITRIDE (G-C_3N_4)-BASED MEMBRANES FOR WASTEWATER TREATMENT

To date, many scientists have been working to develop water and wastewater treatment technologies. However, membrane filtration has been proven as a promising environment-friendly and energy-efficient technology in the current water purification technology due to its ease of preparation and operation.[71] Even though nanofiltration processes have benefits, there are a few drawbacks, such as low-performance water flux and selectivity, along with subordinate fouling/pressure resistance and reliability.[72] To conquer these inadequacies and strengthen the purification efficiency, researchers are trying different methods as well as new materials. To step out, a lot of materials have been tried or being used; one such material is graphic carbon nitride, because of its flexible physicochemical properties such as porous properties, tunable electronic structure, and also physicochemical stability.[73,74] However, herein we summarized an outline of the current progress in the applications of g-C_3N_4 and its composite-based membranes for water purification, as shown in Figure 9.6.[75]

To give you an idea, Chen et al. reported at the very first that polyamide-modified g-C_3N_4 nanosheet-based NF membrane has good permeation and antifouling properties (Figure 9.7). In their work, they have incorporated the g-C_3N_4 nanosheets in the polyamine (PA) active layers. The main goal of incorporating g-C_3N_4 is to

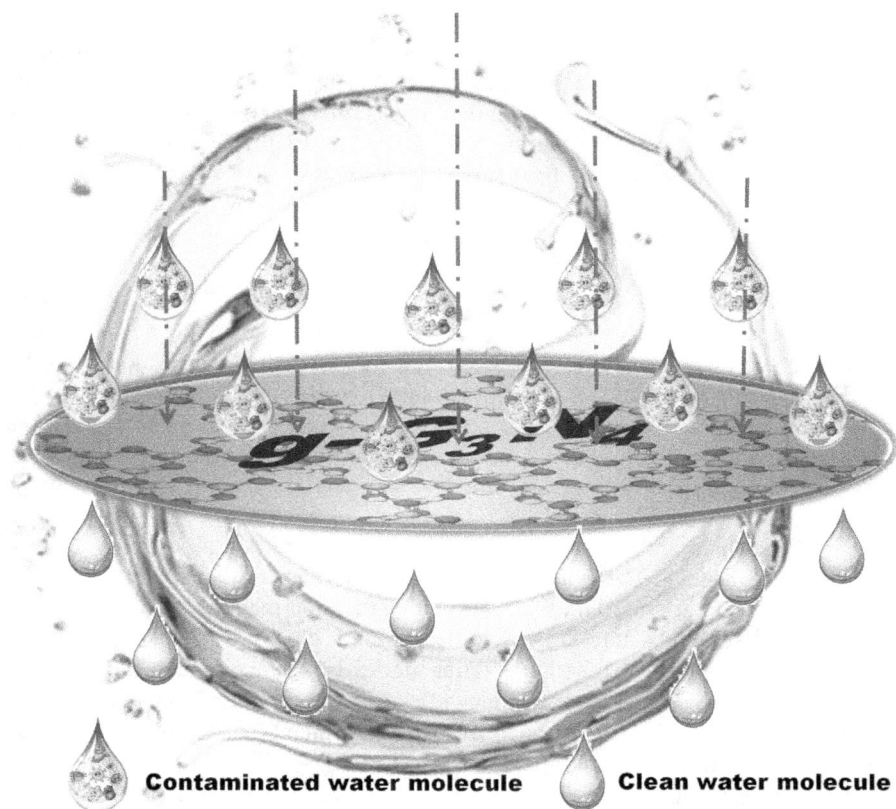

Contaminated water molecule   **Clean water molecule**

FIGURE 9.7   Graphitic nitride (g-C₃N₄)-based membranes for wastewater treatment.

improve the performance of the membrane. They got better results as compared to the unmodified membrane. The water flux increased from 20.9 to 37.6 L/m²h at 2 bar, while the 84.0% of rejection of Na_2SO_4 was maintained.[76]

Madhavi and coworkers made the aforesaid membrane with a slight modification referred to as TFN-CN50. They modified the CN concentration parameter and increased it from 0 to 50 mg/L. As a result, they got superior water flux as compared to the previous one, but the NaCl rejection rate was found to be still constant. The water flux jumps from 25.1 to 45.0 L/m²h under the operating pressure of 16 bars. The steady increase in permeability flow velocity of TFN-CN50 membranes seemed to be primarily due to its extraordinary dispersion of CN with a particulate hierarchical model, where CN material vastly enhanced the hydrophilicity of the hybrid membrane and call for huge nanochannels between both the g-CN and polymer functionality for the diffusion of water.[77]

In addition to this, it has been proven that the surface functional groups have a significant impact on separation performance. Wang and the team have fabricated two types of membranes; the first one is a membrane stacked with highly porous 2D g-C_3N_4 nanosheets one by one for the separation of miscellaneous dyes from direct textile wastewater. Later on, they tried to tune the surface functional groups of the membrane and have been successful in that. It was functionalized with negative ions for the filtration of cationic rhodamine B (RhB) and the methylene blue (MB), maintaining with high permeance of the membrane, near about 90.[78]

Analogously, as it takes less energy and cost to build membranes with low water permeability, the following membranes were considered to be a prospective material for achieving excellent oil-in-water separation. For instance, Shi et al. developed GO incorporated g-C_3N_4 in-situ PVDF membrane (GO/MCU-C_3N_4/PVDF) for oil–water separation. It shows excellent separation performance under visible light along with self-cleaning property. It was fabricated via crosslinking method under vacuum-assisted self-assembly. Moreover, it shows good separation performance due to its high surface area along with high pore volume. The key feature of this work is that it is equipped with high permeation flux, separation performance, and antifouling properties, and stability may indicate potential applications for upcoming water treatment technologies.[79]

Further, Shahabi and their coworkers fabricated polyamide-based multiple graphitic carbon nitride (g-C_3N_4) thin nanosheets membranes functionalized with OH, COOH, and SO_3H groups. The surface functionalization made them highly hydrophilic with high-quality antifouling and desalination properties. Meanwhile, it can be found that the surface functionalization improved the membrane efficiency in terms of water permeation rate without affecting the pollutants rejection performance. More importantly, the surface-functionalized membrane shows better results as compared to the g-C_3N_4 modified membranes.[80]

As we know, the fouling of membrane has always been a serious issue that creates a barrier in the membrane technology used in the large-scale industrial applications of membrane technology. Currently, more research has been done to reduce the fouling rate of the membrane. However, Zheng and the group made CS/PAN@FeOOH/g-C_3N_4 electrospun photo-Fenton catalytic membrane for wastewater treatment. As per the results, it has been found that the photo-Fenton reaction moderated effectively the membrane fouling caused by pollutants along with excellent stability and reusability. Besides, it is found that the OH radical is attributed due to the hydroxyl groups playing a vital role in the more decontaminants. This is the first report based on electrospun membrane combined with photo-Fenton reaction to reduce water contamination, which has provided a new and alternative solution in water purification.[81]

In parallel, ver recently and coworkers made rGO/Ag/M88A nanocomposite membranes for wastewater treatment with self-cleaning properties. Based on the synergistic filtration/photo-Fenton processes, the as-developed p-GAM membrane demonstrated excellent dye removal performance (99.7%) and also significantly improved permeability (189 L/m²h bar) as compared to other membranes. In addition, it is highly reusable and showed similar removal performance of up to ten cycles without destroying its structure, even if there is tremendous pressure on it. The membrane also has antifouling properties and a high water flux recovery rate of more than 90%. Furthermore, when used directly as a photocatalyst in the photo-Fenton system, the as-fabricated membrane demonstrated a high MB degradation efficiency (90%).[82]

In conclusion, the current membrane technologies are well and good; but they have certain limits like poor chemical stability and low mechanical strength, and they also require a high production cost with a complex fabrication process.[74,76,83] Therefore, there is a need to make additional efforts to prepare strong membranes interacting with g-C$_3$N$_4$ laminates that meet the requirements for practical applications.

9.6 CONCLUSION AND OUTLOOK

In conclusion, this chapter has summarized the stepwise progress of g-C$_3$N$_4$ and their hybrid materials in wastewater treatment applications starting with adsorbents, photocatalyst, sensors, and membranes. Because of its controllable physicochemical properties, effortlessness of functionalization has shown tremendous potential in multidisciplinary applications since 1834. Furthermore, 2D polymeric g-C$_3$N$_4$ materials with a moderate bandgap, strong chemical activity, high stability, and outstanding conductivity are available.

Meanwhile, according to the overall literature, a great deal of work has been already done on g-C$_3$N$_4$ and its hybrids, starting from its varieties of synthesis methodologies to their applications. Along with this, according to its past and present performance, it was also found that the whole performance is entirely related to its surfaces, such as defects, working groups, doping, porosity, thickness, and morphology. Although significant results were achieved, satisfactory results have not been achieved.

However, there are still significant challenges, such as porous and high surface area nanostructures, as well as their hybrid composites, for wastewater treatment. In the end, we strongly believe that the proposed work will be helpful in upcoming water purification technology to advance the research progress of g-C$_3$N$_4$ to a large extent in terms of the integration between experimental research and theoretical approaches.

ACKNOWLEDGMENTS

The authors of this work sincerely acknowledge DST, India (DST/BDTD/EAG/2019) and DST, India (DST/TDT/DDP-31/2021) for the financial provision and support. We also sincerely acknowledge Jain University, India for providing infrastructure and various facilities for the work.

REFERENCES

1. Essien, E. E.; Said Abasse, K.; Côté, A.; Mohamed, K. S.; Baig, M. M. F. A.; Habib, M.; Naveed, M.; Yu, X.; Xie, W.; Jinfang, S.; et al. Drinking-water nitrate and cancer risk: A systematic review and meta-analysis. *Archives of Environmental & Occupational Health* **2022**, *77* (1), 51–67. DOI: 10.1080/19338244.2020.1842313.
2. Patil, C. S.; Gunjal, D. B.; Naik, V. M.; Waghmare, R. D.; Dongale, T. D.; Kurkuri, M. D.; Kolekar, G. B.; Gore, A. H. Sustainable conversion of waste tea biomass into versatile activated carbon: Application in quick, continuous, and pressure filtration of miscellaneous pollutants. *Biomass Conversion and Biorefinery* **2022**. DOI: 10.1007/s13399-021-02125-1.

3. Kizil, S.; Sonmez, H. B. Organogels and hydrogels for oil/water separation. In *Oil–Water Mixtures and Emulsions, Volume 2: Advanced Materials for Separation and Treatment*, ACS Symposium Series, Vol. 1408; American Chemical Society, 2022; pp 25–50.

4. Tunesi, M. M.; Soomro, R. A.; Han, X.; Zhu, Q.; Wei, Y.; Xu, B. Application of MXenes in environmental remediation technologies. *Nano Convergence* **2021**, *8* (1), 5. DOI: 10.1186/s40580-021-00255-w.

5. Patil, C. S.; Gunjal, D. B.; Naik, V. M.; Harale, N. S.; Jagadale, S. D.; Kadam, A. N.; Patil, P. S.; Kolekar, G. B.; Gore, A. H. Waste tea residue as a low cost adsorbent for removal of hydralazine hydrochloride pharmaceutical pollutant from aqueous media: An environmental remediation. *Journal of Cleaner Production* **2019**, *206*, 407–418. DOI: 10.1016/j.jclepro.2018.09.140.

6. Bote, P. P.; Vaze, S. R.; Patil, C. S.; Patil, S. A.; Kolekar, G. B.; Kurkuri, M. D.; Gore, A. H. Reutilization of carbon from exhausted water filter cartridges (EWFC) for decontamination of water: An innovative waste management approach. *Environmental Technology & Innovation* **2021**, *24*, 102047. DOI: 10.1016/j.eti.2021.102047.

7. Li, R. Y.; Wang, Z. S.; Yuan, Z. Y.; Van Horne, C.; Freger, V.; Lin, M.; Cai, R. K.; Chen, J. P. A comprehensive review on water stable metal-organic frameworks for large-scale preparation and applications in water quality management based on surveys made since 2015. *Critical Reviews in Environmental Science and Technology* **2021**, 1–34. DOI: 10.1080/10643389.2021.1975444.

8. Moosavi, S.; Lai, C. W.; Gan, S.; Zamiri, G.; Akbarzadeh Pivehzhani, O.; Johan, M. R. Application of efficient magnetic particles and activated carbon for dye removal from wastewater. *ACS Omega* **2020**, *5* (33), 20684–20697. DOI: 10.1021/acsomega.0c01905.

9. Xia, Z.; Zhao, Y.; Darling, S. B. Covalent organic frameworks for water treatment. *Advanced Materials Interfaces* **2021**, *8* (1), 2001507. DOI: 10.1002/admi.202001507 (accessed 2022/06/12).

10. Dongre, R. S.; Sadasivuni, K. K.; Deshmukh, K.; Mehta, A.; Basu, S.; Meshram, J. S.; Al-Maadeed, M. A. A.; Karim, A. Natural polymer based composite membranes for water purification: A review. *Polymer-Plastics Technology and Materials* **2019**, *58* (12), 1295–1310. DOI: 10.1080/25740881.2018.1563116.

11. Thiebault, T. Raw and modified clays and clay minerals for the removal of pharmaceutical products from aqueous solutions: State of the art and future perspectives. *Critical Reviews in Environmental Science and Technology* **2020**, *50* (14), 1451–1514. DOI: 10.1080/10643389.2019.1663065.

12. Emenike, E. C.; Adeniyi, A. G.; Omuku, P. E.; Okwu, K. C.; Iwuozor, K. O. Recent advances in nano-adsorbents for the sequestration of copper from water. *Journal of Water Process Engineering* **2022**, *47*, 102715. DOI: 10.1016/j.jwpe.2022.102715.

13. Constantino, D. S. M.; Dias, M. M.; Silva, A. M. T.; Faria, J. L.; Silva, C. G. Intensification strategies for improving the performance of photocatalytic processes: A review. *Journal of Cleaner Production* **2022**, *340*, 130800. DOI: 10.1016/j.jclepro.2022.130800.

14. Liu, Y.; Chen, H.; Zhu, N.; Zhang, J.; Li, Y.; Xu, D.; Gao, Y.; Zhao, J. Detection and remediation of mercury contaminated environment by nanotechnology: Progress and challenges. *Environmental Pollution* **2022**, *293*, 118557. DOI: 10.1016/j.envpol.2021.118557.

15. Shah, P. N.; Sanghvi, T.; Shah, A.; Saini, B.; Dey, A. Progress in functionalized polymeric membranes for application in waste water treatment. *Polymer-Based Advanced Functional Materials for Energy and Environmental Applications* **2022**, 205–226.

16. Ghafuri, H.; Tajik, Z.; Ghanbari, N.; Hanifehnejad, P. Preparation and characterization of graphitic carbon nitride-supported l-arginine as a highly efficient and recyclable catalyst for the one-pot synthesis of condensation reactions. *Scientific Reports* **2021**, *11* (1), 19792. DOI: 10.1038/s41598-021-97360-x. Li, C.; Wu, X.; Shan, J.; Liu, J.; Huang, X. Preparation, characterization of graphitic carbon nitride photo-catalytic

nanocomposites and their application in wastewater remediation: A review. *Crystals* **2021**, *11* (7), 723. Wang, J.; Wang, S. A critical review on graphitic carbon nitride (g-C3N4)-based materials: Preparation, modification and environmental application. *Coordination Chemistry Reviews* **2022**, *453*, 214338. DOI: 10.1016/j.ccr.2021.214338.

17. Vidyasagar, D.; Bhoyar, T.; Singh, G.; Vinu, A. Recent progress in polymorphs of carbon nitride: Synthesis, properties, and their applications. *Macromolecular Rapid Communications* **2021**, *42* (7), 2000676. DOI: 10.1002/marc.202000676 (accessed 2022/05/18).

18. Miyake, Y.; Seo, G.; Matsuhashi, K.; Takada, N.; Kanai, K. Synthesis of carbon nitride oligomer as a precursor of melon with improved fluorescence quantum yield. *Materials Advances* **2021**, *2* (18), 6083–6093. DOI: 10.1039/d1ma00579k.

19. Zhang, M.; Lai, C.; Li, B.; Huang, D.; Zeng, G.; Xu, P.; Qin, L.; Liu, S.; Liu, X.; Yi, H.; et al. Rational design 2D/2D BiOBr/CDs/g-C3N4 Z-scheme heterojunction photocatalyst with carbon dots as solid-state electron mediators for enhanced visible and NIR photocatalytic activity: Kinetics, intermediates, and mechanism insight. *Journal of Catalysis* **2019**, *369*, 469–481. DOI: 10.1016/j.jcat.2018.11.029.

20. Gaddam, S. K.; Pothu, R.; Boddula, R. Graphitic carbon nitride (g-C3N4) reinforced polymer nanocomposite systems—A review. *Polymer Composites* **2020**, *41* (2), 430–442. DOI: 10.1002/pc.25410 (accessed 2022/05/18).

21. Jiang, W.; Luo, W.; Zong, R.; Yao, W.; Li, Z.; Zhu, Y. Polyaniline/carbon nitride nanosheets composite hydrogel: A separation-free and high-efficient photocatalyst with 3D hierarchical structure. *Small* **2016**, *12* (32), 4370–4378. DOI: 10.1002/smll.201601546 (accessed 2022/05/18). Zhu, B.; Cheng, B.; Fan, J.; Ho, W.; Yu, J. g-C3N4-based 2D/2D composite heterojunction photocatalyst. *Small Structures* **2021**, *2* (12), 2100086. DOI: 10.1002/sstr.202100086 (accessed 2022/05/18).

22. Mittal, S. K.; Goyal, D.; Chauhan, A.; Dang, R. K. Graphene nanoparticles: The super material of future. *Materials Today: Proceedings* **2020**, *28*, 1290–1294. DOI: 10.1016/j.matpr.2020.04.260.

23. Rono, N.; Kibet, J. K.; Martincigh, B. S.; Nyamori, V. O. A review of the current status of graphitic carbon nitride. *Critical Reviews in Solid State and Materials Sciences* **2021**, *46* (3), 189–217. DOI: 10.1080/10408436.2019.1709414. Graphitic carbon nitride: Preparation, properties and applications in energy storage. *Engineered Science* **2020**, *10*, 24–34. DOI: 10.30919/es8d1008.

24. Oseghe, E. O.; Akpotu, S. O.; Mombeshora, E. T.; Oladipo, A. O.; Ombaka, L. M.; Maria, B. B.; Idris, A. O.; Mamba, G.; Ndlwana, L.; Ayanda, O. S.; et al. Multidimensional applications of graphitic carbon nitride nanomaterials – A review. *Journal of Molecular Liquids* **2021**, *344*, 117820. DOI: 10.1016/j.molliq.2021.117820.

25. Das, T. K.; Prusty, S. Review on conducting polymers and their applications. *Polymer-Plastics Technology and Engineering* **2012**, *51* (14), 1487–1500.

26. Song, R.; Murphy, M.; Li, C.; Ting, K.; Soo, C.; Zheng, Z. Current development of biodegradable polymeric materials for biomedical applications. *Drug Design, Development and Therapy* **2018**, *12*, 3117–3145. DOI: 10.2147/dddt.s165440. Kenry; Liu, B. Recent advances in biodegradable conducting polymers and their biomedical applications. *Biomacromolecules* **2018**, *19* (6), 1783–1803. DOI: 10.1021/acs.biomac.8b00275. Guo, B.; Ma, P. X. Conducting polymers for tissue engineering. *Biomacromolecules* **2018**, *19* (6), 1764–1782. DOI: 10.1021/acs.biomac.8b00276. Sikdar, P.; Uddin, M. M.; Dip, T. M.; Islam, S.; Hoque, M. S.; Dhar, A. K.; Wu, S. Recent advances in the synthesis of smart hydrogels. *Materials Advances* **2021**, *2* (14), 4532–4573. DOI: 10.1039/d1ma00193k.

27. Ismael, M.; Wu, Y. A mini-review on the synthesis and structural modification of g-C3N4-based materials, and their applications in solar energy conversion and environmental remediation. *Sustainable Energy & Fuels* **2019**, *3* (11), 2907–2925. DOI: 10.1039/c9se00422j.

28. Miller, T. S.; Jorge, A. B.; Suter, T. M.; Sella, A.; Corà, F.; McMillan, P. F. Carbon nitrides: Synthesis and characterization of a new class of functional materials. *Physical Chemistry Chemical Physics* **2017**, *19* (24), 15613–15638. DOI: 10.1039/c7cp02711g. Ghaemmaghami, M.; Mohammadi, R. Carbon nitride as a new way to facilitate the next generation of carbon-based supercapacitors. *Sustainable Energy & Fuels* **2019**, *3* (9), 2176–2204. DOI: 10.1039/c9se00313d. Li, X.; Masters, A. F.; Maschmeyer, T. Polymeric carbon nitride for solar hydrogen production. *Chemical Communications* **2017**, *53* (54), 7438–7446. DOI: 10.1039/c7cc02532g.

29. Liu, J.; Wang, H.; Antonietti, M. Graphitic carbon nitride "reloaded": Emerging applications beyond (photo)catalysis. *Chemical Society Reviews* **2016**, *45* (8), 2308–2326. DOI: 10.1039/c5cs00767d.

30. Thomas, A.; Fischer, A.; Goettmann, F.; Antonietti, M.; Müller, J.-O.; Schlögl, R.; Carlsson, J. M. Graphitic carbon nitride materials: Variation of structure and morphology and their use as metal-free catalysts. *Journal of Materials Chemistry* **2008**, *18* (41), 4893–4908. DOI: 10.1039/b800274f.

31. Murugan, N.; Chan-Park, M. B.; Sundramoorthy, A. K. Electrochemical detection of uric acid on exfoliated nanosheets of graphitic-like carbon nitride (g-C3N4) based sensor. *Journal of The Electrochemical Society* **2019**, *166* (9), B3163. Chen, L.; Song, J. Tailored graphitic carbon nitride nanostructures: Synthesis, modification, and sensing applications. *Advanced Functional Materials* **2017**, *27* (39), 1702695. Zhang, J.; Chen, Y.; Wang, X. Two-dimensional covalent carbon nitride nanosheets: Synthesis, functionalization, and applications. *Energy & Environmental Science* **2015**, *8* (11), 3092–3108.

32. Dong, G.; Zhang, Y.; Pan, Q.; Qiu, J. A fantastic graphitic carbon nitride (g-C3N4) material: Electronic structure, photocatalytic and photoelectronic properties. *Journal of Photochemistry and Photobiology C: Photochemistry Reviews* **2014**, *20*, 33–50. DOI: 10.1016/j.jphotochemrev.2014.04.002. Cao, Q.; Kumru, B.; Antonietti, M.; Schmidt, B. V. K. J. Graphitic carbon nitride and polymers: A mutual combination for advanced properties. *Materials Horizons* **2020**, *7* (3), 762–786. DOI: 10.1039/c9mh01497g. Zhao, X.; You, Y.; Huang, S.; Wu, Y.; Ma, Y.; Zhang, G.; Zhang, Z. Z-scheme photocatalytic production of hydrogen peroxide over Bi4O5Br2/g-C3N4 heterostructure under visible light. *Applied Catalysis B: Environmental* **2020**, *278*, 119251.

33. Ong, W.-J.; Tan, L.-L.; Ng, Y. H.; Yong, S.-T.; Chai, S.-P. Graphitic carbon nitride (g-C3N4)-based photocatalysts for artificial photosynthesis and environmental remediation: Are we a step closer to achieving sustainability? *Chemical Reviews* **2016**, *116* (12), 7159–7329. DOI: 10.1021/acs.chemrev.6b00075.

34. Darkwah, W. K.; Ao, Y. Mini review on the structure and properties (photocatalysis), and preparation techniques of graphitic carbon nitride nano-based particle, and its applications. *Nanoscale Research Letters* **2018**, *13* (1), 1–15. Mishra, A.; Mehta, A.; Basu, S.; Shetti, N. P.; Reddy, K. R.; Aminabhavi, T. M. Graphitic carbon nitride (g–C3N4)–based metal-free photocatalysts for water splitting: A review. *Carbon* **2019**, *149*, 693–721.

35. Ismael, M.; Wu, Y.; Taffa, D. H.; Bottke, P.; Wark, M. Graphitic carbon nitride synthesized by simple pyrolysis: Role of precursor in photocatalytic hydrogen production. *New Journal of Chemistry* **2019**, *43* (18), 6909–6920. DOI: 10.1039/c9nj00859d. Liu, L.; Qi, Y.; Lu, J.; Lin, S.; An, W.; Liang, Y.; Cui, W. A stable Ag3PO4@g-C3N4 hybrid core@shell composite with enhanced visible light photocatalytic degradation. *Applied Catalysis B: Environmental* **2016**, *183*, 133–141. DOI: 10.1016/j.apcatb.2015.10.035.

36. Liu, J.; Huang, J.; Dontosova, D.; Antonietti, M. Facile synthesis of carbon nitride micro-/nanoclusters with photocatalytic activity for hydrogen evolution. *RSC Advances* **2013**, *3* (45), 22988–22993. DOI: 10.1039/c3ra44490b.

37. Kolesnyk, I.; Kujawa, J.; Bubela, H.; Konovalova, V.; Burban, A.; Cyganiuk, A.; Kujawski, W. Photocatalytic properties of PVDF membranes modified with g-C3N4 in the process of Rhodamines decomposition. *Separation and Purification Technology* **2020**, *250*, 117231. DOI: 10.1016/j.seppur.2020.117231.

38. Sun, K.; Shen, J.; Liu, Q.; Tang, H.; Zhang, M.; Zulfiqar, S.; Lei, C. Synergistic effect of Co(II)-hole and Pt-electron cocatalysts for enhanced photocatalytic hydrogen evolution performance of P-doped g-C3N4. *Chinese Journal of Catalysis* **2020**, *41* (1), 72–81. DOI: 10.1016/S1872-2067(19)63430-3.

39. Bahuguna, A.; Choudhary, P.; Chhabra, T.; Krishnan, V. Ammonia-doped polyaniline–graphitic carbon nitride nanocomposite as a heterogeneous green catalyst for synthesis of indole-substituted 4H-chromenes. *ACS Omega* **2018**, *3* (9), 12163–12178. DOI: 10.1021/acsomega.8b01687.

40. Wang, X.; Blechert, S.; Antonietti, M. Polymeric graphitic carbon nitride for heterogeneous photocatalysis. *ACS Catalysis* **2012**, *2* (8), 1596–1606. DOI: 10.1021/cs300240x. Asadzadeh-Khaneghah, S.; Habibi-Yangjeh, A.; Seifzadeh, D. Graphitic carbon nitride nanosheets coupled with carbon dots and BiOI nanoparticles: Boosting visible-light-driven photocatalytic activity. *Journal of the Taiwan Institute of Chemical Engineers* **2018**, *87*, 98–111. DOI: 10.1016/j.jtice.2018.03.017.

41. Kumar, Y.; Rani, S.; Shabir, J.; Kumar, L. S. Nitrogen-rich and porous graphitic carbon nitride nanosheet-immobilized palladium nanoparticles as highly active and recyclable catalysts for the reduction of nitro compounds and degradation of organic dyes. *ACS Omega* **2020**, *5* (22), 13250–13258. DOI: 10.1021/acsomega.0c01280.

42. Haque, E.; Jun, J. W.; Talapaneni, S. N.; Vinu, A.; Jhung, S. H. Superior adsorption capacity of mesoporous carbon nitride with basic CN framework for phenol. *Journal of Materials Chemistry* **2010**, *20* (48), 10801–10803.

43. Yan, T.; Chen, H.; Wang, X.; Jiang, F. Adsorption of perfluorooctane sulfonate (PFOS) on mesoporous carbon nitride. *RSC Advances* **2013**, *3* (44), 22480–22489. DOI: 10.1039/c3ra43312a. Xin, G.; Xia, Y.; Lv, Y.; Liu, L.; Yu, B. Investigation of mesoporous graphitic carbon nitride as the adsorbent to remove Ni (II) ions. *Water Environment Research* **2016**, *88* (4), 318–324.

44. Yan, T.; Chen, H.; Jiang, F.; Wang, X. Adsorption of perfluorooctane sulfonate and perfluorooctanoic acid on magnetic mesoporous carbon nitride. *Journal of Chemical & Engineering Data* **2014**, *59* (2), 508–515.

45. Kumar, R.; Barakat, M. A.; Alseroury, F. A. Oxidized g-C3N4/polyaniline nanofiber composite for the selective removal of hexavalent chromium. *Scientific Reports* **2017**, *7* (1), 12850. DOI: 10.1038/s41598-017-12850-1. Liu, J.; Chen, Z.; Yu, K.; Liu, Y.; Ge, Y.; Xie, S. Polyaniline/oxidation etching graphitic carbon nitride composites for U (VI) removal from aqueous solutions. *Journal of Radioanalytical and Nuclear Chemistry* **2019**, *321* (3), 1005–1017. Yousefi, M.; Villar-Rodil, S.; Paredes, J. I.; Moshfegh, A. Z. Oxidized graphitic carbon nitride nanosheets as an effective adsorbent for organic dyes and tetracycline for water remediation. *Journal of Alloys and Compounds* **2019**, *809*, 151783.

46. Shen, W.; An, Q.-D.; Xiao, Z.-Y.; Zhai, S.-R.; Hao, J.-A.; Tong, Y. Alginate modified graphitic carbon nitride composite hydrogels for efficient removal of Pb (II), Ni (II) and Cu (II) from water. *International Journal of Biological Macromolecules* **2020**, *148*, 1298–1306.

47. Talukdar, M.; Behera, S. K.; Bhattacharya, K.; Deb, P. Surface modified mesoporous g-C3N4@FeNi3 as prompt and proficient magnetic adsorbent for crude oil recovery. *Applied Surface Science* **2019**, *473*, 275–281. DOI: 10.1016/j.apsusc.2018.12.166.

48. Loeb, S. K.; Alvarez, P. J. J.; Brame, J. A.; Cates, E. L.; Choi, W.; Crittenden, J.; Dionysiou, D. D.; Li, Q.; Li-Puma, G.; Quan, X.; et al. The technology horizon for photocatalytic water treatment: Sunrise or sunset? *Environmental Science & Technology*

2019, *53* (6), 2937–2947. DOI: 10.1021/acs.est.8b05041. Ren, G.; Han, H.; Wang, Y.; Liu, S.; Zhao, J.; Meng, X.; Li, Z. Recent advances of photocatalytic application in water treatment: A review. *Nanomaterials (Basel)* **2021**, *11* (7), 1804. DOI: 10.3390/nano11071804.

49. Kumar, A.; Raizada, P.; Singh, P.; Saini, R. V.; Saini, A. K.; Hosseini-Bandegharaei, A. Perspective and status of polymeric graphitic carbon nitride based Z-scheme photocatalytic systems for sustainable photocatalytic water purification. *Chemical Engineering Journal* **2020**, *391*, 123496. Azhar, U.; Bashir, M. S.; Babar, M.; Arif, M.; Hassan, A.; Riaz, A.; Mujahid, R.; Sagir, M.; Suri, S. U. K.; Show, P. L.; et al. Template-based textural modifications of polymeric graphitic carbon nitrides toward waste water treatment. *Chemosphere* **2022**, *302*, 134792. DOI: 10.1016/j.chemosphere.2022.134792. Rana, A.; Sudhaik, A.; Raizada, P.; Nguyen, V.-H.; Xia, C.; Parwaz Khan, A. A.; Thakur, S.; Nguyen-Tri, P.; Nguyen, C. C.; Kim, S. Y.; et al. Graphitic carbon nitride based immobilized and non-immobilized floating photocatalysts for environmental remediation. *Chemosphere* **2022**, *297*, 134229. DOI: 10.1016/j.chemosphere.2022.134229.

50. Yan, S. C.; Li, Z. S.; Zou, Z. G. Photodegradation of rhodamine B and methyl orange over boron-doped g-C3N4 under visible light irradiation. *Langmuir* **2010**, *26* (6), 3894–3901. DOI: 10.1021/la904023j.

51. Ge, L.; Han, C.; Liu, J. Novel visible light-induced g-C3N4/Bi2WO6 composite photocatalysts for efficient degradation of methyl orange. *Applied Catalysis B: Environmental* **2011**, *108–109*, 100–107. DOI: 10.1016/j.apcatb.2011.08.014.

52. Ge, L.; Han, C.; Liu, J.; Li, Y. Enhanced visible light photocatalytic activity of novel polymeric g-C3N4 loaded with Ag nanoparticles. *Applied Catalysis A: General* **2011**, *409–410*, 215–222. DOI: 10.1016/j.apcata.2011.10.006.

53. Ye, L.; Liu, J.; Jiang, Z.; Peng, T.; Zan, L. Facets coupling of BiOBr-g-C3N4 composite photocatalyst for enhanced visible-light-driven photocatalytic activity. *Applied Catalysis B: Environmental* **2013**, *142–143*, 1–7. DOI: 10.1016/j.apcatb.2013.04.058.

54. Zhou, X.; Jin, B.; Chen, R.; Peng, F.; Fang, Y. Synthesis of porous Fe3O4/g-C3N4 nanospheres as highly efficient and recyclable photocatalysts. *Materials Research Bulletin* **2013**, *48* (4), 1447–1452. DOI: 10.1016/j.materresbull.2012.12.038.

55. Rong, X.; Qiu, F.; Jiang, Z.; Rong, J.; Pan, J.; Zhang, T.; Yang, D. Preparation of ternary combined ZnO-Ag2O/porous g-C3N4 composite photocatalyst and enhanced visible-light photocatalytic activity for degradation of ciprofloxacin. *Chemical Engineering Research and Design* **2016**, *111*, 253–261. DOI: 10.1016/j.cherd.2016.05.010.

56. Reddy, K. R.; Reddy, C. H. V.; Nadagouda, M. N.; Shetti, N. P.; Jaesool, S.; Aminabhavi, T. M. Polymeric graphitic carbon nitride (g-C3N4)-based semiconducting nanostructured materials: Synthesis methods, properties and photocatalytic applications. *Journal of Environmental Management* **2019**, *238*, 25–40. DOI: 10.1016/j.jenvman.2019.02.075. Zhang, J.; Gao, N.; Chen, F.; Zhang, T.; Zhang, G.; Wang, D.; Xie, X.; Cai, D.; Ma, X.; Wu, L.; et al. Improvement of Cr (VI) photoreduction under visible-light by g-C3N4 modified by nano-network structured palygorskite. *Chemical Engineering Journal* **2019**, *358*, 398–407. DOI: 10.1016/j.cej.2018.10.083.

57. Lamkhao, S.; Rujijanagul, G.; Randorn, C. Fabrication of g-C3N4 and a promising charcoal property towards enhanced chromium(VI) reduction and wastewater treatment under visible light. *Chemosphere* **2018**, *193*, 237–243. DOI: 10.1016/j.chemosphere.2017.11.015.

58. Hu, S.; Ma, L.; You, J.; Li, F.; Fan, Z.; Lu, G.; Liu, D.; Gui, J. Enhanced visible light photocatalytic performance of g-C3N4 photocatalysts co-doped with iron and phosphorus. *Applied Surface Science* **2014**, *311*, 164–171. DOI: 10.1016/j.apsusc.2014.05.036. Hu, C.; Hung, W.-Z.; Wang, M.-S.; Lu, P.-J. Phosphorus and sulfur codoped g-C3N4 as an efficient metal-free photocatalyst. *Carbon* **2018**, *127*, 374–383. DOI: 10.1016/j.carbon.2017.11.019.

59. Cao, L.; Wang, R.; Wang, D. Synthesis and characterization of sulfur self-doped g-C3N4 with efficient visible-light photocatalytic activity. *Materials Letters* **2015**, *149*, 50–53. DOI: 10.1016/j.matlet.2015.02.119.
60. Jindal, H.; Kumar, D.; Sillanpaa, M.; Nemiwal, M. Current progress in polymeric graphitic carbon nitride-based photocatalysts for dye degradation. *Inorganic Chemistry Communications* **2021**, *131*, 108786. DOI: 10.1016/j.inoche.2021.108786.
61. Yu, Y.; Xu, W.; Fang, J.; Chen, D.; Pan, T.; Feng, W.; Liang, Y.; Fang, Z. Soft-template assisted construction of superstructure TiO2/SiO2/g-C3N4 hybrid as efficient visible-light photocatalysts to degrade berberine in seawater via an adsorption-photocatalysis synergy and mechanism insight. *Applied Catalysis B: Environmental* **2020**, *268*, 118751. DOI: 10.1016/j.apcatb.2020.118751.
62. Lee, E. Z.; Lee, S. U.; Heo, N.-S.; Stucky, G. D.; Jun, Y.-S.; Hong, W. H. A fluorescent sensor for selective detection of cyanide using mesoporous graphitic carbon(iv) nitride. *Chemical Communications* **2012**, *48* (33), 3942–3944. DOI: 10.1039/c2cc17909a.
63. Barman, S.; Sadhukhan, M. Facile bulk production of highly blue fluorescent graphitic carbon nitride quantum dots and their application as highly selective and sensitive sensors for the detection of mercuric and iodide ions in aqueous media. *Journal of Materials Chemistry* **2012**, *22* (41), 21832–21837. DOI: 10.1039/c2jm35501a. Huang, H.; Chen, R.; Ma, J.; Yan, L.; Zhao, Y.; Wang, Y.; Zhang, W.; Fan, J.; Chen, X. Graphitic carbon nitride solid nanofilms for selective and recyclable sensing of Cu2+ and Ag+ in water and serum. *Chemical Communications* **2014**, *50* (97), 15415–15418. DOI: 10.1039/c4cc06659f.
64. Tian, J.; Liu, Q.; Asiri, A. M.; Qusti, A. H.; Al-Youbi, A. O.; Sun, X. Ultrathin graphitic carbon nitride nanosheets: A novel peroxidase mimetic, Fe doping-mediated catalytic performance enhancement and application to rapid, highly sensitive optical detection of glucose. *Nanoscale* **2013**, *5* (23), 11604–11609. DOI: 10.1039/c3nr03693f.
65. Tian, J.; Liu, Q.; Ge, C.; Xing, Z.; Asiri, A. M.; Al-Youbi, A. O.; Sun, X. Ultrathin graphitic carbon nitride nanosheets: A low-cost, green, and highly efficient electrocatalyst toward the reduction of hydrogen peroxide and its glucose biosensing application. *Nanoscale* **2013**, *5* (19), 8921–8924. DOI: 10.1039/c3nr02031b.
66. Xu, Y.; Niu, X.; Zhang, H.; Xu, L.; Zhao, S.; Chen, H.; Chen, X. Switch-on fluorescence sensing of glutathione in food samples based on a graphitic carbon nitride quantum dot (g-CNQD)–Hg2+ chemosensor. *Journal of Agricultural and Food Chemistry* **2015**, *63* (6), 1747–1755. DOI: 10.1021/jf505759z.
67. Chen, S.; Hao, N.; Jiang, D.; Zhang, X.; Zhou, Z.; Zhang, Y.; Wang, K. Graphitic carbon nitride quantum dots in situ coupling to Bi2MoO6 nanohybrids with enhanced charge transfer performance and photoelectrochemical detection of copper ion. *Journal of Electroanalytical Chemistry* **2017**, *787*, 66–71. DOI: 10.1016/j.jelechem.2017.01.042.
68. Liu, Y.; Jiang, J.; Sun, Y.; Wu, S.; Cao, Y.; Gong, W.; Zou, J. NiO and Co3O4 co-doped g-C3N4 nanocomposites with excellent photoelectrochemical properties under visible light for detection of tetrabromobisphenol-A. *RSC Advances* **2017**, *7* (57), 36015–36020. DOI: 10.1039/c7ra04822j.
69. Feng, Y.; Yan, T.; Wu, T.; Zhang, N.; Yang, Q.; Sun, M.; Yan, L.; Du, B.; Wei, Q. A label-free photoelectrochemical aptasensing platform base on plasmon Au coupling with MOF-derived In2O3@g-C3N4 nanoarchitectures for tetracycline detection. *Sensors and Actuators B: Chemical* **2019**, *298*, 126817. DOI: 10.1016/j.snb.2019.126817.
70. Deiminiat, B.; Rounaghi, G. H. A novel visible light photoelectrochemical aptasensor for determination of bisphenol A based on surface plasmon resonance of gold nanoparticles activated g-C3N4 nanosheets. *Journal of Electroanalytical Chemistry* **2021**, *886*, 115122. DOI: 10.1016/j.jelechem.2021.115122. Xu, L.; Duan, W.; Chen, F.; Zhang, J.; Li, H. A photoelectrochemical aptasensor for the determination of bisphenol A based on the Cu (I) modified graphitic carbon nitride. *Journal of Hazardous Materials* **2020**, *400*, 123162. DOI: 10.1016/j.jhazmat.2020.123162.

71. Kwon, O.; Choi, Y.; Kang, J.; Kim, J. H.; Choi, E.; Woo, Y. C.; Kim, D. W. A comprehensive review of MXene-based water-treatment membranes and technologies: Recent progress and perspectives. *Desalination* **2022**, *522*, 115448. DOI: 10.1016/j.desal.2021.115448.

72. Jia, F.; Xiao, X.; Nashalian, A.; Shen, S.; Yang, L.; Han, Z.; Qu, H.; Wang, T.; Ye, Z.; Zhu, Z.; et al. Advances in graphene oxide membranes for water treatment. *Nano Research* **2022**. DOI: 10.1007/s12274-022-4273-y.

73. Saud, A.; Saleem, H.; Zaidi, S. J. Progress and prospects of nanocellulose-based membranes for desalination and water treatment. *Membranes* **2022**, *12* (5), 462. Zhu, L.; Guo, X.; Chen, Y.; Chen, Z.; Lan, Y.; Hong, Y.; Lan, W. Graphene oxide composite membranes for water purification. *ACS Applied Nano Materials* **2022**, *5* (3), 3643–3653. DOI: 10.1021/acsanm.1c04322.

74. Divya, S.; Oh, T. H. Polymer nanocomposite membrane for wastewater treatment: A critical review. *Polymers* **2022**, *14* (9), 1732.

75. Chen, C.; Chen, L.; Zhu, X.; Chen, B. Graphene nanofiltration membrane intercalated with AgNP@g-C3N4 for efficient water purification and photocatalytic self-cleaning performance. *Chemical Engineering Journal* **2022**, *441*, 136089. DOI: 10.1016/j.cej.2022.136089. Hou, R.; He, Y.; Yu, H.; He, T.; Gao, Y.; Guo, X. A self-cleaning membrane based on NG/g-C3N4 and graphene oxide with enhanced nanofiltration performance. *Journal of Materials Science* **2022**, *57* (20), 9118–9133. DOI: 10.1007/s10853-022-07083-1.

76. Chen, J.; Li, Z.; Wang, C.; Wu, H.; Liu, G. Synthesis and characterization of g-C3N4 nanosheet modified polyamide nanofiltration membranes with good permeation and antifouling properties. *RSC Advances* **2016**, *6* (113), 112148–112157. DOI: 10.1039/c6ra21192e.

77. Mahdavi, M. R.; Delnavaz, M.; Vatanpour, V. Fabrication and water desalination performance of piperazine–polyamide nanocomposite nanofiltration membranes embedded with raw and oxidized MWCNTs. *Journal of the Taiwan Institute of Chemical Engineers* **2017**, *75*, 189–198. DOI: 10.1016/j.jtice.2017.03.039. Yu, H.-Y.; Xu, Z.-K.; Yang, Q.; Hu, M.-X.; Wang, S.-Y. Improvement of the antifouling characteristics for polypropylene microporous membranes by the sequential photoinduced graft polymerization of acrylic acid. *Journal of Membrane Science* **2006**, *281* (1), 658–665. DOI: 10.1016/j.memsci.2006.04.036.

78. Wang, Y.; Li, L.; Wei, Y.; Xue, J.; Chen, H.; Ding, L.; Caro, J.; Wang, H. Water transport with ultralow friction through partially exfoliated g-C3N4 nanosheet membranes with self-supporting spacers. *Angewandte Chemie International Edition* **2017**, *56* (31), 8974–8980. DOI: 10.1002/anie.201701288 (accessed 2022/05/23). Wang, Y.; Wang, X.; Antonietti, M. Polymeric graphitic carbon nitride as a heterogeneous organocatalyst: From photochemistry to multipurpose catalysis to sustainable chemistry. *Angewandte Chemie International Edition* **2012**, *51* (1), 68–89. DOI: 10.1002/anie.201101182 (accessed 2022/05/23). Wang, X.; Wang, G.; Chen, S.; Fan, X.; Quan, X.; Yu, H. Integration of membrane filtration and photoelectrocatalysis on g-C3N4/CNTs/Al2O3 membrane with visible-light response for enhanced water treatment. *Journal of Membrane Science* **2017**, *541*, 153–161. DOI: 10.1016/j.memsci.2017.06.046. Wang, Y.; Liu, L.; Hong, J.; Cao, J.; Deng, C. A novel Fe(OH)3/g-C3N4 composite membrane for high efficiency water purification. *Journal of Membrane Science* **2018**, *564*, 372–381. DOI: 10.1016/j.memsci.2018.07.027. Wang, Y.; Liu, L.; Xue, J.; Hou, J.; Ding, L.; Wang, H. Enhanced water flux through graphitic carbon nitride nanosheets membrane by incorporating polyacrylic acid. *AIChE Journal* **2018**, *64* (6), 2181–2188. DOI: 10.1002/aic.16076 (accessed 2022/05/23).

79. Shi, Y.; Huang, J.; Zeng, G.; Cheng, W.; Hu, J.; Shi, L.; Yi, K. Evaluation of self-cleaning performance of the modified g-C3N4 and GO based PVDF membrane toward oil-in-water separation under visible-light. *Chemosphere* **2019**, *230*, 40–50. DOI: 10.1016/j. chemosphere.2019.05.061. Chen, W.; Ye, T.; Xu, H.; Chen, T.; Geng, N.; Gao, X. An ultrafiltration membrane with enhanced photocatalytic performance from grafted N–TiO2/graphene oxide. *RSC Advances* **2017**, *7* (16), 9880–9887. DOI: 10.1039/ c6ra27666k.
80. Seyyed Shahabi, S.; Azizi, N.; Vatanpour, V.; Yousefimehr, N. Novel functionalized graphitic carbon nitride incorporated thin film nanocomposite membranes for high-performance reverse osmosis desalination. *Separation and Purification Technology* **2020**, *235*, 116134. DOI: 10.1016/j.seppur.2019.116134.
81. Zheng, S.; Chen, H.; Tong, X.; Wang, Z.; Crittenden, J. C.; Huang, M. Integration of a photo-fenton reaction and a membrane filtration using CS/PAN@FeOOH/g-C3N4 electrospun nanofibers: Synthesis, characterization, self-cleaning performance and mechanism. *Applied Catalysis B: Environmental* **2021**, *281*, 119519. DOI: 10.1016/j. apcatb.2020.119519.
82. Yue, R.; Raisi, B.; Rahmatinejad, J.; Ye, Z.; Barbeau, B.; Rahaman, M. S. A photo-Fenton nanocomposite ultrafiltration membrane for enhanced dye removal with self-cleaning properties. *Journal of Colloid and Interface Science* **2021**, *604*, 458–468. DOI: 10.1016/j.jcis.2021.06.157.
83. Papageorgiou, D. G.; Kinloch, I. A.; Young, R. J. Mechanical properties of graphene and graphene-based nanocomposites. *Progress in Materials Science* **2017**, *90*, 75–127. DOI: 10.1016/j.pmatsci.2017.07.004. Liu, Y.; Su, Y.; Guan, J.; Cao, J.; Zhang, R.; He, M.; Gao, K.; Zhou, L.; Jiang, Z. 2D heterostructure membranes with sunlight-driven self-cleaning ability for highly efficient oil–water separation. *Advanced Functional Materials* **2018**, *28* (13), 1706545. DOI: 10.1002/adfm.201706545 (accessed 2022/05/23).

10 Polymer-Activated Carbon Composites for Wastewater Treatment

Dipika Jaspal
Symbiosis Institute of Technology (SIT), Symbiosis
International (Deemed University) (SIU)

Arti Malviya
Lakshmi Narain College of Technology

Smita Jadhav
Bharati Vidyapeeth's College of Engineering for Women

CONTENTS

10.1 INTRODUCTION

Composite material is a combination of different materials with distinct chemical and physical properties. Once they are combined, they provide unique properties to composite material such as strength, stiffness, lightness, and electrical resistance. The constituents sustain their properties within the composite. Composites are used in a variety of applications. In the yester years, composites were used in the form of mud bricks, which were molded by compounding straws and mud bricks giving tensile strength and resistance to the composite. Composites are employed in a majority of products in our daily life. Composite materials are used by human beings since millions of years. In 1500 BC, initial composite materials were prepared by Mesopotamians and Egyptians from straw and mud to construct their shelters. In the middle of the second world war, the use and production of composites was expanded, and correspondingly the fiber-reinforced polymer composites were endorsed by the industrial sector. There are a variety of composites being used in several applications.

This chapter focuses on the use of polymer-activated carbon composites , specifically for the application of waste water treatment.

10.2   TYPES OF COMPOSITES

Xiao et al. used urea, boric acid, and aluminum chloride as starting materials in an aqueous solution, which resulted in a pre-composite gel, which on heat treatment turned into aluminum nitride/boron nitride composite. Synthesized composite was characterized by scanning electron microscope, Fourier transform infrared spectroscopy, and x-ray diffraction techniques, which revealed the hexagonal structure for aluminum nitride phase and turbostatic structure for boron nitride phase (Xiao et al., 1993).

Metin et al. (2013) synthesized chitosan–zeolite composite, which was explored as an adsorbent for the removal of acid black 194 dye. Freundlich adsorption isotherm model was found to be good for studying adsorbent and adsorbate systems. The maximum dye adsorption capacity was found to be 2140 mg/g (Metin et al., 2013). In another study, Lin and Zhan (2012) utilized chitosan–zeolite composite for the removal of humic acid (Lin and Zhan, 2012). Zhang et al. (2015) synthesized crosslinking chitosan–zeolite composite for the elimination of cadmium (Cd). Pseudo-second-order kinetics was followed for the adsorption system with the highest removal capacity of 102.15 mg/g at neutral pH (Zhang et al., 2015). Chitosan modified by zeolite was used as an adsorbent for the eradication of an anionic dye Bezactive orange 16, which revealed the maximum adsorption capacity of 305.8 mg/g. Langmuir adsorption isotherm model proved to be the best fit in the adsorption process (Nešić et al., 2013).

Dhawan et al. (2019) described the composite prepared by using chitosan and graphene oxide, and investigated the use of the mentioned for lead removal (Pb) from aqueous solutions. Freundlich adsorption isotherm was applicable to the adsorption data (Dhawan et al., 2019). In another study, researchers prepared graphene oxide–sand composite and successfully utilized it as an adsorbent in effluent treatment for the elimination of various dyes namely rhodamine B, methyl orange, and methylene blue in different concentration ranges. Results demonstrated that methylene blue dye was effectively eliminated at all levels with 100% efficiency, out of these three dyes (Andrijanto et al., 2018). Various researchers synthesized activated carbon composites, such as zeolite-activated carbon composites, and have used the same for the eradication of methyl–thioninium and nitrogen compounds (Wang et al., 2018), chitosan-activated carbon composite for the degradation of micropollutants of organic nature (Soni et al., 2015). Titanium oxide-activated carbon composite was utilized for the degradation of rhodamine B and methyl orange by He et al. (2009) and Zhang et al. (2012).

10.3   TYPES OF POLYMER COMPOSITES

Polymers have been renowned materials in modern applications since last decade. Though these materials are versatile and easily framed into essential applications

(Afzal and Nawab, 2021), still, using only the polymer cannot meet the requirements of advanced technology. Hence, polymer composites are noticeable across the globe. Polymer composites have a wide range of applications in various fields, especially in aerospace and automotive applications. Tong et al. investigated the different preparation methods to regulate the shape of molecular imprinted polymer composite, using it in biological applications (Tong et al., 2021).

Shojaei and Khasraghi introduced the ideas and knowledges associated with self-healing and self-sensing in polymer composites and review about their potential use, which includes electronics, aerospace, robotics, coating, and transportation (Shojaei and Khasraghi, 2021). Structural polymers are used in high-efficiency engineering applications, for instance, the lightest electrical vehicles, aerospace, and civil engineering constructions. The functional polymer composites comprising of intermittent nanomaterials may be used as innovative and sensitive sensors in several sectors, such as wearable-mountable sensors, medical, electronics, automation, and energy, owing to optical, electrical, and mechanical features. Nanomaterials incorporated in the polymer matrix can improve the mechanical properties of polymers by creating nucleation sites in the polymers, which enhance the mechanical strength by decreasing the mobility of molecules (Nyabadza et al., 2021). Thermosetting polymers are extensively used in polymer composites consisting of vinyl ester, epoxy, unsaturated polyester, etc. Numerous production techniques such as vacuum-assisted resin infusion molding (VARIM), resin transfer molding (RTM), filament winding, pultrusion, hand lay-up processes, and prepreg have been employed for the preparation of thermosetting-based polymer composites (Shojaei and Khasraghi, 2021).

Biswal et al. synthesized polymer nanocomposites to boost the mechanical, chemical, physical, and other properties that increased the suitability of applications in various fields. Different fillers and reinforcements have been used to enhance their overall properties by investigating the different biomedical applications, such as in the treatment of heart, bone, ligament fixing, teeth, and oral tissues, tendon, wound dressing, and skin. Also, polymer composites have been used in advanced medical science to make artificial organs for humans. Though these composite materials have numeral advanced properties, like biocompatibility, biodegradability, and advanced cell adhesiveness, these composite materials also exhibit several limitations, such as generation of acidic by-products having poor cell affinity. Therefore these drawbacks need to be overcomed (Biswal et al., 2020).

10.3.1 POLYMER-ACTIVATED CARBON COMPOSITES

Kumar et al. prepared polyaniline–carbon nanotube composite using various methods like electrophoretic, interfacial polymerization, solution mixing, and in-situ chemical oxidative polymerization. Probable applications of polyaniline/carbon nanotube nanocomposites in various fields, such as sensors, electromagnetic interference shielding, and actuators, have also been discussed. Among several polymer composites, Polyaniline (PANI)/Carbon nanotube (CNT) nanocomposites are widely applied as electromagnetic interface (EMI) shielding and electrostatic discharge (ESD) materials, because of their property like high electrical conductivity.

An enhanced conductivity was depicted by the substance owing to the conducting path being created in the matrix material. CNTs, as a filler, are dispersed homogeneously in the matrix of PANI. Furthermore, interfacial bonding helps in expanding the mechanical properties of nanocomposites, as these transfer significant load throughout the CNT–matrix interface. This process was very useful for proper dispersion of nanotubes without any alteration in the properties. PANI/CNT nanocomposites alter their texture into a fine residue from coarse structure, nanofibers from fibers, and thin film, which affects their properties like mechanical, thermal, and electrical. The imperative properties of PANI/CNT nanocomposites have been extensively used in various applications like sensors, actuators, fuel cells, nanodevices, and supercapacitors (Kumar et al., 2018a).

Lee et al. efficaciously synthesized a composite (represented as CMPEI/CMK-1) using the adsorption method by an ordered mesoporous carbon material CMK-1 and carboxymethylated polyethyleneimine (CMPEI). CMPEI is a chelating agent comprising of carboxylate and amine functional groups. CMPEI/CMK-1 composite material was used to isolate copper ions from aqueous solutions following the Langmuir adsorption isotherm model with an utmost of 41.3 mg adsorption capacity at an acidic pH 5, signifying its use as an efficient adsorbent for the elimination of copper ions from aqueous solutions as well as from the effluents (Lee et al., 2009).

Graphene/polymer composites were synthesized by Sun and Shi (2013), owing to their extensive usage in conductive and high-strength materials, energy-related systems, catalysts, and storage devices specifically in energy alteration. Hence, various techniques have been developed for preparing graphene/polymer composites. Based on contacts between polymers and graphene, the techniques strategies may be categorized into two classes, i.e., covalent and noncovalent (Sun and Shi, 2013). To enhance the properties of composite materials, such as chemical, mechanical, electrochemical properties, and processability, graphene materials were blended with polymers. These materials have been used in a variety of applications such as fuel cells, solar cells, and supercapacitors. Graphene/polymer have shown significant potential in future energy storage and production. Han and Elliott mentioned that a small portion of carbon nanotubes (CNT) by weight when blended into a polymer matrix can augment electrical and mechanical properties of polymer composites. In their study, multiwalled carbon nanotubes (MWCNTs) were disseminated randomly in a liquid epoxy resin through mechanical stirring (Han and Elliott, 2007). Numerical simulation and investigational measurements were directed to know the outcomes of material properties and processing parameters on sensor sensitivity in polymer–carbon nanotube composites (Hu et al., 2010).

The polymer–carbon composite materials were prepared with diverse pore sizes of 2–10 nm, shapes, uniform porosity, and connectivity. The polymer–carbon composites can be utilized as advanced electrode materials, as the electric conductivity of the original carbon is maintained after its synthesis. Furthermore, there is a scope of extension in properties from carbon to porous silica materials. This can be attributed to the properties such as high capacity for metal dispersion, high surface area, facile chemical functionality and regular pore structure (Choi and Ryoo, 2003). Nanoporous composite systems show wide applications in materials science, pollutant removal, ion exchange of selective ions, high performance sensors, and catalysts.

Carbon nanotubes (CNTs) are emerging resources for mixing with polymers with the probability to get light-weight nanocomposites of remarkable thermal, electrical, mechanical properties, and versatile characteristics. Several processing approaches for preparing these nanocomposites are deliberated in specific solution processing, melt mixing, and in-situ polymerization (Breuer and Uttandaraman, 2004).

10.4 SYNTHESIS

In-situ chemical oxidation process involving polymerization was employed to synthesize activated carbon integrated poly-indole. The process was carried out using reagents like benzoyl peroxide and chloroform, sodium dodecylsulfate, and perchloric acid (Begum et al., 2022). A similar synthetic route was used to prepare doped polyaniline using tetrafloroboric acid and persulphate as an oxidant (Vighnesha et al., 2018). Mohan and team prepared polyacrylonitrile-LiI-activated carbon composite gel polymer electrolyte by hot pressing method (Mohan et al., 2013).

Synthesis of Polymer-activated carbon composite was carried out by grafting, where polyglucosamine polymer was grafted over the activated carbon surface, which in turn shaped the oxidation process (Berber, 2020). Elkady et al. synthesized a composite of styrene and acrylonitrile using a solution polymerization process. Electrospinning technique was employed to fabricate the nanofibers for water treatment (Elkady et al., 2016). In the same direction, diethylenetriamine was combined with polyacrylonitrile through the electrospinning method to carry out an amalgamation (Almasian et al., 2015).

Polystyrene-activated carbon composite material was prepared by sulphonation reaction of polystyrene in the calculated amount of toluene followed by the addition of powdered activated carbon (Benregga et al., 2019). A biodegradable membrane production for purification of water was tried by one of the researchers. Polyvinylpyrrolidone and sodium carboxymethyl cellulose were used to prepare aqueous solutions and cross-linked with citric acid along with the activated charcoal, improving its purification and dye adsorption abilities at room temperature, in both dark and light environments (Ramadoss et al., 2020).

Several experiments were carried out to fabricate an ion exchange resin-based sulfonated polystyrene and activated carbon. Matrix properties often have an impact on the mechanical and chemical properties of the composites. Composite materials have high affinity and high capacity to economically remove multiple contaminants from drinking water (Maghchiche et al., 2019).

An activated charcoal-based sorbent was prepared using poly(acrylamide-co-itaconic acid) by polymerization of acrylamide and itaconic acid, and activated charcoal. N, N'-methylene bisacrylamide was used as a crosslinker, while potassium persulfate was employed an initiator (Bajpai and Shrivastava, 2011).

Hypercrosslinked composites of polymer-activated carbon were prepared by Xiong et al. The fabricated composite possessed substantial adsorption capacity and

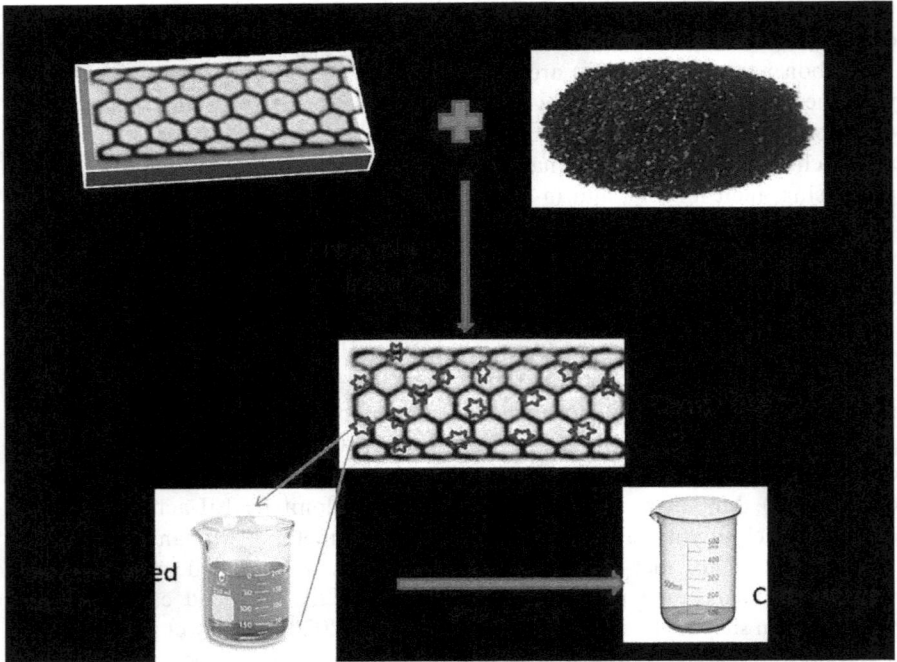

FIGURE 10.1   Preparation of composite for water decontamination.

selectivity (Xiong et al., 2021). Figure 10.1 portrays a general procedure followed during composite fabrication in the water treatment.

10.5   CHARACTERIZATION TECHNIQUES

FTIR was used for functional group analysis and structure elucidation of activated carbon–polyindole composites. The crystalline behavior of the composites was studied using an X-ray diffractometer, while Energy Dispersive X-ray (EDAX) method was used for elemental analysis. The morphological analysis was carried out by scanning electron microscopy. The study demonstrated that the greater the amount of activated carbon in the polyindole, the greater the adsorption characteristics. The existence of elements such as carbon, nitrogen, oxygen, and sulfur exhibited successful doping of the polymer (Begum et al., 2022). Zeng and co-researchers developed a composite with the potential of removing chromium heavy metal ions from water. The surface characterization results indicated a mesoporous nature of the adsorbent with a specific area of 293.4 m^2/g (Zeng et al., 2020).

Frontal polymerization technique was utilized to prepare a series of polyacrylic acid-co-acrylamide and activated carbon composites. Scanning electron microscopy and Fourier transform infrared spectroscopy were used for characterization of the prepared composite, confirming the presence of activated carbon particles in the hydrogel network of the composite (Li et al., 2013).

10.6   APPLICATIONS

The interest in polyaniline has increased considerably during the last few decades, owing to its cost effectiveness, simple fabrication, mechanical suppleness, environmental substantiality, and extraordinary doping chemistry. Polyaniline is used as adsorbent for water purification due to imine and amine functional groups, which associate with most of the organic compounds and inorganic ions, including mercury, chromium, drugs, dyes, etc..

A combination of polyaniline and activated carbon enhances the adsorption ability. The maximum adsorption of polymer-activated carbon from wood sawdust was determined to be 3.4 mg/g at optimal pH of 5 (Ansari and Fahim, 2007).

The synergetic effect for metal ion cadmium removal efficiency was comprehensively studied in composite structures. The adsorption ability of cadmium ions was found to be 35.0 mg/g at a pH of 5.5, 200 mg composite dose, and contact duration of 2 hours. A kinetic and thermodynamic investigation revealed the adsorption process as exothermic chemisorption and spontaneous. They also reported that 89% of the initial adsorption capacity of the composite was maintained even after the three adsorption–desorption cycles (Berber, 2020).

Polyaniline-doped composites were prepared with the use of agricultural waste-activated carbon by a chemical oxidative polymerization process. The fabricated composites were characterized by X-ray diffraction, Fourier transform infrared spectroscopy, scanning electron microscopy, Brunauer–Emmett–Teller, and transmission electron microscopy for the removal of lead ions. Thermogravimetric investigations confirmed the thermal stability of the composite. Adsorption results revealed that the maximum adsorption was obtained at pH 4.0 with an equilibrium attainment in 90 minutes at a lead concentration of 20 mg/L. Many fabricated composites were compared for their adsorption capacities, and it was found that out of activated carbon from polyaniline-peach stones, polyaniline-apricot stones, polyaniline-orange peels, polyaniline-banana peels, and polyaniline-pomegranate peels, the removal efficiency of polyaniline-orange peels-activated carbon nanocomposite was superior (6.81 mg/g) than the other composites. The Freundlich isotherm gave better fit data than the Langmuir isotherm with a pseudo-second-order kinetic model. Desorption experiments using 0.1 M HCl was most effective for the removal of the metal (Somaia et al., 2019).

In an investigation, the maximum adsorption for basic violet dye was 67.11 mg/g within 30 minutes after the polymer composite fabrication (Elkady et al., 2016).

Cyclodextrin polymer having cross-linked polydopamine was developed as an eco-friendly composite for high removal efficiency of various dyes due to novel structural traits of the polymer composite (Chen et al., 2020). Ramadoss et al. studied the dye adsorption efficiency of the polymer-activated carbon composite for the elimination of toxic textile dyes, and reported the efficiency to be 100% for methyl orange and 57% for rhodamine B dyes within 180 minutes. Acidic, neutral, and alkaline pH conditions were all appropriate for dye removal (Ramadoss et al., 2020). Agricultural waste like cellulose was doped with 2-acrylamido-2-methylpropane sulfonic acid and acrylic acid, along with a crosslinker for fabricating polymer composites having dye removal characteristics. The grafted copolymer was flaky and highly thick fibrous structure representing the surface functionalization of agrowaste. The adsorption behavior of the composite was explored for the eradication of cationic and anionic dyes. The pH of 7 and 2.2 was optimum for removal of the cationic and anionic dye, respectively. The adsorption data fitted well the pseudo-second-order kinetic and Langmuir isotherm model (Kumar et al., 2018b). In another experiment, polystyrene was combined with activated carbon of a large surface area. The composite had polymer as the binder of the matrix. The composite was found to have a greater dye (methyl orange) removal tendency than activated carbon due to the adjunction of the polymer (Lakhal et al., 2018).

Polymer composites have also been fabricated for the elimination of salts from saline water. Polymer composites with activated carbon showed outstanding traits in desalination of groundwater as well as the brackish water, along with remarkable antimicrobial performance (Ashraf et al., 2013).

Polypropylene–sawdust composites were fabricated using in-situ ferric chloride oxidation method and showed the potential to remove chromium, zinc, dyes like carmoisine, methylene blue, etc. Spontaneous and the pseudo-second-order rate kinetics was followed in the adsorption process. The adsorption of Cr(VI), Zn(II), and nitrate followed the Freundlich isotherm model that well fitted the adsorption of metals, while Langmuir model was appropriate with dyes (Huang et al., 2014). In comparison to the use of sawdust as an adsorbent, the composites of poly(3-methyl thiophene)-sawdust showed an adsorption of 191 mg/g for removal of methylene blue dye, which was seven folds elevated than sawdust. The surface morphology studies reflected that an increased surface area was due to heterogeneous surface and porosity, leading to the higher adsorption capacity of the dye. The acid ethanol solution acted as a desorption agent, and adsorption efficiency remained at 93% even after three recycles. The same combination was tested for the removal of silver from wastewater. pH, concentration, composite dosage, and contact time influenced the adsorption efficiency. Both the Langmuir isotherm and Freundlich isotherms described the adsorption process. The maximum metal recovery obtained was 40% and 42% by 0.5 nitric acid and 1.0 M ammonia solutions, respectively (Huang et al., 2014).

A cationic dye methylene blue was adsorbed over polymer-activated charcoal composites. The maximum sorption capacity reported was 909.0, 312.5, and 192.3 mg/g at 296 K, 310 K and 323 K, respectively. The pseudo-second-order equation was found to fairly fit the uptake data with a regression value greater than 0.99, confirming the applicability of pseudo-second-order equation. The colorant uptake increased with pH, and a spontaneous and an exothermic process was observed (Bajpai and Shrivastava, 2011).

Chitosan-based novel polymeric activated carbon composite showed an enhanced removal tendency for chromium metal. This was attributed to the presence of high metal-binding active sites and porous structures. Researchers also suggested the use of *Sargassum horneri* as a potential pioneer for the composite for heavy metal eradication (Zeng et al., 2020).

Impregnation of chitosan-based polyvinyl alcohol was done with granular activated carbon in different ratios to investigate the adsorption ability of methylene blue removal over the adsorbent. Pseudo-first-order model was operational for the experimental data, as confirmed by regression values. The composite showed a higher adsorption tendency around 33 mg/g more than activated carbon (Salehi and Farahani, 2017).

10.7   CONCLUSION

Activated carbon-based composites act as successful decontamination agents. Different composites fabricated from a range of polymeric matrix and activated materials exhibit different selectivity. The composites have shown the capacity to adsorb a variety of contaminants, including heavy metals, dyes, ions, drugs, organics, and inorganic and also a substantial desalination capacity. Exploration is under process to unveil a broad spectrum of composites that are cost-effective, efficient as well as eco-friendly, and reusable. The entire fabrication process of the composites needs to be improved, and novel carbon-based composites should be fabricated with multiple microstructures and applicability to a broad range of contaminants eradication from wastewaters. Future research for controlled fabrication of composites is enviable. Adsorption mechanism and kinetics impart valuable direction to compose and optimize composite fabrication.

REFERENCES

Afzal A., Nawab Y. (2021) Polymer composites, In *Woodhead Publishing Series in Composites Science and Engineering, Composite Solutions for Ballistics,* Woodhead Publishing, 139–152, https://doi.org/10.1016/B978-0-12-821984-3.00003-6.

Almasian A., Olya M. E., Mahmoodi N. M. (2015) Preparation and adsorption behavior of diethylenetriamine/polyacrylonitrile composite nanofibers for a direct dye removal, *Fibers and Polymers,* 16(9), 1925–1934.

Andrijanto E., Subiyanto G., Marlina N., Citra H., Lintang C. (2018) Preparation of graphene oxide sand composites as super adsorbent for water purification application MATEC web conf. 156 05019, DOI: 10.1051/matecconf/201815605019.

Ansari R., Fahim N. K. (2007) Application of polypyrrole coated on wood sawdust for removal of Cr(VI) ion from aqueous solutions, *Reactive and Functional Polymer,* 67, 367–374, https://doi.org/10.1016/j.reactfunctpolym.2007.02.001.

Ashraf M. A., Maah M. J., Qureshi A. K., Gharibreza M., Yusoff I. (2013) Synthetic polymer composite membrane for the desalination of saline water, *Desalination and Water Treatment,* 51(16–18), 3650–3661, https://doi.org/10.1080/19443994.2012.751152.

Bajpai S. K., Shrivastava S. (2011) Sorptive removal of methylene blue from aqueous solutions by polymer/activated charcoal composites, 119(5), 2525–2532, DOI: 10.1002/app.31073.

Begum B., Ijaz S., Khattak R., Qazi R. A., Khan M. S., Mahmoud K. H. (2022) Preparation and characterization of a novel activated carbon@polyindole composite for the effective removal of ionic dye from water. *Polymers,* 14, 3, https://doi.org/10.3390/polym14010003.

Benregga F. Z., Maghchiche A., Nasri R., Haouam A. (2019) Preparation of composite materials from activated carbon and waste plastic for water treatment, *International Journal of Environmental Chemistry,* 5(2), 38–48.

Berber M. R. (2020) Surface-functionalization of activated carbon with polyglucosamine polymer for efficient removal of cadmium ions, *Polymer Composites,* https://doi.org/10.1002/pc.25599.

Biswal T., BadJena S.K., Pradhan D. (2020) Synthesis of polymer composite materials and their biomedical applications, *Materials Today: Proceedings*, 30, 305–315, https://doi.org/10.1016/j.matpr.2020.01.567Breuer O., Uttandaraman S. (2004) *Polymer Composites*, 25(6).

Chen H., Zhou Y., Wang J., Lu J., Zhou Y. (2020) Polydopamine modified cyclodextrin polymer as efficient adsorbent for removing cationic dyes and Cu2+, *Journal of Hazardous Materials*, 389, Article ID 121897, https://doi.org/10.1016/j.jhazmat.2019.121897.

Choi M., Ryoo R. (2003) Ordered nanoporous polymer-carbon composites, *Nature Materials*, 2(7), 473–476, https://doi.org/10.1038/nmat923.

Dhawan A. K., Seyler J. W., Bohrerbcbohre B. C. (2019) Preparation of a core-double shell chitosan-graphene oxide composite and investigation of Pb (II) absorption, *Heliyon*, 5, e01177, DOI: 10.1016/j.heliyon.2019.e01177.

Elkady M., El-Aassar M., Hassan H. (2016) Adsorption profile of basic dye onto novel fabricated carboxylated functionalized co-polymer nanofibers, *Polymers*, 8(5), 177, https://doi.org/10.3390/polym8050177.

Han Y., Elliott J. (2007) Molecular dynamics simulations of the elastic properties of polymer/carbon nanotube composites, *Computational Materials Science*, 39, 315–323, https://doi.org/10.1016/j.commatsci.2006.06.011.

He Z., Yang S., Ju Y., Sun C. (2009) Microwave photocatalytic degradation of rhodamine B using TiO2 supported on activated carbon: Mechanism implication, *Journal of Environmental Sciences*, 21(2), 268–272, https://doi.org/10.1016/S1001–0742(08)62262–7.

Hu N., Karube Y., Arai M., Watanabe T., Yan C., Li Y., Liu Y., Fukunaga, H. (2010) Investigation on sensitivity of a polymer/carbon nanotube composite strain sensor, *Carbon*, 48(3), 680–687, https://doi.org/10.1016/j.carbon.2009.10.012.

Huang Y., Li J., Chen X., Wang X. (2014) Applications of conjugated polymer based composites in wastewater purification, *RSC Advances*, 4, 62160–62178, DOI: 10.1039/C4RA11496E.

Kumar A., Kumar V., Awasthi K. (2018a) Polyaniline-carbon nanotube composites: Preparation methods, properties, and applications, *Polymer-Plastics Technology and Engineering*, 57(2), 70–97, DOI: 10.1080/03602559.2017.1300817.

Kumar R., Sharma R. K., Singh A. P. (2018b) Removal of organic dyes and metal ions by cross-linked graft copolymers of cellulose obtained from the agricultural residue, *Journal of Environmental Chemical Engineering*, 6(5), 6037–6048, https://doi.org/10.1016/j.jece.2018.09.021.

Lakhal F. Z., Maghchiche A., Nasri R., Haouam A. (2018) Composite material polystyrene activated carbon for water purification, *Journal of Materials and Environmental Science*, 9(8), 2411–2417.

Lee H. I., Jung Y., Kim S., Yoon J. A., Kim J. H., Hwang J. S., Yun M. H., Yeon J. W., Hong C. S., Kim J. M. (2009) Preparation and application of chelating polymer–mesoporous carbon composite for copper-ion adsorption, *Carbon*, 47(4), 1043–1049, ISSN 0008–6223, https://doi.org/10.1016/j.carbon.2008.12.024.

Li S., Huang H., Tao M., Liu X., Cheng T. (2013) Frontal polymerization preparation of poly(acrylamide-co -acrylic acid)/activated carbon composite hydrogels for dye removal, *Journal of Applied Polymer Science*, 129(6), DOI: 10.1002/app.39139.

Lin J., Zhan Y. (2012) Adsorption of humic acid from aqueous solution onto unmodified and surfactant-modified chitosan/zeolite composites, *Chemical Engineering Journal*, 200–202, 202–213, https://doi.org/10.1016/j.cej.2012.06.039.

Maghchiche A., Benregga F. Z., Nasri R., Haouam A. (2019) Preparation of a composite materials from activated carbon and waste plastic for water treatment, *International Journal of Environmental Chemistry*, 5 (2).

Metin A. U., Çiftçi H., Alver E. (2013) Efficient removal of acidic dye using low-cost biocomposite beads, *Industrial and Engineering Chemistry Research*, 52, 10569–10581.

Mohan V. M., Murakami K., Kono A., Shimomura M. (2013) Poly(acrylonitrile)/activated carbon composite polymer gel electrolyte for high efficiency dye sensitized solar cells, *Journal of Material Chemistry A*, 1, 7399–7407, https://doi.org/10.1039/C3TA10392G.

Nesic A. R., Velickovic S. J., Antonovic D. G. (2013) Modification of chitosan by zeolite A and adsorption of Bezactive Orange 16 from aqueous solution, *Composites Part B: Engineering*, 53, 145–151, https://doi.org/10.1016/j.compositesb.2013.04.053.

Nyabadza A., Vázquez M., Coyle S., Fitzpatrick B., Brabazon D. (2021) Review of materials and fabrication methods for flexible nano and micro-scale physical and chemical property sensors, *Applied Sciences*, 11(18), 8563, https://doi.org/10.3390/app11188563

Ramadoss P, Thankam R., Mohammed I. R., Arivuoli D. (2020) Low-cost and biodegradable cellulose/PVP/activated carbon composite membrane for brackish water treatment, 137(22), 48746, https://doi.org/10.1002/app.48746.

Salehi E., Farahani A. (2017) Macroporous chitosan/polyvinyl alcohol composite adsorbents based on activated carbon substrate, *Journal of Porous Materials*, 24, 1197–1207, https://doi.org/10.1007/s10934-016-0359-9.

Shojaei, A., Khasraghi, S. S. (2021). Self-healing and self-sensing smart polymer composites. In *Composite Materials*, 307–357. Elsevier.

Somaia G., Mohammada D. E. Abulyazied S., Ahmed M. (2019) Application of polyaniline/activated carbon nanocomposites derived from different agriculture wastes for the removal of Pb(II) from aqueous media, *Desalination and Water Treatment*, 170, 199–210, https://doi.org/10.5004/dwt.2019.24694.

Soni, U., Bajpai J., Bajpai A. K. (2015) Chitosan-activated carbon nanocomposites as potential biosorbent for removal of nitrophenol from aqueous solutions, *International Journal of Nanomaterials and Biostructure*, 5(4), 53–61, https://www.scopus.com/inward/record.uri?eid=2-s2.0-.

Sun Y., Shi G. (2013) Graphene/polymer composites for energy applications, *Journal of Polymer Science Part B: Polymer Physics*, 51, 231–253, https://doi.org/10.1002/polb.23226.

Tong, P., Li, M. et al. (2021)Molecularly imprinted polymer composites in biological analysis, In *Woodhead Publishing Series in Composites Science and Engineering, Molecularly Imprinted Polymer Composites*, Woodhead Publishing, 143–172, https://doi.org/10.1016/B978-0-12-819952-7.00001-9.

Vighnesha K. M., Sandhya S. et al. (2018) Synthesis and characterization of activated carbon/conducting polymer composite electrode for supercapacitor applications, *Journal of Materials Science: Materials in Electronics*, 29, 914–921, https://doi.org/10.1007/s10854-017-7988-x.

Wang M., Xie R., Chen Y., Pu X., Jiang W., Yao L. (2018) A novel mesoporous zeolite-activated carbon composite as an effective adsorbent for removal of ammonia-nitrogen and methylene blue from aqueous solution, *Bioresource Technology*, 268, 726–732, https://doi.org/10.1016/j.biortech.2018.08.037.

Xiao T.D., Gonsalves K.E., Strutt, P.R. (1993) Synthesis of Aluminum Nitride/Boron Nitride composite materials, *Journal of the American Ceramic Society*, 76, 987–992. https://doi.org/10.1111/j.1151-2916.1993.tb05323.x

Xiong Y., Robert T., Woodward D. D., Arwyn E., Tian T., Hassan A., Ardakani M., Petit C. (2021) Understanding trade-offs in adsorption capacity, selectivity and kinetics for propylene/propane separation using composites of activated carbon and hypercrosslinked polymer, *Chemical Engineering Journal*, 426, 131628, https://doi.org/10.1016/j.cej.2021.131628.

Zeng G., Hong C., Zhang Y. et al. (2020) Adsorptive removal of Cr(VI) by Sargassum horneri–based activated carbon coated with chitosan, *Water Air and Soil Pollution*, 231, 77, https://doi.org/10.1007/s11270-020-4440-2.

Zhang F., Wang M., Zhou L., Ma X., Zhou Y. (2015) Removal of Cd (II) from aqueous solution using cross-linked chitosan-zeolite composite, *Desalination and Water Treatment*, 54, 2546–2556, https://doi.org/10.1080/19443994.2014.901190.

Zhang, Z. H., Xua, Y. Ma X. P. et al. (2012) Microwave degradation of methyl orange dye in aqueous solution in the presence of nano TiO2-supported activated carbon (supported-TiO2/AC/MW), *Journal of Hazardous Materials*, 209–210, 271–277, https://doi.org/10.1016/j.jhazmat.2012.01.021.

11 Polymer–MXene Composites for Wastewater Treatment

Himadri Tanaya Das, Swapnamoy Dutt,
Elango Balaji T., Payaswini Das,
and Nigamananda Das
Utkal University

CONTENTS

11.1 INTRODUCTION

In last few decades, water consumption across the globe has augmented owing to industrialisation and urbanisation. Mostly due to improper management of natural water resources and everyday human activities, the water pollution has surged to a risky level. For instance, several recent reports have demonstrated that toxic contaminants of emerging concern can be found at very low levels (sub 1 µg/L) in wastewater influents/effluents and raw/finished drinking water around the world. So, to prevent this calamitous situation, various conventional as well as advanced treatment techniques have been utilised in water and wastewater treatment plants to treat both existing and emerging pollutants, including coagulation/ flocculation/sedimentation/filtration, chlorination, ozonation, carbon/nanomaterial adsorption, membranes and ultrasonication [1,2].

Owing to their promising energy storage, environmental and industrial applications, MXenes are widely explored, which is why they are been increasingly recognised in current research and hopefully will bring tremendous research scope in the upcoming times. Especially, recent studies have shown that the effective fabrication of various MXenes has resulted in high potential for applications of MXene-based materials in the area of water purification and organic solvent filtration. In particular, excellent stability, superior surface area, high electrical/thermal conductivity, excellent oxidation resistance and hydrophilicity-like favourable properties have

DOI: 10.1201/9781003328094-11

made them attractive as suitable material for applications. On the other hand, for eliminating toxic compounds from water, utilisation of polymers as water treatment coagulants and flocculants have been broadly investigated due to their physicochemical characteristics. For instance, cationic polymers are used in the water treatment industry as primary coagulants, whereas non-ionic and anionic polymers are used as flocculants or filter aids, and are usually used in conjunction with inorganic coagulants. On the other hand, polymers or polymeric materials are also known for their capability in water treatment, especially the role of conducting polymers have been examined in several studies. Owing to benefits such as chemical inertness, thermal stability and high surface areas, polymeric photocatalysts exhibit excellent activity [3–5]. By selecting different combinations and ratios of building blocks and the degree of polymerisation, different optoelectronic and photophysical properties can be administered in the polymer, such as high absorption area, tunable band gaps, stable excited states, etc. [6,7]. Also, charge migration in 2D polymers can be improved by pi stacking, which provides a percolation channel within the layers [8]. Particularly polymers like poly(3-hexylthiophene) (P3HT) and poly(diphenylbutadiyne) (PDPB) nanofibres have shown superior photocatalytic performances [9,10]. Other than that, polymeric or composite polymeric membrane have also proved to be considerable solution for water treatment [11–13]. For instance, electrospun polymeric nanofibrous membranes are widely studied where several benefits are observed, such as, mild operating conditions, and low energy consumption for synthesis and small footprint. Several nanofibrous membranes were fabricated with PAN, PSU, PES and nylon-6, which exhibited great outcomes for liquid separation and particle removal. Owing to unique characteristics such as specific surface area and high porosity, polymer along with MXenes are mixed to form composite materials, which exhibited excellent adsorption, catalysis and filtration activities. For example, $PPy/Ti_3C_2T_x$ composites, AgNP-loaded $MXene/Fe_3O_4$/polymer nanocomposite, PVA/PAA/MXene@PdNPs functional nanocomposites, etc. are developed that have shown superior activities [14–17]. Like these, in the upcoming sections we will illustrate more about the synthesis and the performance of polymer- and MXene-based materials for wastewater purification.

11.2   SYNTHESIS OF POLYMER–MXENE COMPOSITES

For the synthesis of MXene, polymer composite electrospinning strategy is mainly adapted in reports, as it has the unique ability to develop tailored membranes or structures, especially many different polymers due to its ability to prepare highly porous membranes with narrow pore size distributions and low tortuosity. Particularly for the fabrication of composite, core sheath, hollow nanofibres and ceramic nanofibres electrospinning is employed. In a study, PVA/poly(acrylic acid)/Fe_3O_4/MXene@ AgNP (PVA/PAA/Fe_3O_4/MXene@AgNP) functional nanocomposites were synthesised through the electrospinning process. For the synthesis, firstly, the layered MAX (Ti_3AlC_2) powders were exfoliated through the treatment of 40% HF solution at a temperature of 25°C, as shown in Figure 11.1a [16]. Selective HF etching originated from only "A" layers from the MAX phase. Afterwards, MAX became a multilayer MXene phase. In order to intercalate MXene, DMSO is utilised at room

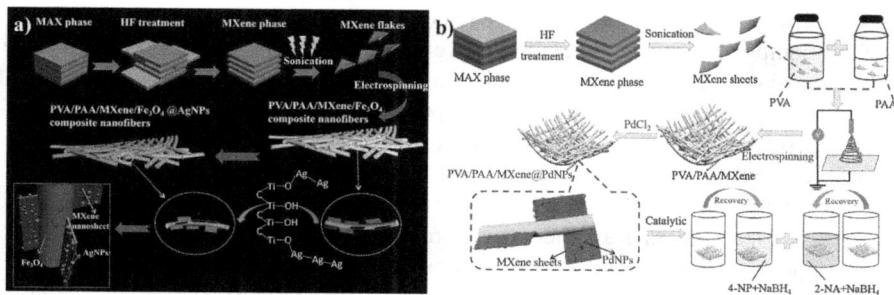

FIGURE 11.1 (a) Schematic illustration of the preparation process of the present composite fibres, and (b) schematic illustration of preparation process of the obtained composite fibres. (Reproduced from Refs. [14,16]. Copyrights (ACS, 2019) (Elsevier, 2019).)

temperature, as it could improve the c-lattice parameter of MXene from 19.5 ± 0.1 to 35.04 ± 0.02 Å. The DMSO-intercalated MXene sonicated in water led to themselves being delaminated into separate flakes similar to "paper" and formation of nanosheet colloidal solution. Multilayer MXene was intercalated with DMSO and ultrasonically dispersed to obtain MXene nanosheet colloidal solution for the next step of electro-spinning. PVA, PAA, Fe_3O_4 and MXene nanosheet colloidal solution were mixed in proportion to form an electrospinning precursor solution. Through electrospinning, the composite nanofibres with MXene flake "wings" were obtained. When the com-posite nanofibres were placed in a slowly stirred $AgNO_3$ solution, a [Ag^+-DMSO] complex monomer was first formed. The rapidly transferred electrons form oxygen lone pair electrons ([$\ddot{O}S$–$(CH_3)_2$]), leading to the formation of Ag^+–[DMSO]. The charge was transferred between Ag–MXene complexes to initialise the dimerised MXene–Ag DMSO. The MXene–Ag dimeric complexes were bonded to −OH and reduced to stable Ag^0 nanoclusters. As the nanoclusters were further nucleated and grown on the surface of the MXene nanosheets, spherical AgNPs were formed. With the increment of self-reduction time, they may have a small amount of AgNPs drifted/anchored on the surface of polymer fibres. In this way, the AgNP-loaded composite nanofibres were synthesised, which exhibited large specific surface area; the sur-face has a reduction/nucleation site and superior catalytic activity. Another minute observation has been mentioned in the report that the layered MXene nanosheets were firmly "locked" on the fibre surface by composite nanofibres to achieve high dispersibility. The synthesis of the composite was further confirmed with the TEM, XPS and EDX, where elements of C, O, Fe, Ti and Ag were seemed to be present. Along with that, strong affiliation in composite nanofibres and addition of the Fe_3O_4 NPs and MXene flakes in the composite nanofibre layers were confirmed through XRD. In another study, electrospinning strategy (Figure 11.1b) was adapted to pre-pare polyvinyl alcohol (PVA)/polyacrylic acid (PAA)/MXene fibre membrane hav-ing structural characteristics (such as higher specific surface, uniform distribution of MXene sheets on the spun fibres that provided application advantages) and excellent catalytic reduction activity [14]. Here, a solution of the MXene flake dispersion was separately added to the PVA and PAA solutions. Then, both the solutions of PVA and PAA were uniformly mixed with each other to be spun. At the end, PVA/PAA/ MXene composite nanofibre membranes were synthesised via electrospinning. After the composite fibre membrane was dried, it was placed in a $PdCl_2$ solution with an appropriate time. Finally, ethanol and ultrapure water was utilised to wash the membrane for several times, and dried to acquire PVA/PAA/MXene@PdNPs composite nanofibre membrane. Successful loading of PdNPs and existence of carbon, oxygen, titanium and palladium elements was confirmed through XRD, SEM and XPS.

Tape-casting process is another process recommended to synthesise MXene polymer composite. This process is utilised in order to gain accurate control of the layer thickness in a wide range, easy to scale up and the possibility to cast large area structures. Other than that, sequential tape casting can be used to develop multi-layered structures via directly casting different layers one on top of the other. This technical tactic restricts as much as possible the number of manufacturing stages, with significant benefits in terms of the costs and preparation duration of the production process. Yang et al. [18] used tape-casting method to develop MXene/PVB/ $Ba_3Co_2Fe_{24}O_{41}$/Ti_3C_2 composite for where at first Co_2Z powder were prepared via the conventional solid-state reaction process. Afterwards, $BaCO_3$, CoO and Fe_2O_3 powders were mixed following the stoichiometry of $Ba_3Co_2Fe_{24}O_{41}$ (Co_2Z). After the successful etching of Al to prepare Ti_3C_2 MXene in HF at room temperature, Co_2Z powders were added to 40 vol% methyl–ethyl–ketone, and 60 vol% ethanol, triethyl phosphate dispersant and MEK were ball milled for 4 hours. With respect to the mass of random Co_2Z powders, 10 wt% (001) plate-like Co_2Z powders were then added into the above mixture, which was ball milled with a slow rotation for another 2 hours to form a slurry for tape casting. The slurry of PVB/Co_2Z was tape cast to form a sheet with a thickness of ~200 µm on a polyethylene film by a doctor blade apparatus. After drying, the first layer sheet was formed. The slurry of PVB/Ti_3C_2 MXene was prepared by the same method and was tape cast to form the second sheet with a thickness of ~230 µm on the first layer by a doctor blade apparatus. After drying, a PVB/Co_2Z/Ti_3C_2 MXene composite was obtained. Using the similar tape-casting process, MXene/polymer films was prepared where Ti_3AlC_2 MAX powders were used as raw material for preparing $Ti_3C_2T_x$ MXene, and anhydrous ethanol, triethyl phosphate, polyvinyl butyral, glycerol and dioctylphthalate were utilised as the solvent, dispersing agent, binder, lubricant and plasticiser, respectively [19]. The as-prepared films exhibited great electromagnetic and mechanical characteristics.

Chitosan, cellulose like natural polymers are also physically blended with MXenes to fabricate nanocomposites. For instance, acetylcholinesterase/chitosan/MXene is prepared Liya Zhou and team via drop casting process of the $Ti_3C_2T_x$ nanosheets circulated in 0.20% CS solution onto the GCE and was further coated by AChE solution to create AChE/CS-$Ti_3C_2T_x$/GCE biosensor [20]. The prepared CS-$Ti_3C_2T_x$ composite has excellent conductivity and high electro catalytic activity. In another study, vacuum-assisted filtration self-assembly process was employed to fabricate ultrathin and highly flexible $Ti_3C_2T_x$ (d-$Ti_3C_2T_x$, MXene)/cellulose nanofibre (CNF) composite paper with a nacre-like lamellar structure, which exhibited superior electrical conductivity, excellent integration of the mechanical strength and toughness [21]. Vacuum-assisted filtration is also employed for employed for MXene/PVDF membrane, where presence of functional groups, lateral size, number of layers, and surface and edge properties were investigated, as they can influence in applicative purpose [7]. Using the same method, $Ti_3C_2T_x$ MXene/polystyrene composite was prepared [22].

11.3 APPLICATIONS OF POLYMER AND MXENE COMPOSITES

11.3.1 Dye and Photocatalyst

Photocatalyst has a significant role in decreasing the environmental pollution caused by industrial wastewater. The preparation of catalyst and adsorbents through simple methods and obtaining high activity is still a great challenge [23,24].

Yin et al. developed a novel method to achieve high activity with simple method sandwich like $Cu_2O/TiO_2/Ti_3C_2$ composite prepared by a simple solvent reduction method. The composite exhibited excellent catalysis towards 2-nitroaniline and 4-nitrophenol, the pseudo-first-order rate constants are 0.163 and 0.114 min^{-1}. After eight cycles, the conversion rates are as high as 95% and 92%, which shows that the catalyst has high reutilisation rate. The good performance is due to the in-situ-formed Cu_2O, which minimises the aggregation during the formation [25]. Khadidja et al. synthesised ZnO microrods/MXene via facile hydrothermal method, which showed high photocurrent intensity and showed excellent separation efficiency of photogenerated charges. Rhodamine B was degraded under UV radiation by the composite within 18 minutes and retained more than 90% of photocatalytic efficiency after seven cycles, which makes the composite an efficient candidate for application in wastewater treatment [26]. 2D a-Fe_2O_3-doped Ti_3C_2 MXene composite was reported by Zhang et al. through a facile ultrasonic-assisted self-assembly method; the as-synthesised composite showed high degradation rate of 98% (Figure 11.2) [27].

Liu et al. reported $BiOBr/Ti_3C_2$. Othman et al. developed $AgNPs/TiO_2/Ti_3C_2T_x$ MXene composites as efficient photocatalyst for Methylene Blue(MB) and Rhodamine

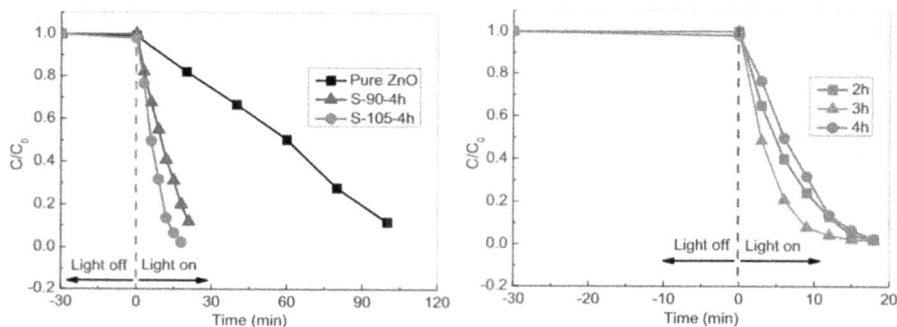

FIGURE 11.2 Typical degradation rate of RhB with pure ZnO and ZnO/MXene composites: (a) comparison of degradation rate of RhB for pure ZnO, S-90–4 h and S-105-4 h; (b) comparison of degradation rate of RhB for S-105-2 h, S-105-3 h and S-105-4 h. (Reproduced from Ref. [26]. Copyrights (Elsevier, 2021).)

Blue (RhB) degradation under UV and solar light. The composite showed rate constants of 0.162 and 0.143 min^{-1} for MB and RhB under UV light, and it showed rate constants of 0.028 and 0.020 min^{-1} for MB and RhB under simulated solar light [28]. Li et al. reported 1D/2D CdS-Ti$_3$C$_2$T$_x$ composite prepared via one-step electrostatic self-assembly, which was further studied for photocatalytic hydrogen evolution; the results showed that 10% CdS-Ti$_3$C$_2$T$_x$ composite shows better performance [29]. Low et al. prepared TiO$_2$ *in-situ* grown on Ti$_3$C$_2$ through calcination method; the composite exhibited excellent photocatalytic CO$_2$ reduction; the composite exhibited 3.7 times higher photocatalytic CO$_2$ reduction for CH$_4$ production; and the excellent performance is due to the unique morphology and the features of individual components in the composite. The ultrahigh conductivity of Ti$_3$C$_2$ enhanced the photogenerated electron transfer [30]. Xu et al. reported g-C$_3$N$_4$/Ti$_3$C$_2$ as photocatalyst for hydrogen production. It showed high hydrogen production rate of 2181 μmol/g, which is higher than the bulk and protonated g-C$_3$N$_4$, which has hydrogen evolution rate of 393 and 816 μmol/g [31]. Iqbal et al. reported La- and Mn-codoped bismuth ferrite embedded on MXene, which were prepared by low-cost solvent sol–gel method. The composite degraded 93% of the organic pollutants [32]. Yang et al. designed PtO decorated on Ti$_3$C$_2$/TiO$_2$ obtained via in-situ conversion and following PtO nanodots photodeposition. The composite showed high photocatalytic hydrogen evolution activity with a rate high as 2.54 mmol/g h. The high activity is due to the photogenerated electrons and holes that flow in opposite direction to make separation efficiency high and suppress undesirable hydrogen black reaction, which resulted in higher photogenerated carrier separation and transmission efficiency [33]. Ding et al. investigated charge transfer in 2D CdS/2D MXene Schottky heterojunctions for high efficiency photocatalytic hydrogen production; the composite exhibits five times higher activity than the pristine CdS with 3226 μmol/g h [34]. Liu et al. reported a simple hydrothermal method for the synthesis of ZnO@Ti$_3$C$_2$, which exhibited excellent photocatalytic degradation performance of 94.84%. Compositing of ZnO with Ti$_3$C$_2$ forms a Schotky junction, which prevents the photogenerated electrons returning to ZnO and facilitates the separation of carriers [35]. Shi et al. studied the photocatalytic activity of bismuth oxyhalides composited with Ti$_3$C$_2$T$_x$ for enhanced photocatalytic degrading property of RhB and phenol. The experimental results indicate that the material has excellent chemical stability. The composite removed almost 50% of phenol after 5 hour and almost all RhB after 40 minutes under visible light irradiation [36]. Zhou et al. reported hierarchical ZnO/MXene formed by electrostatic self-assembly; the composite exhibited excellent photocatalytic activity in degrading MB. The composite achieved high degradation rate of 99.53% after irradiation with ultraviolet light for 120 minutes; also, the degradation rate constants are 0.04317 min^{-1}, which is 16.23 times higher than those of pure ZnO. Also, the composite showed high stability of 97.3% after four cycles, while the ZnO only maintained 22% after four cycles. The improved performance is due to promoted electron transfer in the heterostructures [37]. Zhang et al. reported α-Fe$_2$O$_3$/ZnFe$_2$O$_4$@Ti$_3$C$_2$ MXene composite, which can be easily obtained by ultrasonic-assisted self-assembly approach. The composite showed high-rate constant of 0.02686 min^{-1} for the 10% composite. It showed excellent photocatalytic activity and reusability in degradation of RhB and reduction of Cr(VI) under visible light [38].

11.3.2 MEMBRANE

Wastewater generation is unavoidable due to the fact that it forms an integral part in all sectors of life. Membrane technology has grabbed much attention in the recent years due to the advantages it offers in water and wastewater treatment. Reducing the size of the equipment, energy requirement and low capital cost, membrane technology gives many prospects in wastewater treatment [39]. Zeng et al. developed a self-cleaning photocatalytic composite membrane based on g-C_3N_4@MXene for the removal of dyes and antibiotics. The composite showed very good removal ratio; they are 98% for Congo Red, 96% for Trypan Blue (TB), and 86% for tetracycline hydrochloride and the membrane could perform three consecutive cycles without any performance degradation [40]. Lin et al. reported a bionics inspired, modified two-dimensional MXene composite membrane that showed a rejection ratio of direct red reached 88.9% and direct black 38 reached 88.6%; and the composite membrane showed a pure water flux of 271.2 L/m²h, which is 277% more than the unmodified membrane [41]. Zeng et al. reported a composite membrane from functionalised metal organic framework integrated MXene for oil/water emulsion separation. The composite membrane exhibited excellent water flux of 5347 L m²/h; the rejection ratio for three different oil/water emulsions reached up to 98%. Further, it showed 84% flux recovery ratio during multicycle oil/water emulsion filtration [42]. Yang et al. developed Fe_3O_4@MXene composite nanofiltration membrane for heavy metal ions removal from wastewater, the membrane could achieve the maximum removal of 63.2% for Cu^{2+}, 64.1% for Cd^{2+} and 70.2% for Cr^{6+}. The pure water flux of the membrane is 125.1 L/m h, which is 44.1% higher when compared to pure MXene membrane [42]. Feng et al. reported dual-layered covalent organic framework/MXene membranes for fast water treatment, which showed high permeance of 563 L/(m²h bar) with high rejection to Congo Red of 99.6% [43]. Hu et al. reported MXene nanosheets on porous polyvinylidene fluoride substrate and in-situ mineralisation of β-FeOOH on the membrane surface. The composite showed fast permeation flux of 1022.7 L/m²h and favourable separation efficiency of 99.72% [44]. Zhao et al. reported PEI modified GO/MXene composite membrane, which shows high water permeability of 9.5 ± 0.4 L m²/h bar. The composite membrane had a rejection of over 70% for Ca^{2+} and Mg^{2+}, which is seven times higher than GO membrane [45]. Ma et al. reported GO/MXene/PPS composite membrane for dye wastewater treatment, which showed high water flux of 45.78 L/m²h bar with a high rejection rate of 99.99% for many organic dyes like MB [46]. Sun et al. reported $Ti_3C_2T_x$ and carbon nanotube composite as membrane for municipal wastewater treatment; the composite showed four times higher water flux, and it also exhibited good concentration efficiency [47]. Ajibade et al. reported PAN UF membrane modified with 3D-MXene/O-MWCNT for the removal of oil and dyes from industrial wastewater. The composite showed a high water flux of 301 L m²/h and high flux recovery ratio of 98%, as shown in Figure 11.3a and b [48]. Han et al. reported 2D MXene/PES composite membrane, which exhibited excellent flux 115 L/m²h (as shown in Figure 11.3c) and good rejection to Congo Red dye (92.3%) [49].

Feng et al. reported Ti_3C_2Tx MXene/poly(arylene ether nitrile) fibrous composite membrane for the purification of emulsified oil and degradation of organic

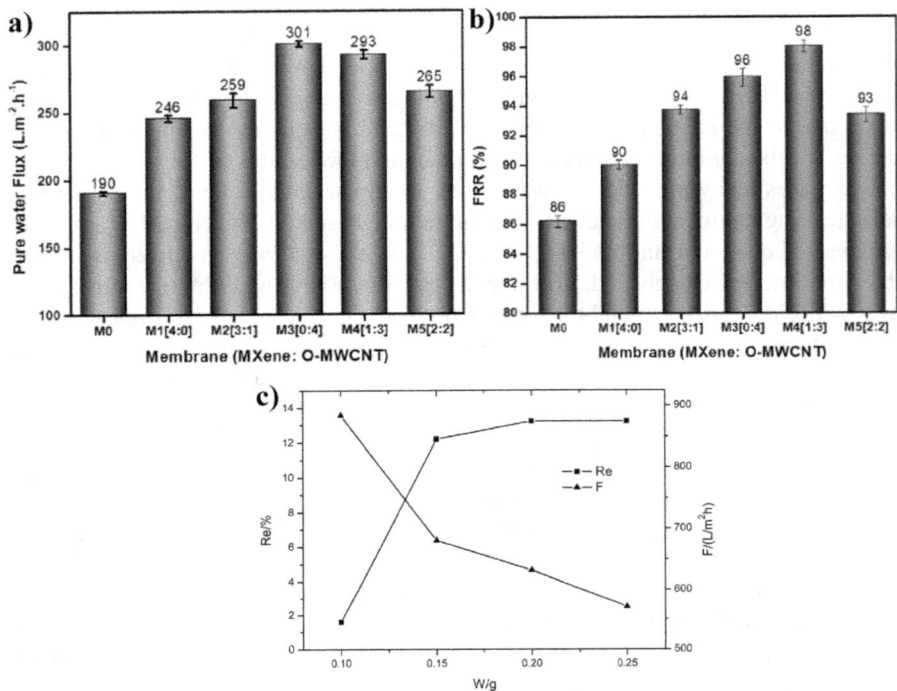

FIGURE 11.3   (a) Pure water flux for M0-M5. (b) Flux recovery ratio of M0-M5 after stage 5 filtration, (c) the performance of the MXene membrane with different MXene (1 g L1 Na2SO4 solution, 0.1 MPa). (Reproduced from Refs. [48,49].)

compounds; the membrane achieved permeation flux of 08–1003 L/m h; also the material exhibited good photocatalytic property with 92.31% degradation of MB within 60 minutes [50]. Cheng et al. reported BiOCl-PPy composited with MXene for removal of contaminants from wastewater; the membrane displayed high permeation performance with pure water flux of 3680.2 L/m²h. Also, the composite membrane showed high removal rate of 99.9% for three different dyes [51].

11.4   FUTURE PERSPECTIVE AND CONCLUSION

The discovery of MXene has opened up greater possibilities in investigating several environmental and industrial applications, where many of them have exhibited beneficial outcome owing to beneficial characteristics (surface area, excellent conductivity) of MXene. However, in search of further precise advancements, several composites are been recommended to combine with MXene to develop favourable structures to improve its applicability. Polymers or especially conductive polymers are one such set of materials that have given promising outputs. In order to modify material properties, these versatile variety of long chains of carbon molecules (polymers) are been efficiently incorporated with MXenes. Apart from easy manufacturing and low cost, compositing with polymer have enhanced several characteristics: resistance to abrasion, resistance to corrosion, resistance to fatigue, resistance to impact, great strength, resistance to fracture and great stiffness. Like this, we have illustrated several parts indicating the impact of utilising MXene and polymer composites. We have elaborately mentioned several synthesising as well as applicative references to understand the workability of the composite in very detailed manner. In this chapter, we have specifically discussed the composite's usability in wastewater treatment. Although the composite has utilisation in other applications such as photothermal conversion, flexible supercapacitors, modified yarn energy storage, etc. Various composites such as $Ti_3C_2T_x$/poly(vinyl alcohol)/poly(acrylic acid), $Ti_3C_2T_x$/polyether sulphone, Ti_3C_2/PVA/PAA, etc. are found to be effective for wastewater treatment, which is elaborately discussed in the review. Not only particular mixture of polymers and MXenes, but also more complex structures such as AgNP-loaded MXene/Fe_3O_4/polymer, layered PVB/Co_2Z/Ti_3C_2 MXene composite, etc., are also been explored to achieve higher performance. In upcoming times, more versatile as well as complex composite must be approached to record better applicative activity, keeping in mind the factors such as cost of manufacturing, availability, toxicity and viability.

REFERENCES

1. Y.A.J. Al-Hamadani, B.-M. Jun, M. Yoon, N. Taheri-Qazvini, S.A. Snyder, M. Jang, J. Heo, Y. Yoon, Applications of MXene-based membranes in water purification: A review, *Chemosphere*, 254 (2020) 126821. https://doi.org/10.1016/j.chemosphere.2020.126821.
2. H.T. Das, T.E. Balaji, S. Dutta, N. Das, T. Maiyalagan, Recent advances in MXene as electrocatalysts for sustainable energy generation: A review on surface engineering and compositing of MXene, *Int. J. Energy Res.* https://doi.org/10.1002/er.7847.
3. V.S. Vyas, B.V. Lotsch, Organic polymers form fuel from water, *Nature*, 521 (2015) 41–42. DOI: 10.1038/521041a.
4. S. Kumar, M.Y. Wani, C.T. Arranja, J.d.A. e Silva, B. Avula, A.J.F.N. Sobral, Porphyrins as nanoreactors in the carbon dioxide capture and conversion: A review, *J. Mater. Chem. A*, 3 (2015) 19615–19637. DOI: 10.1039/C5TA05082K.
5. H.T. Das, K. Mahendraprabhu, T. Maiyalagan, P. Elumalai, Performance of solid-state hybrid energy-storage device using reduced graphene-oxide anchored sol-gel derived Ni/NiO nanocomposite, *Sci. Rep.*, 7 (2017) 1–14. DOI: 10.1038/s41598-017-15444-z.
6. V.S. Vyas, V.W.-h. Lau, B.V. Lotsch, Soft photocatalysis: Organic polymers for solar fuel production, *Chem. Mater.*, 28 (2016) 5191–5204. DOI: 10.1021/acs.chemmater.6b01894.
7. H.T. Das, S. Vinoth, M. Thirumoorthi, T. Alshahrani, H.H. Hegazy, H.H. Somaily, M. Shkir, S. Aifaify, Tuning the optical, electrical, and optoelectronic properties of CuO thin films fabricated by facile SILAR dip-coating technique for photosensing applications, *J. Inorg. Organometal. Polym. Mater.*, 31 (2021) 2606–2614. DOI: 10.1007/s10904-021-01928-z.
8. M. Calik, F. Auras, L.M. Salonen, K. Bader, I. Grill, M. Handloser, D.D. Medina, M. Dogru, F. Löbermann, D. Trauner, A. Hartschuh, T. Bein, Extraction of photogenerated electrons and holes from a covalent organic framework integrated heterojunction, *J. Am. Chem. Soc.*, 136 (2014) 17802–17807. DOI: 10.1021/ja509551m.
9. S. Saha, N. Chaudhary, A. Kumar, M. Khanuja, Polymeric nanostructures for photocatalytic dye degradation: Polyaniline for photocatalysis, *SN Appl. Sci.*, 2 (2020) 1115. DOI: 10.1007/s42452-020-2928-4.

10. K.K. Jaiswal, S. Dutta, C.B. Pohrmen, I. Banerjee, H.T. Das, N. Jha, B. Jha, M. Verma, W. Ahmad, Bio-fabrication of selenium nanoparticles/micro-rods using cabbage leaves extract for photocatalytic dye degradation under natural sunlight irradiation, *Int. J. Environ. Anal. Chem.*, (2021) 1–18.

11. Y. Liao, C.-H. Loh, M. Tian, R. Wang, A.G. Fane, Progress in electrospun polymeric nanofibrous membranes for water treatment: Fabrication, modification and applications, *Prog. Polym. Sci.*, 77 (2018) 69–94. https://doi.org/10.1016/j.progpolymsci.2017.10.003.

12. M.S. Uddin, H.T. Das, T. Maiyalagan, P. Elumalai, Influence of designed electrode surfaces on double layer capacitance in aqueous electrolyte: Insights from standard models, *Appl. Surf. Sci.*, 449 (2018) 445–453.

13. T.E. Balaji, H. Tanaya Das, T. Maiyalagan, Recent trends in bimetallic oxides and their composites as electrode materials for supercapacitor applications, *ChemElectroChem*, 8 (2021) 1723–1746. https://doi.org/10.1002/celc.202100098.

14. J. Yin, F. Zhan, T. Jiao, H. Deng, G. Zou, Z. Bai, Q. Zhang, Q. Peng, Highly efficient catalytic performances of nitro compounds via hierarchical PdNPs-loaded MXene/polymer nanocomposites synthesized through electrospinning strategy for wastewater treatment, *Chin. Chem. Lett.*, 31 (2020) 992–995. https://doi.org/10.1016/j.cclet.2019.08.047.

15. M. Boota, B. Anasori, C. Voigt, M.-Q. Zhao, M.W. Barsoum, Y. Gogotsi, Pseudocapacitive electrodes produced by oxidant-free polymerization of pyrrole between the layers of 2D titanium carbide (MXene), *Adv. Mater.*, 28 (2016) 1517–1522. https://doi.org/10.1002/adma.201504705.

16. X. Huang, R. Wang, T. Jiao, G. Zou, F. Zhan, J. Yin, L. Zhang, J. Zhou, Q. Peng, Facile preparation of hierarchical AgNP-loaded MXene/Fe3O4/polymer nanocomposites by electrospinning with enhanced catalytic performance for wastewater treatment, *ACS Omega*, 4 (2019) 1897–1906. DOI: 10.1021/acsomega.8b03615.

17. H.T. Das, P. Barai, S. Dutta, N. Das, P. Das, M. Roy, M. Alauddin, H.R. Barai, Polymer composites with quantum dots as potential electrode materials for supercapacitors application: A review, *Polymers*, 14 (2022) 1053.

18. H. Yang, J. Dai, X. Liu, Y. Lin, J. Wang, L. Wang, F. Wang, Layered PVB/Ba3Co2Fe24O41/Ti3C2 Mxene composite: Enhanced electromagnetic wave absorption properties with high impedance match in a wide frequency range, *Mater. Chem. Phys.*, 200 (2017) 179–186. https://doi.org/10.1016/j.matchemphys.2017.05.057.

19. B. Ji, S. Fan, S. Kou, X. Xia, J. Deng, L. Cheng, L. Zhang, Microwave absorption properties of multilayer impedance gradient absorber consisting of Ti3C2TX MXene/polymer films, *Carbon*, 181 (2021) 130–142. https://doi.org/10.1016/j.carbon.2021.05.018.

20. L. Zhou, X. Zhang, L. Ma, J. Gao, Y. Jiang, Acetylcholinesterase/chitosan-transition metal carbides nanocomposites-based biosensor for the organophosphate pesticides detection, *Biochem. Eng. J.*, 128 (2017) 243–249. https://doi.org/10.1016/j.bej.2017.10.008.

21. W.-T. Cao, F.-F. Chen, Y.-J. Zhu, Y.-G. Zhang, Y.-Y. Jiang, M.-G. Ma, F. Chen, Binary strengthening and toughening of MXene/cellulose nanofiber composite paper with nacre-inspired structure and superior electromagnetic interference shielding properties, *ACS Nano*, 12 (2018) 4583–4593. DOI: 10.1021/acsnano.8b00997.

22. K. Rasool, K.A. Mahmoud, D.J. Johnson, M. Helal, G.R. Berdiyorov, Y. Gogotsi, Efficient antibacterial membrane based on two-dimensional Ti3C2Tx (MXene) nanosheets, *Sci. Rep.*, 7 (2017) 1598. DOI: 10.1038/s41598-017-01714-3.

23. H.T. Das, S. Dutta, R. Beura, N. Das, Role of polyaniline in accomplishing a sustainable environment: Recent trends in polyaniline for eradicating hazardous pollutants, *Environ. Sci. Pollut. Res.*, 29 (2022) 49598–49631. DOI: 10.1007/s11356-022-20916-5.

24. J. Theerthagiri, J. Park, H.T. Das, N. Rahamathulla, E.S.F. Cardoso, A.P. Murthy, G. Maia, D.V.N. Vo, M.Y. Choi, Electrocatalytic conversion of nitrate waste into ammonia: A review, *Environ. Chem. Lett.*, (2022) DOI: 10.1007/s10311-022-01469-y.

25. J. Yin, B. Ge, T. Jiao, Z. Qin, M. Yu, L. Zhang, Q. Zhang, Q. Peng, Self-assembled sandwich-like MXene-derived composites as highly efficient and sustainable catalysts for wastewater treatment, *Langmuir*, 37 (2021) 1267–1278. DOI: 10.1021/acs.langmuir.0c03297.

26. M.F. Khadidja, J. Fan, S. Li, S. Li, K. Cui, J. Wu, W. Zeng, H. Wei, H.-G. Jin, N. Naik, Z. Chao, D. Pan, Z. Guo, Hierarchical ZnO/MXene composites and their photocatalytic performances, *Colloids Surf. A: Physicochem. Eng. Aspects*, 628 (2021) 127230. https://doi.org/10.1016/j.colsurfa.2021.127230.

27. H. Zhang, M. Li, J. Cao, Q. Tang, P. Kang, C. Zhu, M. Ma, 2D a-Fe2O3 doped Ti3C2 MXene composite with enhanced visible light photocatalytic activity for degradation of Rhodamine B, *Ceram. Int.*, 44 (2018) 19958–19962. https://doi.org/10.1016/j.ceramint.2018.07.262.

28. Z. Othman, A. Sinopoli, H.R. Mackey, K.A. Mahmoud, Efficient photocatalytic degradation of organic dyes by AgNPs/TiO2/Ti3C2Tx MXene composites under UV and solar light, *ACS Omega*, 6 (2021) 33325–33338. DOI: 10.1021/acsomega.1c03189.

29. J.-Y. Li, Y.-H. Li, F. Zhang, Z.-R. Tang, Y.-J. Xu, Visible-light-driven integrated organic synthesis and hydrogen evolution over 1D/2D CdS-Ti3C2Tx MXene composites, *Appl. Catal. B: Environ.*, 269 (2020) 118783. https://doi.org/10.1016/j.apcatb.2020.118783.

30. J. Low, L. Zhang, T. Tong, B. Shen, J. Yu, TiO2/MXene Ti3C2 composite with excellent photocatalytic CO$_2$ reduction activity, *J. Catal.*, 361 (2018) 255–266. https://doi.org/10.1016/j.jcat.2018.03.009.

31. H. Xu, R. Xiao, J. Huang, Y. Jiang, C. Zhao, X. Yang, In situ construction of protonated g-C3N4/Ti3C2 MXene Schottky heterojunctions for efficient photocatalytic hydrogen production, *Chinese J. Catal.*, 42 (2021) 107–114. https://doi.org/10.1016/S1872-2067(20)63559-8.

32. M.A. Iqbal, S.I. Ali, F. Amin, A. Tariq, M.Z. Iqbal, S. Rizwan, La- and Mn-Co doped bismuth ferrite/Ti3C2 MXene composites for efficient photocatalytic degradation of Congo Red dye, *ACS Omega*, 4 (2019) 8661–8668. DOI: 10.1021/acsomega.9b00493.

33. J.-X. Yang, W.-B. Yu, C.-F. Li, W.-D. Dong, L.-Q. Jiang, N. Zhou, Z.-P. Zhuang, J. Liu, Z.-Y. Hu, H. Zhao, Y. Li, L. Chen, J. Hu, B.-L. Su, PtO nanodots promoting Ti3C2 MXene in-situ converted Ti3C2/TiO2 composites for photocatalytic hydrogen production, *Chem. Eng. J.*, 420 (2021) 129695. https://doi.org/10.1016/j.cej.2021.129695.

34. M. Ding, R. Xiao, C. Zhao, D. Bukhvalov, Z. Chen, H. Xu, H. Tang, J. Xu, X. Yang, Evidencing interfacial charge transfer in 2D CdS/2D MXene Schottky heterojunctions toward high-efficiency photocatalytic hydrogen production, *Solar RRL*, 5 (2021) 2000414. https://doi.org/10.1002/solr.202000414.

35. M. Liu, J. Li, R. Bian, X. Wang, Y. Ji, X. Zhang, J. Tian, F. Shi, H. Cui, ZnO@Ti3C2 MXene interfacial Schottky junction for boosting spatial charge separation in photocatalytic degradation, *J. Alloys Compd.*, 905 (2022) 164025. https://doi.org/10.1016/j.jallcom.2022.164025.

36. X. Shi, P. Wang, L. Lan, S. Jia, Z. Wei, Construction of BiOBrxI1−x/MXene Ti3C2Tx composite for improved photocatalytic degradability, *J. Mater. Sci.: Mater. Electron.*, 30 (2019) 19804–19812. DOI: 10.1007/s10854-019-02346-1.

37. W. Zhou, B. Yu, J. Zhu, K. Li, S. Tian, Hierarchical ZnO/MXene (Nb2C and V2C) heterostructure with efficient electron transfer for enhanced photocatalytic activity, *Appl. Surf. Sci.*, 590 (2022) 153095. https://doi.org/10.1016/j.apsusc.2022.153095.

38. H. Zhang, M. Li, C. Zhu, Q. Tang, P. Kang, J. Cao, Preparation of magnetic α-Fe2O3/ZnFe2O4@Ti3C2 MXene with excellent photocatalytic performance, *Ceram. Int.*, 46 (2020) 81–88. https://doi.org/10.1016/j.ceramint.2019.08.236.

39. E. Obotey Ezugbe, S. Rathilal, Membrane technologies in wastewater treatment: A review, *Membranes*, 10 (2020) 89.

40. G. Zeng, Z. He, T. Wan, T. Wang, Z. Yang, Y. Liu, Q. Lin, Y. Wang, A. Sengupta, S. Pu, A self-cleaning photocatalytic composite membrane based on g-C3N4@MXene nanosheets for the removal of dyes and antibiotics from wastewater, *Sep. Purif. Technol.*, 292 (2022) 121037. https://doi.org/10.1016/j.seppur.2022.121037.

41. Q. Lin, Y. Liu, G. Zeng, X. Li, B. Wang, X. Cheng, A. Sengupta, X. Yang, Z. Feng, Bionics inspired modified two-dimensional MXene composite membrane for high-throughput dye separation, *J. Environ. Chem. Eng.*, 9 (2021) 105711. https://doi.org/10.1016/j.jece.2021.105711.

42. G. Zeng, Y. Liu, Q. Lin, S. Pu, S. Zheng, M.B.M.Y. Ang, Y.-H. Chiao, Constructing composite membranes from functionalized metal organic frameworks integrated MXene intended for ultrafast oil/water emulsion separation, *Sep. Purif. Technol.*, 293 (2022) 121052. https://doi.org/10.1016/j.seppur.2022.121052.

43. C. Feng, K. Ou, Z. Zhang, Y. Liu, Y. Huang, Z. Wang, Y. Lv, Y.-E. Miao, Y. Wang, Q. Lan, T. Liu, Dual-layered covalent organic framework/MXene membranes with short paths for fast water treatment, *J. Membr. Sci.*, 658 (2022) 120761. https://doi.org/10.1016/j.memsci.2022.120761.

44. J. Hu, Y. Zhan, G. Zhang, Q. Feng, W. Yang, Y.-H. Chiao, S. Zhang, A. Sun, Durable and super-hydrophilic/underwater super-oleophobic two-dimensional MXene composite lamellar membrane with photocatalytic self-cleaning property for efficient oil/water separation in harsh environments, *J. Membr. Sci.*, 637 (2021) 119627. https://doi.org/10.1016/j.memsci.2021.119627.

45. X. Zhao, Y. Che, Y. Mo, W. Huang, C. Wang, Fabrication of PEI modified GO/MXene composite membrane and its application in removing metal cations from water, *J. Membr. Sci.*, 640 (2021) 119847. https://doi.org/10.1016/j.memsci.2021.119847.

46. X.-Y. Ma, T.-T. Fan, G. Wang, Z.-H. Li, J.-H. Lin, Y.-Z. Long, High performance GO/MXene/PPS composite filtration membrane for dye wastewater treatment under harsh environmental conditions, *Compos. Commun.*, 29 (2022) 101017. https://doi.org/10.1016/j.coco.2021.101017.

47. P.-F. Sun, Z. Yang, X. Song, J.H. Lee, C.Y. Tang, H.-D. Park, Interlayered forward osmosis membranes with Ti3C2Tx MXene and carbon nanotubes for enhanced municipal wastewater concentration, *Environ. Sci. Technol.*, 55 (2021) 13219–13230. DOI: 10.1021/acs.est.1c01968.

48. T.F. Ajibade, H. Tian, K. Hassan Lasisi, Q. Xue, W. Yao, K. Zhang, Multifunctional PAN UF membrane modified with 3D-MXene/O-MWCNT nanostructures for the removal of complex oil and dyes from industrial wastewater, *Sep. Purif. Technol.*, 275 (2021) 119135. https://doi.org/10.1016/j.seppur.2021.119135.

49. R. Han, X. Ma, Y. Xie, D. Teng, S. Zhang, Preparation of a new 2D MXene/PES composite membrane with excellent hydrophilicity and high flux, *RSC Adv.*, 7 (2017) 56204–56210. DOI: 10.1039/C7RA10318B.

50. Q. Feng, Y. Zhan, W. Yang, A. Sun, H. Dong, Y.-H. Chiao, Y. Liu, X. Chen, Y. Chen, Bi-functional super-hydrophilic/underwater super-oleophobic 2D lamellar Ti3C2Tx MXene/poly (arylene ether nitrile) fibrous composite membrane for the fast purification of emulsified oil and photodegradation of hazardous organics, *J. Colloid Interface Sci.*, 612 (2022) 156–170. https://doi.org/10.1016/j.jcis.2021.12.160.

51. X. Cheng, J. Liao, Y. Xue, Q. Lin, Z. Yang, G. Yan, G. Zeng, A. Sengupta, Ultrahigh-flux and self-cleaning composite membrane based on BiOCl-PPy modified MXene nanosheets for contaminants removal from wastewater, *J. Membr. Sci.*, 644 (2022) 120188. https://doi.org/10.1016/j.memsci.2021.120188.

12 Polymer–Carbon Dots Composites for Waste Water Treatment

Madhusmita Bhuyan and Priyanka Sahu
Sambalpur University

Santosh Ganguly
Bharat Pharmaceutical Technology

Dibakar Sahoo
Sambalpur University

Priyatosh Sarkar
Bharat Pharmaceutical Technology

CONTENTS

DOI: 10.1201/9781003328094-12

12.1 INTRODUCTION

Water is our earth's most precious natural resource, and its quality availability is essential for the survival of living creatures on this planet. But due to increased population and rapid industrialisation, water quality is degrading daily due to the addition of water pollutants into the water bodies.[1,2] In today's date, we are heading towards renewable energy resources as we are already at the edge of depleting non-renewable resources. Every human being requires clean and pure water to live a healthy existence, yet excessive and injudicious water use in many parts of the world has made drinking water a scarce resource.[3] The growth of novel materials and processes has been fuelled by the global challenge of water quality and conservation. As a result, there is a huge demand for effective water filtration solutions to combat pollution. The approach must be holistic, and science, technology, and environmental management must coexist. The fundamental wastewater remediation is required with appropriate material with high separation capacity, low cost, porosity, and reusability. In this regard, developing cost-effective and stable materials and methods for adequate fresh water is a challenge for the water industry. So it's essential to design a low-cost adsorbents for effective removal of hazardous pollutants from wastewater, which can be further used for primary purposes of mankind.

Due to rising water demand, traditional wastewater treatment methods and techniques are inefficient in supplying clean drinking water. As a result, nanotechnology allows for developing improved materials for effective water filtration by enhancing their capabilities. For this purpose, carbon-based materials are the most used adsorbents for heavy metals removal from wastewater. Carbon dots (CDs) are quantum dots with particle sizes smaller than 10 nm. They have emerged among the most fluorescent nanomaterials due to their easy preparation, thermal stability, and relatively nontoxic nature. Recently, a broad range of synthesis methods has been developed to produce CDs, typically classified as top-down and bottom-up approaches. Although it has a large surface area, its aggregation limits its application; nevertheless, aggregation can be reduced by turning nanomaterials into nanocomposites. So composites are generally a combination of constituents (two or more) differing in properties, made up of single material with an integrated property set. The unique advantages of using composites are that it combines the properties of two materials into one for specific applications. A feasible pathway is to fabricate CD-based nanocomposites materials by integrating CDs into compatible solid matrices through intermolecular interaction or covalent bonding.[4] Polymer-based nanocomposites (PNCs)[5] have piqued the interest of current researchers, because carbon has a unique electronic structure and may make covalent bonds with a wide range of metals and non-metals. Carbon has numerous unique qualities at the nanoscale compared to other adsorbent materials, including strong mechanical properties, high thermal stability, and high electrical conductivity.[6] As a result, carbon dot polymer composites have become popular in a variety of applications, including wastewater treatment and desalination.[1] Hence the use of solid materials in wastewater treatment is the main focus of researchers. Compared with other types of carbon/polymer nanocomposites materials such as graphene, graphene oxide,[7] and carbon nanotube–polymer composites,[8] the advantages of selecting CD-polymer nanocomposites lie in the much broader choices of CD precursors. This review, which focuses on carbon dot polymer nanocomposites, intends to present a comprehensive view of the water quality issues to the methods for treating wastewater. Additionally, a full presentation of the various CD-polymer composite synthesis methods and their use in wastewater treatment was made.[9]

12.2 WATER QUALITY CHALLENGES

According to a report, progress on sanitation and drinking water (2013) by the World Health Organization and UNICEF mentions that approximately 768 million people do not have access to clean water. As per the report [climate change: impacts and adaption and vulnerability (2014) by the inter-governmental panel on climate change (IPCC)], a warning has been issued that states that 80% of the world is under severe threat of water scarcity. Thus, many states are facing a shortage of water, as only 10% of the water available globally is for domestic use. Considering the condition of India as per NITI Aayog in 2018, total 21 significant cities will hardly have any ground water by 2020. As stated by the central water commission, the reservoirs of India, which almost compromise 2/3 of the total reservoirs, are running below average water levels. The rising scarcity is mainly due to the increase in population and the advent of the era of industrialisation, which ultimately leads to widespread pollution of the water bodies. Water quality is one of our societies' main challenges during the 21st century. Water quality degradation translates directly into social, economic, and environmental problems.

The pollution of freshwater resources is caused by the disposal of large quantities of insufficiently treated or untreated wastewater into rivers, lakes, aquifers, and coastal water. According to UNESCO facts, it shows that:

I. 90% of sewage in developing countries is discharged untreated directly into water bodies.[10]
II. Industry releases an estimated 300–400 megatons of waste into water bodies every year.[10]
III. Nitrates from agriculture are the most common chemical containment in the world's groundwater aquifers.[10]
IV. One in nine people worldwide use drinking water from unimproved and unsafe sources.

As a result, in such a difficult situation, it is crucial to use water wisely and convert waste water to a useable form. Increased contamination in water bodies necessitates the development of materials that are highly efficient, cost-effective, and adaptable to a variety of contaminants.[11] Furthermore, better water treatment technologies such as adsorption, RO purification, UV purification, and others can help.[12] Regarding the amount of pollution produced, the functioning principles of various technologies differ. Due to their unique qualities, carbon dots can be used to clean water, which works wonders when combined with polymers to make nanocomposites.

12.3   WASTEWATER TREATMENT APPROACHES

Water is very crucial for the survival of all human beings.[1] It is used for various things, including drinking, irrigation, washing, cooking, and numerous industrial processes. The World Health Organization's (WHO) minimal quality standards should be strictly adhered to. These water restrictions aren't being followed, though, because of water pollution. As a result, industrial or municipal wastewater causes more water pollution daily (Table 12.1).

This wastewater can be converted to pure and clean water by some methods, and approaches so that the water can be used further for domestic purposes without wasting the water, which will work constructively for humankind.

TABLE 12.1
Illustration of Industrial and Municipal Wastewater Sources

Industrial Wastewater	Municipal Wastewater
Automobile	Households
Refinery	Hospitals
Mining	Commercial
Electronics	Institutions

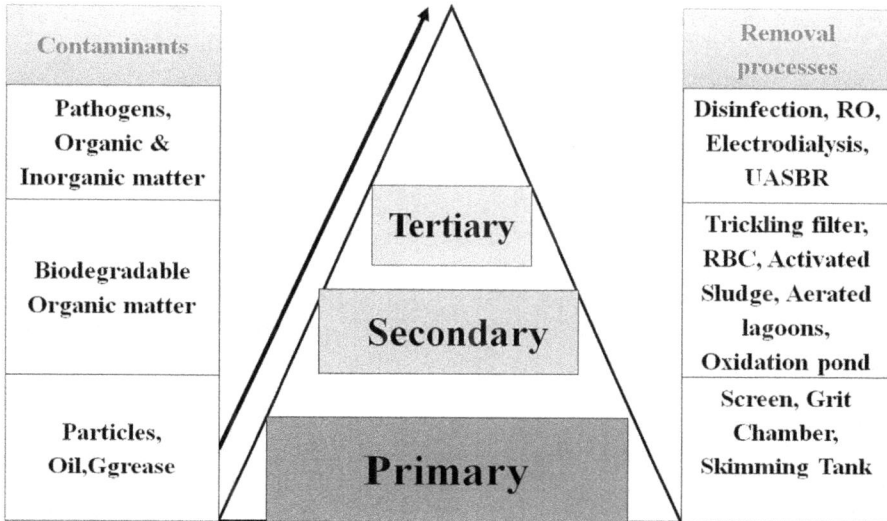

FIGURE 12.1 (a) Flow diagram of primary, secondary, and tertiary treatment process; (b) flow chart for levels of waste water treatment.

12.3.1 CONVENTIONAL APPROACHES

This method includes three stages: primary, secondary and tertiary treatments (Figure 12.1).

12.3.1.1 Primary Treatment

It consists of a preliminary treatment as well as the sedimentation process. Large particles and debris from wastewater, as well as oil and lipids, are removed in this procedure. Sedimentation or chemical precipitation[13] removes organic solids and colloidal particles from the waste water.[14]

12.3.1.2 Secondary Treatment

In this process, suitable microorganisms are fed into the wastewater so that the biodegradable soluble organic compounds are degraded through microorganisms. Here the BOD (biochemical oxygen demand)[15] is checked as a measuring parameter for wastewater;[16] as organic matter is removed, the BOD level decreases.

12.3.1.3 Tertiary Treatment

In tertiary treatment, a remarkable quantity of phosphorous, nitrogen, various organic matter, heavy metals, and pathogenic bacteria are removed. This step is essential for treating wastewater. However, there are many advantages and benefits of treating wastewater using conventional methods. But it also has some disadvantages; for instance, a large area is required for installation, and a huge amount of sludge is generated; also, we need to provide a good climate for proper functioning, corrosive in nature, etc. Thus, gradually, researchers are now approaching some advance methods for the purification of water that must be eco-friendly and low cost. Till now, some of the advanced techniques for water purification are activated carbon filters,[17] ozonation,[18] UV radiation,[19] nanocomposites, and many more.[20]

12.3.2   ADVANCED APPROACHES

12.3.2.1   Activated Carbon Filters

Carbon filters remove the contaminants through adsorption; it soaks up the particles like a sponge.

The excellent absorption capacity of carbon is a result of its enormous surface area, and water's assistance in interacting with different-sized particles accelerates the purifying process. The process is therefore effective at treating wastewater.

12.3.2.2   Ozonation

It is an advanced oxidation process, including forming highly reactive oxygen species (ROS).[21] This ROS is very much reactive and is also short-lived. So they are capable of extracting pathogens as well as the organic and inorganic pollutants present in the water. This process is efficient for the disinfection and degradation of organic and inorganic pollutants.

12.3.2.3   UV Radiation

This is used for disinfection. Here the UV radiation penetrates the cell wall of the membrane and helps to destroy the genetic mutant of those tiny microorganisms and prohibit further growth of the cell.[22]

12.3.2.4   Nanocomposites

Nanoscale composite materials have a lot of surface area, a lot of chemical reactivity, a lot of mechanical strength, and a lot of cost effectiveness;[23] thus, they have a lot of promise to clean water in different ways. Nanocomposites are suitable in removing various microorganisms, organic and inorganic pollutants from wastewater due to their precise binding action (chelation,[24] absorption, ion exchange).[25] These Nanocomposites can be used to purify wastewater, including metal nanocomposite, metal oxide nanocomposite, carbon nanocomposite, polymer nanocomposite, and membrane nanocomposite.[26] Carbon-based nanocomposites are the most widely employed of all nanocomposites, because their large surface area, zeolites, and carbon-based compounds are the most effective adsorbents for removing contaminants.[27] Recently, carbon dots (CDs) have been exploited as effective adsorbents for removing heavy metals from wastewater. Carbon dots[28] are fluorescent and discrete nanoparticles in nature with sp^2 hybridization[29] diameter of less than 10 nm, and it was first obtained in 2004 during the purification of single-walled carbon nanotubes.[30] CD shows the graphitic structure, which is water soluble, and distinguished photoluminescence (PL) properties.[29] Recently, several studies have been done to understand their structural and PL behaviour, in the recent years, biowastes have been turned into the most captivating sources for the synthesis of CDs. Since biowastes offer economic nature and added advantage of green synthesis, they are less harmful to the environment.[31] So CDs have evolved as a promising photocatalyst due to wide light-absorbance capacity,[32] high photoluminescence properties, and electron transfer ability.[33]

12.4 CARBON DOTS PHOTOPHYSICAL PROPERTIES

Different synthetic methods are there to obtain CDs; they exhibit good optical properties and have high absorption, photoluminescence (PL), chemiluminescence (CL), and phosphorescence spectra.[34]

12.4.1 ABSORPTION SPECTROSCOPY

CDs exhibit high optical absorption maxima in the ultraviolet (UV) range (250–350 nm), and the resultant absorption tail is generally the modest in the visible region.[35] Carbon dot absorption peaks are most likely found around 240 nm. Because of the change in the $\pi-\pi^*$ energy level,[36] which can be controlled by surface engineering with heteroatom doping, CD absorption spectra can be regulated. Surface flaws embedded in CDs cause broad spectral characteristics in the absorption spectra.[37]

12.4.2 PHOTOLUMINESCENCE SPECTROSCOPY (PL)

CDs appearing in the spectrum's blue and green regions have the highest PL.[34,38] Variations in the initial precursors, synthetic methodologies, and surface engineering[39] can influence CDs' PL behaviour. CDs with an average particle size of 1.54 nm have PL spectra with the highest peaks are seen at 460, 540, and 620 nm, respectively, with excitation wavelengths of 380, 460, and 540 nm.[40] N-CDs (nitrogen-doped CDs)[41] produced from m-amino benzoic acid show exceptionally powerful PL maxima peaking at 415 nm, according to Wang et al.[42]

12.4.3 PHOSPHORESCENCE SPECTROSCOPY

CDs also have phosphorescence spectra caused by the excited triplet state of carbonyl groups on their spherical surface. Deng et al. reported a room temperature phosphorescence (RTP) of CDs.[43] By departing the sample at 325 nm,[43] they achieved a phosphorescence peak at 500 nm with a 380 ms lifespan. Li et al. used a hydrothermal technique to prepare CDs from citric acid and urea, exhibiting good PL behaviour. They also made CD–CA powder combining CD and cyanuric acid (CA).[44] At standard temperature, the aqueous solution of CD–CA powder emits green phosphorescent light, which can be seen with the naked eye.[45]

12.4.4 CHEMILUMINESCENCE SPECTROSCOPY

Another feature of CDs is chemiluminescence[46] (CL), which is still not fully explored by the researchers. CL of CDs was first reported by Lin et al.[47] in the presence of $KMnO_4$ and cerium (IV) ion. When different oxidiser amalgamates with electrons, they release energy as CL and form holes within CDs.[47] On the other hand Zhao et al. find CL in CD derived from glucose and PEG1500 as a surface-passivating agent.[48]

An increase in CL intensity for the luminol-K3Fe(CN)6 system was observed by Zhang et al., when CDs were present at alkaline pH recently reported. The CD composite material as-fabricated is used to detect the anticancer drug 2-methoxy estradiol.[49]

12.4.5   Quantum Yield (QY)

QY parameter is mainly used for light-emitting systems, leading to use in photo devices in the future. It has been reported that surface modification on CDs can achieve high QY.[50] The lower QY precursors, having electron-withdrawing groups (EWGs) such as carboxylic and epoxy, can be engineered to convert electron-donating groups (EDGs) by doping CDs and producing new CDs having high QY. Initially, CDs made from candle soot, graphite, and citric acid had very low QY maxima of up to 10%. Recently, researchers achieved extremely high QY. Zhuo et al. used citric acid and glutathione as starting materials to synthesise CDs with a high QY of 80% in an aqueous medium.[51]

12.4.6   Spectroscopic Origin of CDs

Although several works have been done to explain CDs' photophysical properties, the origin of PL is still unclear. The PL behaviours of CDs are mainly explained by (i) the bandgap transition and (ii) the surface defect model.

   a. **Bandgap Transition Model.**
      The quantum confinement effect and the size of the CD affect the PL in CDs.[52] Li et al. reported that CDs made with alkali-assisted electrochemical sizes ranging from 1.2 to 3.8 nm displayed blue PL having an emission maximum near 450 nm.[53] The size change can also influence the colour of the emission. The theoretical model shows that the PL spectrum peak positions and intensities vary with particle size.[53]
   b. **Surface Defect Model.**
      Surface defects[54] originating from surface oxidation, functionalities, and heteroatom doping can form inside CDs and affect PL-exhibited CDs.[37] Increased surface defects come from surface oxidation, resulting in more emissive sites inside CDs.[55] Doping of heteroatoms increases the electron density within the cyclic structure of CDs, resulting in a multiphoton PL mechanism with anti-Stoke transitions, further enhancing the photoluminescent feature. Chen et al., Hu et al., and other groups observed major changes in the PL spectrum of CDs, which contributed to surface defects resulting bandgap energies to change.[56] Ding et al. observed that red-shifted PL of CDs from 440 to 625 nm by increasing the oxygen percentage in the CD matrix.[57]

Carbon dots' photophysical parameters include the radioactive decay rate with its average decay time.[58] Dopping different materials and the supply of resources affect its photophysical performance. Some researchers have already synthesised n-type and p-type doped carbon dots by doping phosphorus, boron, and nitrogen.

Researchers found that photophysical property is enhanced when phosphorus is dopped with carbon dot due to extra electron addition from n-type doping[59] to carbon dots, making it suitable for the radioactive relaxation pathways; but when boron is doped with carbon dot containing nitrogen, it forms the nonradiative electron–hole recombination pathways; as a result, QY, radiative rate, and average decay time are reduced. Researchers investigated the ensemble-averaged state/bulk state and single-particle level photophysical properties of CDs, which are passivated with electron-accepting (CD-A) and electron-donating (CD-D) molecules on its surface in 2012, which led to the conclusion that CD-A dominates the overall photophysical properties, developing at least two associated geometries dependent on time, concentration, intramolecular electron, and hydrogen bonding. Based upon the result of instantaneous intensity, bleaching kinetics, and photoblinking,[60] single-particle studies do not reveal an "acceptor dominating" scenario, showing that the direct interaction of these CDs may affect photophysical properties in the bulk state due to the formation of hierarchical structural assemblies. The most noticeable property of a carbon dot is its excitation-dependent photoluminescence. A carbon dot has a multicolour property, which means that the emission of the carbon dot can be tuned depending on the excitation wavelength. Many experiments show that modulation in the surface states of carbon dots provides us a source of excitation-dependent luminescence properties. By modifying the surface of a carbon dot, it can also exhibit single emission wavelengths.[58]

12.5 POLYMER–CD COMPOSITES

Polymers are mainly classified into an organic polymer and inorganic polymer.[61] The main difference between these two polymers is that organic polymers contain carbon atoms in their backbone, but inorganic polymers do not contain carbon atoms in their backbone.[62] Examples of organic polymers are polypropylene (PP), PVC, PVA, PMMA, PVDF, etc. and inorganic polymers are polysilanes, polysiloxanes, polyphosphazenes, etc. Polymers exhibit many characteristics like selective functionality, thermal resistance, binding affinity, and processability,[63] making them excellent material for wastewater treatment and environmental remediation. Fabricating natural polymers such as chitosan, cellulose, plant gums, and synthetic polymer such as PVA, PLA, PVP, etc., with carbon dots improve different properties of C-dot. The polymer acts as a support for the CDs and prevents their aggregation. The performance of polymer nanocomposites for water purification is influenced by surface-active catalytic action, adsorption ability, and physical properties.[64]

A polymer nanocomposite is a composite material comprising a polymer matrix and an inorganic dispersive phase with one nanometric dimension in scale.[65] These are made up of fibres embedded with organic polymer matrix. These fibres are introduced to enhance the properties of the material.

Though the polymer matrices are of different categories, two variants are widely used for the PNCs (polymer nanocomposites) fabrication; they are

a. Thermosetting polymer
b. Thermoplastic polymer

When heat is applied to thermoplastic polymers, they become soft or molten and pliable; on the other hand, when heat is applied to thermosetting, they bring about degradation without going through the fluid phase. Both these polymers can be employed for wastewater treatment. Thermosetting polymer[66] possesses high thermal and structural stability. These properties largely depend on their chemical structure crosslinking density[67] and processing conditions such as temperature and pressure.[68] Some examples of these varieties of polymers are polyimides, phenolics, etc. Similarly, thermoplastic polymer[69] is very useful for the fabrication of PNCs because of their good mechanical properties, durability, and versatility in processing, allowing them to use numerous forms.[70] Some of the examples of thermoplastics are polyethylene,[71] styrene acrylonitrile, polypropylene, etc.[72]

12.5.1   Inorganic Nanomaterials for PNCs

Basically the incorporation of inorganic nanoparticles into polymeric matrices helps to improve some sought of unique properties. These inorganic nanoparticles tend to provide some mechanical and thermal stability to the polymer matrix[73] and other functionalities, depending on structure, shape, crystallinity, and chemical nature of that inorganic nanoparticles. Some of the example are graphene oxide magnetite,[74] silica, titanium dioxide,[75] etc.[76]

12.5.2   Synthesis of Polymer–Carbon Dots Nanocomposites

There are several methods for synthesis of carbon dot polymer composites such as hydrothermal treatment, sol–gel procedure, solvothermal treatment, microwave treatment, thermal decomposition, etc.[77]

12.5.2.1   Hydrothermal Treatment

The hydrothermal method is the most commonly used method for creating carbon dots and CD/polymer composites. Issa and Abidin used this method to create a composite film of polyvinyl alcohol (PVA)/nitrogen-doped CD (PVA/CD).[78] The CDs were produced by branching polyethyleneimin (PEI) and carboxy methyl cellulose (CMC) obtained from the waste of discarded oil palm fruit bunches. CMC was employed as an essential and affordable carbon precursor, while PEI was doped with nitrogen to make nitrogen-doped CDs (N-CDs). CMC and PEI were first dissolved in deionised water to produce CDs. Then the mixture was cooked for 2 hours at 260°C in a Teflon-lined autoclave. The exact carbonaceous product was obtained by centrifuging the supernatant. To eliminate the undesired particles, precipitation and vacuum filtration were done. The impure ions were then removed by using a dialysis membrane. Then, CDs were encapsulated with PVA polymer by stirring N-CDs continuously into the PVA solution. The hydrogel was then placed on a glass substrate and cured at 80°C for 1 hour. At last, the PVA/CD film was peeled away from the glass substrate surface. Bulk thermoplastic polyurethane elastomer (TPU)/CD composites were also developed by Bai et al. They use citric acid and 2-aminothiophenol (2AT) as the carbon precursor and nitrogen dopant, respectively,[79] in the hydrothermal process used to make the CDs. Citric acid and 2AT were distributed in ultrapure water and heated for 3 hours at 170°C. The supernatant was dialysed after centrifuging the solution. A light-yellow powder was obtained after freeze drying and stored below 5°C. A well-known polymer is obtained, which is polyurethane. To create TPU/CD composites via in-situ polymerisation, CDs were disseminated in N, N-dimethyl formamide (DMF) and kept in a flask with nitrogen environment. Poly(tetramethyleneglycol (PTMG) was added while stirring to obtain a homogenous mixture. Methylene-bis(4-cyclohexylisocyanate)(HMDI) was then added to the reactor. The dibutyltindilaurate (DBTDL) reaction was carried out at 80°C for 2 hours, until the production of isocyanate groups (NCO). Finally, 1,4-butanediol (BDO) was added to the mixture. While the viscosity was rapidly raised, the agitator was accelerated. A heated mould was set up at 140°C and pressed for 10 minutes at 10 MPa to handle the resulting mixture. The finished product was a nearly 1 mm thick sheet. TPU/CD composites were created after curing for 24 hours at 100°C. Zheng et al. synthesised CD/polymer composite via hydrothermal treatment using hexamethylenetetramine (HMT) as carbon precursor and polycarbonate as polymer precursors.[80] The CD/polymer composite was prepared by single-step method, placing HMT solution and polycarbonate into a Teflon-line autoclave and heating it at 150°C for 12 hour. After that, when the Teflon-line autoclave was cooled down naturally, the solution was dialysed in distilled water for a week.[81] G. Liu developed bisphenol A (BPA) composite using CDs and molecularly imprinted polymer (MIP).[82] For CD synthesis, 0.5 g citric acid was first mixed with N-(β-aminoethyl)-γ-aminopropyl methyl dimethoxysilane (AEAPMS), and put into an autoclave and consecutively degassed with nitrogen. After 2 hours at 240°C inside an autoclave, the solution was taken out, filtered, and washed with petroleum ether three times. The obtained products were stored in cooled places after being dispersed in ethanol for further use. And the MIP-coated CDs composite was fabricated via a sol–gel polymerisation; for that a suitable amount of BPA was mixed with 40 mL CDs anhydrous ethanol solution by using ultrasound. Then, 350 μL (3-aminopropyl) triethosysilane (APTES) (functional monomer), 1.5 mL tetraethoxysilane (TEOS) (cross-linker), 200 μL ammonia solution, and 800 μL ultrapure water were added with the solution with constant stirring. Finally, the mixture was kept in dark place for 1 day. The product was centrifuged and washed properly to get rid of reagent. The non-imprinted NIP-coated CDs composite can also be prepared using the same method without usingany template molecules. Then the materials were dried at 60°C overnight in a vacuum. The obtained MIP-coated CDs composite was kept in the refrigerator as a water solution. K. Jlassi synthesised chitosan (CH)–CDs hybrid hydrogel nanocomposite film by hydrothermal mixing 4 g of petroleum in 90 mL of H_2SO_4 and 30 mL of HNO_3 and keeping under stirring at 120°C for 12 hours.[1] Then the mixture was diluted ten times with H_2O and neutralised with ammonia. Then the solution shifted to a hydrothermal Teflon-lined autoclave was heated at 180°C for 12 hours. The resultant supernatant was filtered and dialysed with 3500 Da MWCO (molecular-weight cut-off) and kept for further analysis. Then chitosan-CDs membrane (CH–CDs membrane) was prepared by mixing CDs (3 wt%) with chitosan (10 wt%) in 0.1 M acetic acid, which was further stirred at 30°C. Then, the resulting hydrogel was placed on a glass substrate and dried at 80°C for 24 hours to make it film, and it was neutralised by using NaOH (3 M) and washed with H_2O water several times before being dried and kept for further utilisation.[1]

12.5.2.2 Solvothermal Treatment

Several reports have been published on using the solvothermal method to synthesise CD/polymer composites. Wang et al., created solid white-light-emitting phosphors (WCDs@PS) by self-assembling blue emissive CDs (B-CD) and orange emissive CDs (O-CDs) with polystyrene (PS).[83] PS displaces solid matrices and block certain specific agents to prevent intermolecular fluorescence resonance energy transmission and multicoloured CD aggregation-induced quenching. The blue and orange emissive CDs were made using the solvothermal process. In a nutshell, urea and citric acid were disseminated in N, N-dimethyl formamide (DMF). The mixed solution was then cooked at 180°C for 6 hours. After being cooled down, the B-CDs and O-CDs were then separated using silica column chromatography; the B-CDs and O-CDs were then freeze-dried to get powdered CDs. After that, the CDs were stirred for 4 hours to prepare a mixed phosphors (PS) solution. Then, the remaining CDs were removed from the mixture by centrifuging. And the precipitates were separated and dried using the freeze-drying method. The white emissive CDs@PSpowderand silica gel was mixed and placed on a GaN-LED (gallium nitride light-emitting diode) chip to generate white-light-emitting diodes (WLEDs). Then the light-emitting diode (LED) pedestal's bottom was bonded to the chip. The item was then dried for 1 hour at 150°C. By solvothermal treatment followed by electrospinning, Safaei and coworkers generated CD/PAN (polyacrylonitrile) composite nanofibers.[84] CDs made from chitosan were made using the solvothermal process. Chitosan was commonly dissolved in 2% acetic acid and cooked for 16 hours at 180°C in a Teflon autoclave. It was centrifuged after chilling to remove the larger particles from the dark-brown result. Then the CD solution went through microporous filters to get pure CD solution. And The CDs were spread out using DMF. At room temperature, the mixture was stirred for 2 hours at 25°C; the mixture was sonicated. The dispersion was given a 2-hour stir after adding PAN powder. Afterward, the sample was injected into a syringe needle attached to a high-voltage generator. In the electrospinning process, the needle was used as an electrode. A syringe pump fed the dispersions at a rate of 0.11 mL/h for 3 hours.

Bhattacharya et al. synthesised self-healing fluorescence gels by reacting CDs with polyethyleneimin (PEI) by solvothermal treatment, and the CDs are prepared from aldehyde precursors.[85] The self-healing gel was prepared by Schiff base reaction between the amine residues in the PEI network and the aldehyde units on the CDs surface. The gels' viscoelasticity depends on the CD and polymer ratios. To produce three different colour CDs, the other composites should be synthesised under different synthesis conditions; these are glutaraldehyde/ethanol, benzaldehyde/ethanol, and cyclooctadiene-aldehyde. That polymer/chloroform was heated to varying temperatures for extra time period by solvothermal method. In particular, dialdehyde glutaraldehyde produced green-FL CDs (G-CDs), benzaldehyde produced blue-FL CDs (B-CDs), and cyclooctadiene-aldehyde produced yellow-FL CDs (Y-CDs), respectively.[81]

Hess et al. fabricated CDs/PVA (polyvinyl alcohol) aqueous solutions using solvothermal treatment.[86] For solvothermal synthesis, a microwave was used; they stirred polyvinyl alcohol, citric acid, and polyethylene imine together with water to prepare a precursor solution, which was further followed by microwave synthesis at 135°C at various periods.

12.5.2.3 Microwave Treatment

In some publications, CD/polymer composites are also prepared with the help of the microwave treatment method. Li et al., for example, create molecularly imprinted polymers (HMIP) with a single hole with CDs (HMIP@CDs). They first placed ammonium citrate and cysteine in a microwave for 2.5 minutes at 750 W to create sulphur CDs (S-CDs) with carboxyl groups. A homogenous crosslinking polymer shell was generated at the polystyrene particle's surface by separating microphase of the polymer layer when carboxyl functional polystyrene particle was produced and swelled. HMIP@CDs were created by dissolving the polystyrene particle in dichloromethane. The precursor (3-aminopropyl) triethosysilane (APTES) and tetraethoxysilane (TEOS) were used to incorporate AEAPMS-CDs into HMIP@CDs via the dehydration synthesis reactions, and the polymerisation process lasted 4 hours.[82] Tian et al. also created CDs@PVA via post-synthetic thermal annealing at 200°C.[87] Microwave treatments were used to develop CDs-1 from citric acid and NH_3 solution. CDs-2 were made by heating CDs-1 at 200°C under N_2 protection. As a result of annealing CDs at 200°C, dehydration and carbonisation occurred. In a nutshell, the mixture of citric acid and NH_3 solution was heated in a 650 W microwave for 5 minutes, yielding dark-brown liquid. The centrifugation was done to remove the agglomerated particles.[77] To obtain CDs-1, the supernatant was lyophilised. The CDs-1 was then heated in nitrogen for 0.5 hours at 200°C. The product was centrifuged after being diluted in ethanol. To obtain CDs-2, the supernatant was freeze-dried. CDs-1 and CDs-2 have dissolved in polyvinyl alcohol (PVA) solution before being it for spin coating onto quartz plates and annealed at different temperatures for 0.5 hour. Then citric acid was dissolved in a mixture of water and glycerol. The solution was then treated with diethylenetriamine (DETA). The mixture was sonicated for 5 minutes before being heated in a 750 W microwave for 5 minutes. During this process, the colour of the solution changed from yellow to black. After centrifugation, the solution was preserved in water solution. The CDs were then created by freeze drying them for 48 hours. Then, to create composite, the polyacrylonitrile (PAN) support was altered into a disc and dispersed in NaOH solution for 1 hour at 50°C. The remaining NaOH was removed by neutralising it with water. Water was used to keep the hydrolysed PAN support. During stirring a certain number of CDs were dissolved in water, and a polyethyleneimine (PEI) solution was also stirred. The CDs and PEI solutions were mixed by stirring to create a PEI-CD solution. After that, the casting of the mixture was done onto PAN support layer and left it for 10 minutes. After that, n-hexane and trimesoyl chloride (TMC) solution were cast onto the PEI-CD surface and allowed to cross-link for 2 minutes; then the membrane was kept in air for 20 minutes to dry and followed by heating at 60°C for 2 hours to remove the remaining n-hexane and complete the crosslinking reaction. The yield PAN/PEI-CD composite membrane was then freeze-dried to obtain powdered CDs. Wang et al. synthesised CDs/CS (chitosan) composites by microwave treatment, where CS is used as both polymer matrix and CDs precursor.[88] CDs were prepared by sonicating 0.5 g of chitosan and water, and then the solution was dried at 300°C in the muffle furnace for 2 hours to get the resultant CD powder. Then the composite was prepared by dissolving 0.34 g of chitosan with 98 mL of distilled water and 2 mL of acetic acid. Then, 5 wt% carbon dots solution was added to the solution and stirred at 80°C for 30 minutes. Then the residue was washed and dried at 60°C to get CDs/CS composite.[89]

Yi Liu prepared diethylene glycol (DEG)-CDs by one-step microwave-assisted method by using sucrose as a carbon source and diethylene glycol (DEG)[90] as the reaction medium to prepare this composite; they took 3 mL DEG, 1 mL sucrose solution, and 200 mL of concentrated H_2SO_4 in a glass tube. The mixture was sonicated and then heated in a microwave oven for some time. Then it is centrifuged to remove the unwanted particles. Then to get pure DEG-CDs, the solution was then dialysed against double-distilled water.[91]

12.5.2.4 Pyrolysis/Thermal Decomposition

The pyrolysis method was used to create CD-modified polyethersulfone (PES) membranes. Furthermore, V. Rimal et al. synthesised CDs by using this method from oleic acid, which they then incorporated into a cellulose acetate polymer matrix to create a composite.[92] In a nutshell, NaOH was mixed with double-distilled water and stirred together until it became a clear solution. After that, oleic acid was taken in another beaker and kept at 230°C–260°C. The obtained solution was then filtered, followed by sonication. The CD solution was then evaporated to separate anhydrous CDs from the solution. The dried solid CDs were dispersed in ethanol. Then to synthesise CD/polymer composite, they fused cellulose acetate with dimethylformamide by stirring magnetically. The CD solution was then dissolved in DMF by stirring and sonicating. And the mixture was then transferred to a Petri dish and dried at normal temperature. Finally, the composite was produced by heating the sample at 60°C for 2.5 hours.

Koulivand et al. also created CD-modified PES using pyrolysis-assisted method.[93] By heating citric acid at 160°C for 55 minutes while stirring and neutralising it with NaOH solution, CDs were made using the pyrolysis technique.[94] The solution was then freeze-dried to produce CDs' powder. Subsequently, the CD was modified by PES membranes using the nonsolvent-induced phase inversion method. For that, first, the CDs were added into dimethylacetamide[95] (DMAc) as a polymer solvent. Then, polyvinyl pyrrolidone (PVP) and PES were added into the solution under stirring. The solutions were heated at 50°C for 2 hour in an hot air oven. After cooling, thin films were made by drop cast onto the clean glass. Then the films were immersed in a water bath to produce thicker membranes. Finally, they were stored in water overnight and dried at 25°C.

12.5.2.5 Ultrasonic Treatment

This method is also used to create CD/polymer composites. Kumar et al., for example, created polyvinyl alcohol (PVA)-(N@CD) nitrogen-containing carbon dot nanocomposites by embedding N@CDs in PVA polymer.[88] He made PVAN@CD composites using ultrasonic treatment and drop-casting method. PVA-N@CDs were synthesised by mixing N@CDs aqueous solution with PVA solution. In this mixing process, the pH of the N@CDs solution and the PVA solution should be between 6 and 8 to get more stable composites. Then the thin film was created using the drop-casting method, and the solvent was evaporated naturally at normal temperature.[96] Furthermore, Li et al. fabricated a pyrophosphate ion (PPi)-responsive alginate hydrogel using Cu^{2+} as a crosslinking agent and CDs as a visual indicator.[97] CDs were incorporated with Cu/Alg gel to yield the composite of CDs@Cu/Alg. The CDs were prepared using an ultrasonic method. In brief, NaOH was added to etylpyridinium chloride (CPC) aqueous solution. Then, the mixture was ultrasonicated for 30 minutes at 25°C. During this process, the pH of the solution should be neutral; after that, dialysis was done to remove impurities. Finally, the CDs were obtained by drying the purified product at 65°C. The Alg/Cu gel was prepared by mixing an aqueous alginate solution with $CuCl_2$. Then, washing of hydrogel was conducted to remove the unreacted reagent. For the preparation of CDs@Alg/Cu, CDs were first dissolved in an aqueous alginate solution. Then, alginate gelation was triggered by mixing the $CuCl_2$ aqueous solution. After that, centrifugation and washing were done with water.[81] H. Huang prepared fluorescent carbon nanoparticles (FCNs), by ultrasonicating cigarette ash (1 g), PEG-SH (700 mg), and N, N-dimethyl formamide (DMF) mixer e solution for 30 minutes. Then to get the final product the supernatant was centrifuged, and the solution was dialysed against distilled water and alcohol for 3 days to remove the unwanted particles.[98]

12.5.3 Physics and Chemistry of Polymer–CD Composite

Carbon dots may be nanocrystallites or amorphous nanoparticles with sp^2 bonding.[29] Generally, the carbon dot shows fringe spacing around 0.34 nm, which corresponds to (002) interlayer spacing of graphite. CDs' surface contains different types of functional groups, including amine (–NH_2), hydroxyl (–OH), and carboxylic (–COOH) groups.[84] Due to large sp^2 π-conjugated structure, it shows different characteristics like good photo stability, high surface area, etc., with an increase in the size of carbon dot, the highest occupied molecular orbital (HOMO) and the lowest unoccupied molecular orbital (LUMO) shift to higher and lower energy by reducing HOMO-LUMO gap.[4] When polymer composites with carbon dots (PCDs) through condensation and crosslinking process, the quantum yield of the composite increases. Typically, PCDs are asymmetric particles with many different structures with a low degree of carbonisation. PCDs contain both the form of fluorescent polymers and CDs. It contains more organic polymer chains and less inorganic carbon core.[99] PCDs exhibit mainly two different properties; the first is due to low carbonisation, low molecular weight chain, and many functional groups. The second is shows polydispersity on the PCDs structure due to different types of reaction condition PCDs.

12.5.4 Application of Polymer–CD Composites for Wastewater Treatment

The principle of PCN-based waste water treatment was based on adsorption, photocatalysis, membrane filtration and disinfection (Figure 12.2). Due to the enormous surface area of nanoparticles in polymer composites, these materials exhibit highly modifiable adsorption behaviour.[100] Nanocomposites' optimised adsorption behaviour makes them perfect for various technical applications such as chemical sensors, water purification, drug delivery, and fuel cell technology. PNCs have been widely used for the adsorptive removal of toxic heavy metal ions, organic and inorganic dyes, and microorganisms from wastewater.[101] Singh et al. examined various types

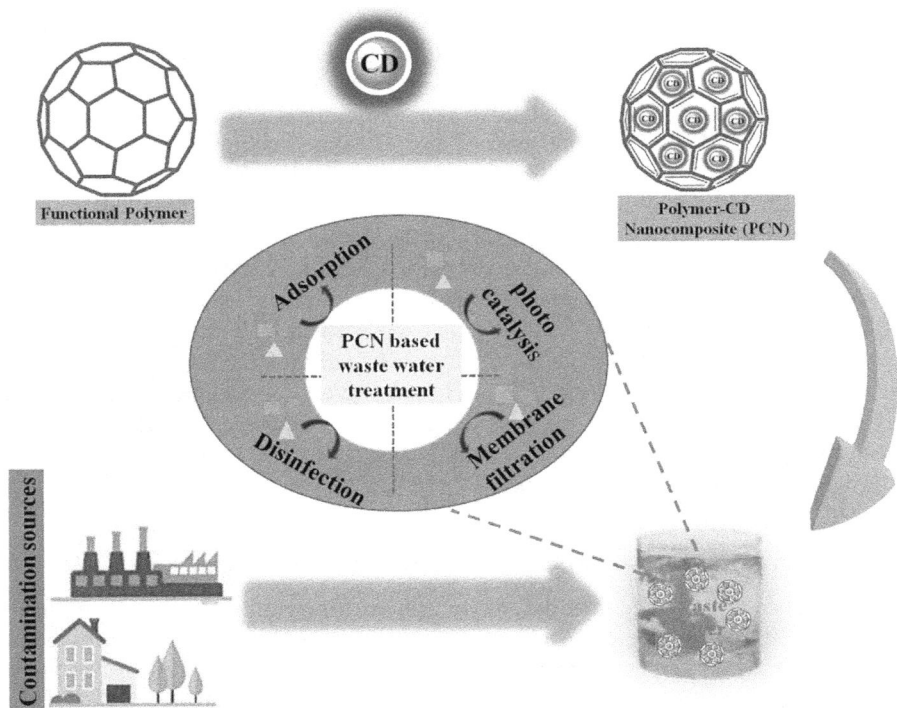

FIGURE 12.2 Principle of PCN-based waste water treatment via adsorption, photo catalysis, membrane filtration, and disinfection mechanism.

of water purification methods.[102] These methods are primarily classified according to their separation techniques, such as chemical degradation,[103] biological treatment, and physical adsorption. All of these methods use a combination of processes to improve water quality. According to the findings, numerous techniques, such as adsorption, biological treatments, can improve dye's natural degradation and less sludge formation.[104,105] Some of the polymer- and CD-based nanocomposite is enlisted in Table 12.2.

12.5.4.1 Removal of Heavy Metal Ions

Nowadays, ground water is mainly contaminated by heavy metal ions. These metal ions are deposited in natural waters through the discharge of industrial wastewater.[106] And these have a direct and indirect impact on the ecosystem[107] and the human body via the drinking of water, skin contact with this impure water, and the food chain. Many heavy metals like Cr, Pb, Hg, Cu, As, etc., are harmful to our body. One of them is Cr^{3+}, which is very much toxic in nature. Detection methods for Cr^{3+} in water include photo detection, electrochemical radiation[108] and spectral detection. CQDs fluorescent probe detection[109] is a type of optical detection with low biological toxicity, environmental friendliness, and higher precision. As a result, CQDs have gained

TABLE 12.2
CD-Polymer-Based Nanocomposite and Their Applications

Synthesis Process	Carbon Dot	Polymer	Application	Reference
Hydrothermal	Nitrogen-CD	Polyvinyl alcohol (PVA)	Removal of cadmium ions, heavy metals	78
Hydrothermal	Carbon dots (CDs)	Molecularly imprinted polymer (MIP)	Detection of BPA in river water samples	82
Hydrothermal	CD	Chitosan (CH)	Dyes in waste water	151
Solvothermal	CD	Polyvinyl alcohol	Removal of cadmium ions, heavy metals	78
Utlrasonic treatment	CDs@Cu/Alg	Pyrophosphate ion (PPi)	Detection of alkaline phospate	97
Hydrothermal	CD	Nanofibrillated cellulose	Detection of heavy metal ions (Cr3+)	110
Microwave	CD	Chitosan	Heavy metal absorption	114
In-situ carbonisation	CD	Thenoyltrifluoroacetole (TTA)	Radioactive separation	152

a lot of attention and are marked as a new type of heavy metal ion detector. Studies have shown that CDs-based polymer composites have a higher adsorption capacity of heavy metals from an aqueous solution. Zihui Song reported a polymer aerogel,[110] a porous material with extremely low density, low thermal conductivity, and large surface area when compositing with CQDs. They discovered that CQDs/aerogel composite could be used productively for adsorption and detecting Cr^{3+} from contaminated water through different experiments and analyses. To improve its adsorption capacity, Khare et al. prepared a nanocompositefilm[111] by using chitosan, carbon nanofibers,[112] iron (Fe)-oxide nanoparticles (NPs), and polyvinyl alcohol[113] with the combination (80 mg per g of chitosan/Fe-CNF composite). These materials have a high metal absorption capacity and are being investigated for systematic removal of Cr (VI) from waste water.[114] K. Jlassi investigated the selectivity towards removal of heavy metals in their work, and he prepared CH–CDs composite by hydrothermal treatment and tested it in the presence of Zn^{2+}, Cd^{2+} and Pb^{2+} at pH around 8.0. They found that the removal efficiency of CH–CDs towards Cd^{2+} is highest, which is 93%, and the removal efficiency of CH–CDs towards Zn^{2+} and Pb^{2+} were approximately 55% and 40%, respectively.[1]

12.5.4.2 Removal of Dyes

The tuneable surface and catalytic properties of polymer nanocomposites[115] have enormous potential to remove dyes from wastewater. Dyes remain in water for a long time, prohibiting the photosynthetic process and not allowing sunlight to pass through the water. It also dissolves oxygen from the water and reduces the stream's recreation value, resulting in the retardation of the aquatic biota expansion and creating water-borne diseases. Most dyes are harmful to the environment and aquatic life, and they cause allergies, dermatitis,[116] skin irritation, and mutations in humans. There are many other methods like adsorption,[117] filtration, coagulation,[118] biochemical degradation, and photo catalytic method for removing dyes from wastewater. Many researchers use polymer nanocomposites for efficient dye removal in this regard. PNC exhibits various properties, including high adsorption capacity, good photocatalytic behaviour, and magnetic behaviour for dye removal. A.G. El-Shamy reported a nanocomposite film made up of zinc peroxide (ZnO_2)[119] and carbon dots (CDs) embedded with polyvinyl alcohol, i.e. PVA/$CZnO_2$, which is very much efficient for pollutant dye removal from wastewater. From BET analysis, it is obtained that PVA/$CZnO_2$ is a suitable catalyst. PVA/$CZnO_2$ composite displays ultra-high adsorption capacity against methylene blue (MB),[120] which is 98% at 60 minutes.[121] S. Nayak repoted PVP-CD composite, which can remove cationic dyes like Malachite Green[122] and Crystal Violet and anionic dye like Eosin Y[123] from wastewater, and the dye adsorption capacity of PVP-CD hydrogel was the highest compared to only carboxylated-PVP[124] or PVP hydrogel mixed with a free CD.[114,125]

12.5.4.2.1 Brewing Waste for Wastewater Treatment

Every year, the brewery industry produces huge amounts of derivates,[126] constituting a valuable source for bio-based compounds. Two major beer wastes can be converted into carbon dots using a simple, scalable, and environment-friendly hydrothermal treatment (CDs). S. Cailotto synthesised and characterised CDs chemically, morphologically, and optically,[127] resulting in photoluminescence emission at 420 nm, whose lifetimes range from 5.5–7.5 ns. In order to create a reusable catalytic system for wastewater treatment, he also prepared a composite of CDs with a polyvinyl alcohol matrix[128] and tested examined it for dye removal ability, which leads to the conclusion that methylene blue can be efficiently adsorbed from water solutions into the composite hydrogel and then fully degraded by UV irradiation. A. T. N. Keumaleu recycled microbrewery wastes to produce carbon dots (CDs).[129] Microbrewery wastes are mainly composed of an organic component containing phosphorous and nitrogen. They synthesised CDs by using the microwave method. They found that CDs were spherical with an average diameter of 5.3 nm, and the N-doped CDs, containing many functional groups (hydroxyl, ethers, esters, carboxyl and amino groups), exhibited good photoluminescence with a quantum yield of 14%. Finally, the interaction between carbon dots and metal ions was investigated towards developing CDs as a sensing technology for water treatment, food quality, and safety detection. And they found that the synthesised CDs are sensitive towards Cu^{2+}, Fe^{3+}, Zn^{2+} and Al^{2+} ion sensing.

12.5.4.3 Removal of Radioactive Ions

Radioactive substances are widely incorporated in many industrial sectors, including nuclear power plants, biomedical engineering, etc. With the rapid increase of nuclear technology, more radioactive wastewater has been produced through various channels. These radioactive ions have a high heat-releasing capacity and an extensively long half-life, which has a severe and long-lasting environmental impact. They may enter the food chain and cause diseases such as anaemia, bone cancer, leukaemia, metabolic disorders, and even can cause death. As a result, it is essential to develop efficient materials and technology for detecting and removing radioactive ions from the water. However, the most challenging part of radioactive wastewater decontamination is its effective stabilisation and the solidification of soluble radioactive nuclides present in wastewater, which is critical for final disposal. As a result, researchers have developed various decontamination technologies (such as membrane separation, adsorption, and biosorption) and cutting-edge technology (such as chemical precipitation, evaporation, adsorption, ion exchange, membrane separation, and so on) for radioactive wastewater treatment. Carbon dots and their composite materials have been extensively used for detecting radioactive ions due to their easy availability of raw materials, simple synthesis and its various functional process, unique optical properties, and abundance of functional groups.[130] Uranium is the most toxic radioactive element. So it is essential to detect uranium from wastewater. In 2015, Z. Wang prepared novel CDs by microplasma method.[131] These CDs are quenched by uranyl ions in water; this property makes CD a good potential sensor in uranium detection. Z. Zhang reported a 3,4-hydroxy pyridinone-functionalised CDs (HOPO-CDs) to detect uranium in water.[132] He found that HOPO-CDs could detect uranium in 30 seconds with excellent sensitivity and high selectivity. Wang fabricated the amphiphilic CDs (A-CDs) with hydrophilic and hydrophobic properties.[133] In the presence of iodine, the A-CDs were quenched and showed a detection limit of 3.5 nmol/L in water. S. Dolai uses thenoyltrifluoroacetone (TTA) embedded with CDs into silica aerogel to prepare TTA-CDs composite,[134] quenched dramatically after adding Sm^{3+}, and Eu^{3+} ions. X. Guo composite zeolite imidazolate framework (ZIF-8) and CDs prepare the magnetic composite adsorbents that can easily absorb uranium from waste water.[135]

12.5.4.4   Waste Chimney Oil Modification

Nowadays, from households and restaurants, much wastage has been produced; kitchen chimney oil is one of them, which is very much toxic. Still, researchers are cycling up the waste kitchen chimney oil to prepare fluorescent multifunctional carbon quantum dots by simple ultrasonic method.[136] Then they characterised and tested it against intercellular Fe^{3+} ion as a sensing probe in the waste water.[136] Thus, they found that in the practical field, the C-dots can sense Fe^{3+} ion in a wide range of concentrations (1 nM to 600 μM) with a detection limit of 0.18 nM, which can be called as 'tracer metal' 'chemosensor'. Then researchers doped carbon dot with PVA polymer to increase its sensing efficiency.[137]

12.5.4.5   Bacterial Elimination from Wastewater

The antibacterial activity was studied against pure cultures of gram-positive bacteria *Staphylococcus aureus* (*S. aureus*) and gram-negative bacteria *Escherichia coli* (*E. coli*) under sunlight exposure in sunlight the presence of PVP-CD hydrogel and free CD. This investigation observed that the bacterial killing capacity increases from 3.7 to 4.7 log by dopping. However, the photo-antibacterial activity of free CD was better than that of PVP-CD because of the large surface contact of CD with bacterial cells than the swollen PVP-CD hydrogel. Gram-positive bacteria, such as *Staphylococcus epidermidis*, are killed by sunlight in less time than gram-negative bacteria, such as *Escherichia coli*. In contrast, Bacillus subtilis endospores take more time.[138]

Experiments show that when a certain amount of dye in water solution is photo degraded in sunlight using PVP-CD composite; the percolating water is effectively bacteria-free for further applications. Because of the high-level contact of the CD with bacterial cells, the free CD in solution demonstrated higher antibacterial activity. However, when CD is present in the three-dimensional network with PVP, such interconnection is impossible, leading to a decrease in antibacterial efficacy.[125]

12.5.4.6 Low-Dimension Carbon–Polymer Nanocomposite Membrane for Wastewater Treatment

Carbon–polymer nanocomposite membranes are created by mixing different nano-carbon (filler)[139] with polymer matrix. These membranes are used in a wide range of applications; the 0D carbon nanoforms incorporated into polymer matrices have demonstrated high capability and potential to eliminate a wide range of water pollutants,[140] such as pathogens, heavy metal ions, and recalcitrant organic compounds from wastewater. Thus the 0D carbon–polymer nanocomposites can replace the traditional ones for wastewater remediation. Brunet et al. prepared hydrophilic functionalised C60 species[141] to kill pathogenic microorganisms[142] and heavy metal ion detection from wastewater. Conventional materials have low metal adsorption capacity and removal efficiency for wastewater treatment. So, the use of C60 resulted in the development of a porous structure with an increase in the hydrophobicity.[143] For example, using 0.001%–0.004% C60 in activated carbon increases the absorption capacity of heavy metals from wastewater. Following a Langmuir model, Alekseeva et al. reported a C60-based nanocomposite-PS film with improved effectivity for removing Cu^{2+} ions.[144] The fabrication of C60-based polymer films increases their hydrophobicity, improving their adsorption capacity and making recycling cling easy. The asymmetric ultrafiltration (UF) membranes were created using a phase inversion technique on poly(phenyleneisophtalamide)-C60 composite membranes.[145] Compared to pristine polymer membranes, C60 incorporates hydrophobic polymers like C60-PPO(poly(phenyleneisophtalamide) nanocomposites 95% efficient. Plisko and colleagues created novel polyamide-C60(OH)22–24 thin-film nanocomposite (TFN) hollow fibre membranes by incorporating polysulfone substrate in C60(OH)22–24.[146] They observed that the TFN membrane with 0.5 wt% C60(OH)22–24 had the best antifouling performance for organic matter removal. Shen et al. created a new TFN membrane using interfacial polymerisation (IP)[147] to load fullerenol. The membrane exhibited high antifouling ability, stability, and high efficiency in Mg^{2+}/Li^+ separation with a high separating factor containing 0.01% (w/v) fullerenol. These membranes were developed to recover Li+ from seawater. To remove contaminants from wastewater, Wang et al. prepared a composite of mesoporous organosilica incorporated with CDs, which are used as an adsorbent to remove toxic organic pollutants (2, 4-dichlorophenol) and inorganic metal ions [mercury (II), copper (II), and lead (II)] from wastewater.[148] CQDs show better water treatment performance due to their hydrophilicity, desirable size, tuneable surface functional properties, and favourable polymer affinity.[148] To prepare composite membranes, a 5 nm CQD was embedded with a polyethyleneimine matrix[149] and dip-coated on a polyacrylonitrile. The prepared composites were used for organic solvent separation, solute rejection, solvent resistance, and solvent flux. The membranes prepared with highly carbonated CQD are more efficient then low carbonated CQD, which act as a non-polar solvent and are accelerated through their hydrophobic domains, restricting polar solvent permeation.[150]

12.6 CONCLUSION AND FUTURE PERSPECTIVE

Worldwide population expansion and climate change are closely related to our global development and water situation. The need for fresh water is rapidly increasing. Of late, nanomaterials, especially carbon dots, are hugely exploited for water cleaning. Nanomaterials enable to create advanced materials such as membranes, nanocatalysts, functionalised surfaces, etc. for effective water treatment. However, nanocomposites, specifically polymer–carbon dots nanocomposites, are now more effective than nanomaterials due to aggregation and instability. Due to their high potentiality[95] nanocomposites, it attracts increasing investment from governments and businesses worldwide in various research fields. The current write-up indicates that versatile conjugated polymer-based composites can successfully remove various pollutants from aqueous solutions. Although there are several reports of recent polymer–CD composites, we cannot ignore their disadvantages. Different composites have different capacities to absorb various organic and inorganic pollutants. And also, the current composites can remove a limited number of pollutants, so it is urgently required to manufacture a universal composite capable of absorbing a large number of pollutants, as water contains many pollutants. The recent composites should be replaced by new ones that are more efficient, cost-effective, impede practical applications, and are environmentally friendly. Most of the nanoparticles have a certain level of toxicity. So composites made of nanoparticles may emit into the environment and cause concern to the environment over a longer time. Although there are several assessment guidelines for nanomaterials, they are grossly inadequate. As a result, a thorough assessment of nanomaterial toxicity is necessary before their real field applications. However, to make polymer nanocomposites, especially CD/polymer composites, as an efficient and cost-effective material for water purifications, much more work needs to be done, in particular, the selection of nanomaterials and polymers, the polymer nanocomposites' interaction, pollutants' interactions, and the optimisation of water purification conditions. The ideas mentioned above have the potential to pave the road to a more sustainable future. To make a bridge between laboratory research and industrial applications, more extensive technical research is required.

REFERENCES

1. K. Jlassi, K. Eid, M. H. Sliem, A. M. Abdullah, M. M. Chehimi and I. Krupa, *Environ. Sci. Eur.*, 2020, **32**, 12.
2. R. Bhateria and D. Jain, *Sustain. Water Resour. Manag.*, 2016, **2**, 161–173.
3. A. S. Subala, K. V Anand and M. S. A. Sibi, *International Journal of Innovative Science and Research Technology*, 2020, **5**.
4. Z. Feng, K. H. Adolfsson, Y. Xu, H. Fang, M. Hakkarainen and M. Wu, *Sustain. Mater. Technol.*, 2021, **29**.
5. G. Zhao, X. Huang, Z. Tang, Q. Huang, F. Niu and X. Wang, *Polym. Chem.*, 2018, **9**, 3562–3582.

6. X. Wang, Y. Feng, P. Dong and J. Huang, *Front. Chem.*, 2019, **7**, 671.
7. S. C. Ray, *Appl. Graphene Graphene-Oxide Based Nanomater.*, 2015, 39–55.
8. Z. Spitalsky, D. Tasis, K. Papagelis and C. Galiotis, *Prog. Polym. Sci.*, 2010, **35**, 357–401.
9. K. K. Chan, S. H. K. Yap and K. T. Yong, *Biogreen Synthesis of Carbon Dots for Biotechnology and Nanomedicine Applications*, Springer, Berlin Heidelberg, 2018, vol. 10.
10. A. N. Obilonu, C. Chijioke, W. E. Igwegbe, O. I. Ibearugbulem and Y. F. Abubakar, *Int. Lett. Nat. Sci.*, 2013, **4**, 44–53.
11. A. S. Ayangbenro and O. O. Babalola, *Int. J. Environ. Res. Public Health*, 2017, **14**, 94.
12. A. Ahmad and T. Azam, *Water Purification Technologies*, Elsevier Inc., 2019.
13. J. C. L. E. Bell, P. E. R. Stenius and C. Axberg, 1983, **17**, 1073–1080.
14. A. Gottfried, A. D. Shepard, K. Hardiman and M. E. Walsh, *Water Res.*, 2008, **42**, 4683–4691.
15. K. Gregory, I. Simmons, A. Brazel, J. Day, E. Keller, A. Sylvester and A. Yáñez-Arancibia, *Environ. Sci. A Student's Companion*, 2014, 297–297.
16. I. S. A. Abeysiriwardana-Arachchige and N. Nirmalakhandan, *Algal Res.*, 2019, **43**, 101643.
17. B. I. Dvorak, *Univ. Nebraska–Lincoln Extension, Inst. Agric. Nat. Resour.*, 2013, 1–4.
18. U. S. Environmental Protection Agency, *United States Environ. Prot. Agnecy*, 1999, 1–7.
19. A. M. Ashok Paidalwar IshaP Khedikar Tech Student Assistant Professor, *IJSTE-International J. Sci. Technol. Eng.*, 2016, **2**, 9.
20. J. Wang and H. Chen, *Sci. Total Environ.*, 2020, **704**, 135249.
21. K. Das and A. Roychoudhury, *Front. Env. Sci.*, 2014, **2**, 53.
22. T. Maisch, *Mini-Rev. Med. Chem.*, 2012, **9**, 974–983.
23. D. L. Huang, R. Z. Wang, Y. G. Liu, G. M. Zeng, C. Lai, P. Xu, B. A. Lu, J. J. Xu, C. Wang and C. Huang, *Environ. Sci. Pollut. Res.*, 2015, **22**, 963–977.
24. L. Deng, Y. Su, H. Su and Wang, *Adsorption*, 2002, **56**, 35–99.
25. O. A. Dar, M. A. Malik, M. I. A. Talukdar and A. A. Hashmi, *Bionanocomposites Green Synth. Appl.*, 2020, 505–518.
26. R. Pandiyan, S. Dharmaraj, S. Ayyaru, A. Sugumaran, J. Somasundaram, A. S. Kazi, S. C. Samiappan, V. Ashokkumar and C. Ngamcharussrivichai, *J. Hazard. Mater.*, 2022, **421**, 126734.
27. R. Gusain, N. Kumar and S. S. Ray, *Coord. Chem. Rev.*, 2020, **405**, 213111.
28. I. Singh, R. Arora, H. Dhiman and R. Pahwa, *Turkish J. Pharm. Sci.*, 2018, **15**, 219–230.
29. J. Manioudakis, F. Victoria, C. A. Thompson, L. Brown, M. Movsum, R. Lucifero and R. Naccache, *J. Mater. Chem. C*, 2019, **7**, 853–862.
30. N. Saran, K. Parikh, D. S. Suh, E. Muñoz, H. Kolla and S. K. Manohar, *J. Am. Chem. Soc.*, 2004, **126**, 4462–4463.
31. L. A. Pfaltzgraff and J. H. Clark, *Green Chemistry, Biorefineries and Second Generation Strategies for Re-use of Waste: An Overview*, 2014.
32. M. N. Egorova, A. E. Tomskaya, A. N. Kapitonov, A. A. Alekseev and S. A. Smagulova, *AIP Conf. Proc.*, 2018, **2041**, 020029.
33. U. A. Rani, L. Y. Ng, C. Y. Ng and E. Mahmoudi, *Adv. Colloid Interface Sci.*, 2020, **278**, 102124.
34. R. Jelinek, *Carbon Nanostructures*, Springer International Publishing, 2017, vol. 10, pp. 29–46.
35. A. R. Blaustein and C. Searle, *Encycl. Biodivers. Second Ed.*, 2013, 296–303.
36. X. Li, S. P. Lau, L. Tang, R. Ji and P. Yang, *Nanoscale*, 2014, **6**, 5323–5328.
37. L. Bao, Z. L. Zhang, Z. Q. Tian, L. Zhang, C. Liu, Y. Lin, B. Qi and D. W. Pang, *Adv. Mater.*, 2011, **23**, 5801–5806.

38. O. Adedokun, A. Roy, A. O. Awodugba and P. S. Devi, *Luminescence*, 2017, **32**, 62–70.
39. S. Zhu, Q. Meng, L. Wang, J. Zhang, Y. Song, H. Jin, K. Zhang, H. Sun, H. Wang and B. Yang, *Angew. Chemie*, 2013, **125**, 4045–4049.
40. C. Sun, Y. Zhang, S. Kalytchuk, Y. Wang, X. Zhang, W. Gao, J. Zhao, K. Cepe, R. Zboril, W. W. Yu and A. L. Rogach, *J. Mater. Chem. C*, 2015, **3**, 6613–6615.
41. A. Ghanem, R. Al-Qassar Bani Al-Marjeh and Y. Atassi, *Heliyon*, 2020, **6**, e03750.
42. R. Wang, X. Wang and Y. Sun, *Sensors Actuators, B Chem.*, 2017, **241**, 73–79.
43. Y. Deng, D. Zhao, X. Chen, F. Wang, H. Song and D. Shen, *Chem. Commun.*, 2013, **49**, 5751–5753.
44. A. K. Bera, K. G. Aukema, M. Elias and L. P. Wackett, *Sci. Rep.*, 2017, **7**, 1–9.
45. Q. Li, M. Zhou, M. Yang, Q. Yang, Z. Zhang and J. Shi, *Nat. Commun.*, 2018, **9**, 734,
46. C. L. Shen, Q. Lou, K. K. Liu, L. Dong and C. X. Shan, *Nano Today*, 2020, **35**, 100954.
47. Z. Lin, W. Xue, H. Chen and J. M. Lin, *Chem. Commun.*, 2012, **48**, 1051–1053.
48. L. Zhao and F. Di, *Nanoscale*, 2013, **5**, 2655
49. M. Zhang, Y. Jia, J. Cao, G. Li, H. Ren, H. Li and H. Yao, *Green Chem. Lett. Rev.*, 2018, **11**, 379–386.
50. S. C. Ray, A. Saha, N. R. Jana and R. Sarkar, *J. Phys. Chem. C*, 2009, **113**, 18546–18551.
51. Y. Zhuo, H. Miao, D. Zhong, S. Zhu and X. Yang, *Mater. Lett.*, 2015, **139**, 197–200.
52. W. Kwon, Y. H. Kim, C. L. Lee, M. Lee, H. C. Choi, T. W. Lee and S. W. Rhee, *Nano Lett.*, 2014, **14**, 1306–1311.
53. G. Eda, Y. Y. Lin, C. Mattevi, H. Yamaguchi, H. A. Chen, I. S. Chen, C. W. Chen and M. Chhowalla, *Adv. Mater.*, 2010, **22**, 505–509.
54. O. R. De La Fuente, M. A. González-Barrio, V. Navarro, B. M. Pabón, I. Palacio and A. Mascaraque, *J. Phys. Condens. Matter*, 2013, **25**, 484008
55. H. Ding, S. B. Yu, J. S. Wei and H. M. Xiong, *ACS Nano*, 2016, **10**, 484–491.
56. D. Huang and C. Persson, *Thin Solid Films*, 2013, **535**, 265–269.
57. B. Gayen, S. Palchoudhury and J. Chowdhury, *J. Nanomater.*, 2019, Article ID 3451307.
58. M. K. Barman, B. Jana, S. Bhattacharyya and A. Patra, *J. Phys. Chem. C*, 2014, **118**, 20034–20041.
59. A. Barman, M. Kumar, J. Bikash, B. Santanu, and Patra, *J. Phys. Chem. C*, 2014, **118**, 20034–20041.
60. I. Rodrigues and J. Sanches, *2010 7th IEEE Int. Symp. Biomed. Imaging* From *Nano to Macro, ISBI 2010- Proc.*, 2010, pp. 1265–1268.
61. T. Kaliyappan and P. Kannan, *Prog. Polym. Sci.*, 2000, **25**, 343–370.
62. H. R. Allcock, *Adv. Mater.*, 1994, **6**, 106–115.
63. Y. Chujo and K. Tanaka, *Bull. Chem. Soc. Jpn.*, 2015, **88**, 633–643.
64. L. P. Lingamdinne, J. R. Koduru and R. R. Karri, *J. Environ. Manage.*, 2019, **231**, 622–634.
65. F. R. Passador and L. A. Pessan, *Matrices and Lamellar Clays*, Elsevier Inc., 2017.
66. A. Kausar, *Issue 1 Kausar A. Am. J. Polym. Sci. Eng.*, 2017, **5**, 1–12.
67. B. Sbr, *Egypt. J. Solids*, 2007, **30**, 157–173.
68. M. O. Akharame, O. S. Fatoki, B. O. Opeolu, D. I. Olorunfemi and O. U. Oputu, *Polym. - Plast. Technol. Eng.*, 2018, **57**, 1801–1827.
69. G. R. E. Mărieş and A. M. Abrudan, *IOP Conf. Ser. Mater. Sci. Eng.*, 2018, **393**, 11.
70. H. Wu, W. P. Fahy, S. Kim, H. Kim, N. Zhao, L. Pilato, A. Kafi, S. Bateman and J. H. Koo, *Prog. Mater. Sci.*, 2020, **111**, 100638.
71. X. Zhong, X. Zhao, Y. Qian and Y. Zou, *Insight – Mater. Sci.*, 2018, **1**, 1.
72. N. V. Tsarevsky, T. Sarbu, B. Göbelt and K. Matyjaszewski, *Macromolecules*, 2002, **35**, 6142–6148.
73. W. Caseri, *Chem. Eng. Commun.*, 2009, **196**, 549–572.
74. S. Tanwar and D. Mathur, *Mater. Today Proc.*, 2020, **30**, 17–22.
75. X. Yan and X. Chen, *Titanium Dioxide Nanomaterials*, 2012, vol. 1352.

76. M. M. Adnan, A. R. M. Dalod, M. H. Balci, J. Glaum and M. A. Einarsrud, *Polymers (Basel).*, 2018, **10**, 15.
77. L. Ndlwana, N. Raleie, K. M. Dimpe, H. F. Ogutu, E. O. Oseghe, M. M. Motsa, T. A. M. Msagati and B. B. Mamba, *Materials (Basel).*, 2021, **14**, 5094.
78. M. A. Issa and Z. Z. Abidin, *Molecules*, 2020, **25**, 3541.
79. J. Bai, W. Ren, Y. Wang, X. Li, C. Zhang, Z. Li and Z. Xie, *High Perform. Polym.*, 2020, **32**, 857–867.
80. X. Zheng, G. Ding, H. Wang, G. Cui and P. Zhang, *Mater. Lett.*, 2019, **238**, 22–25.
81. M. Zulfajri, S. Sudewi, S. Ismulyati, A. Rasool, M. Adlim and G. G. Huang, *Coatings*, 2021, 11.
82. G. Liu, Z. Chen, X. Jiang, D. Q. Feng, J. Zhao, D. Fan and W. Wang, *Sens. Actuat., B Chem.*, 2016, **228**, 302–307.
83. C. Wang, T. Hu, Y. Chen, Y. Xu and Q. Song, *ACS Appl. Mater. Interfaces*, 2019, **11**, 22332.
84. B. Safaei, M. Youssefi, B. Rezaei and N. Irannejad, *Smart Sci.*, 2018, **6**, 117–124.
85. S. Bhattacharya, R. S. Phatake, S. Nabha Barnea, N. Zerby, J. J. Zhu, R. Shikler, N. G. Lemcoff and R. Jelinek, *ACS Nano*, 2019, **13**, 1433–1442.
86. S. C. Hess, F. A. Permatasari, H. Fukazawa, E. M. Schneider, R. Balgis, T. Ogi, K. Okuyama and W. J. Stark, *J. Mater. Chem. A*, 2017, **5**, 5187–5194.
87. Z. Tian, D. Li, E. V. Ushakova, V. G. Maslov, D. Zhou, P. Jing, D. Shen, S. Qu and A. L. Rogach, *Adv. Sci.*, 2018, **5**, 1800795.
88. X. Y. Du, C. F. Wang, G. Wu and S. Chen, *Angew. Chemie - Int. Ed.*, 2021, **60**, 8585–8595.
89. S. A. Mathew, P. Praveena, Y. S. Hubert, V. Narayanan and A. Stephen, *AIP Conf. Proc.*, 2019, **2115**, 030152.
90. Y. Liu, N. Xiao, N. Gong, H. Wang, X. Shi, W. Gu and L. Ye, *Carbon N. Y.*, 2014, **68**, 258–264.
91. Y. Liu, N. Xiao, N. Gong, H. Wang, X. Shi, W. Gu and L. Ye, *Carbon N. Y.*, 2014, **68**, 258–264.
92. V. Rimal, S. Shishodia and P. K. Srivastava, *Appl. Nanosci.*, 2020, **10**, 455–464.
93. H. Koulivand, A. Shahbazi, V. Vatanpour and M. Rahmandoust, *Sep. Purif. Technol.*, 2020, **230**, 115895.
94. H. Al-Haj Ibrahim, *Recent Adv. Pyrolysis*, 2020, January 2.
95. F. Benkessou and I. El Serafi, *J. Anal. Bioanal. Tech.*, 2016, August 15.
96. C. Xia, S. Zhu, T. Feng, M. Yang and B. Yang, *Adv. Sci.*, 2019, **6**, 1901316.
97. Y. Li, Z. Z. Huang, Y. Weng and H. Tan, *Chem. Commun.*, 2019, **55**, 11450–11453.
98. H. Huang, Y. Cui, M. Liu, J. Chen, Q. Wan, Y. Wen, F. Deng, N. Zhou, X. Zhang and Y. Wei, *J. Colloid Interface Sci.*, 2018, **532**, 767–773.
99. Y. Song, S. Zhu, J. Shao and B. Yang, *J. Polym. Sci. Part A Polym. Chem.*, 2017, **55**, 610–615.
100. C. Niu, T. Lan, D. Wang, J. Pan, J. Chu, C. Wang, H. Yuan, A. Yang, X. Wang and M. Rong, *Appl. Surf. Sci.*, 2020, **520**, 146257.
101. M. Barrera, M. Mehrvar, K. A. Gilbride, L. H. McCarthy, A. E. Laursen, V. Bostan and R. Pushchak, *Chem. Eng. Res. Des.*, 2012, **90**, 1335–1350.
102. A. K. Singh and J. K. Singh, *New J. Chem.*, 2017, **41**, 4618–4628.
103. J. A. Matthews, *Encycl. Environ. Chang.*, 2014, July 3.
104. A. Alharbi, M. I. Nelson, A. Worthy and H. Sidhu, *ANZIAM J.*, 2014, **55**, 348.
105. M. Samer, *Biological and Chemical Wastewater Treatment Processes*, 2015, October 14.
106. Y. C. Ho, K. Y. Show, X. X. Guo, I. Norli, F. M. Alkarkhi and N. Mor, *Ind. Waste*, 2012, March 7.
107. A. Balasubramanian, *Earth Sci.*, 2008, **12**, 1–7.

108. D. C. Prabhakaran, P. C. Ramamurthy, Y. Sivry and S. Subramanian, *Int. J. Environ. Anal. Chem.*, 2020, **00**, 1–21.
109. M. Pan, X. Xie, K. Liu, J. Yang, L. Hong and S. Wang, *Nanomaterials*, 2020, **10**, 1–25.
110. Z. Song, X. Chen, X. Gong, X. Gao, Q. Dai, T. T. Nguyen and M. Guo, Opt. Mater., 2020, **100**, 109642.
111. S. Kumari, R. P. Singh, N. N. Chavan, S. V. Sahi and N. Sharma, *Nanomaterials*, 2021, **11**, 1275.
112. L. Feng, N. Xie and J. Zhong, *Materials (Basel).*, 2014, **7**, 3919–3945.
113. M. Aslam, M. A. Kalyar and Z. A. Raza, *Polym. Eng. Sci.*, 2018, **58**, 2119–2132.
114. N. Pandey, S. K. Shukla and N. B. Singh, *Nanocomposites*, 2017, **3**, 47–66.
115. V. Melinte, L. Stroea and A. L. Chibac-Scutaru, *Catalysts*, 2019, **9**, 986.
116. T. Woolfson, *Minist. Def.*, 2008, 1–19.
117. M. Alaqarbeh, *Green and Appl. Chem.*, 2021, **13**, 12,
118. S. Palta, R. Saroa and A. Palta, *Indian J. Anaesth.*, 2014, **58**, 515–523.
119. S. Verma and S. L. Jain, *Inorg. Chem. Front.*, 2014, **1**, 534–539.
120. A. G. El-Shamy, *Polymer (Guildf).*, 2020, **202**, 122565.
121. A. G. El-Shamy and H. S. S. Zayied, *Synth. Met.*, 2020, **259**, 116218.
122. N. P. Raval, P. U. Shah and N. K. Shah, *Appl. Water Sci.*, 2017, **7**, 3407–3445.
123. C. Identification, *SubStance*, 1910, **13**, 1–8.
124. Kamaruddin, D. Edikresnha, I. Sriyanti, M. M. Munir and Khairurrijal, *IOP Conf. Ser. Mater. Sci. Eng.*, 2017, **202**, 012043.
125. S. Nayak, S. R. Prasad, D. Mandal and P. Das, *J. Hazard. Mater.*, 2020, **392**, 122287,
126. K. Rachwał, A. Waśko, K. Gustaw and M. Polak-Berecka, *PeerJ*, 2020, **8**, 1–28.
127. S. Cailotto, D. Massari, M. Gigli, C. Campalani, M. Bonini, S. You, A. Vomiero, M. Selva, A. Perosa and C. Crestini, *ACS Omega*, 2022, **7**, 4052–4061.
128. A. Al-Taie, X. Han, C. M. Williams, M. Abdulwhhab, A. P. Abbott, A. Goddard, M. Wegrzyn, N. J. Garton, M. R. Barer and J. Pan, *Microbiol. Res.*, 2020, **241**, 126587.
129. A. T. Nkeumaleu, D. Benetti, I. Haddadou, M. Di Mare, C. M. Ouellet-Plamondon and F. Rosei, *RSC Adv.*, 2022, **12**, 11621–11627.
130. X. Zhang, P. Gu and Y. Liu, *Chemosphere*, 2019, **215**, 543–553.
131. S. Zhang, J. Wang, Y. Zhang, J. Ma, L. Huang, S. Yu, L. Chen, G. Song, M. Qiu and X. Wang, *Environ. Pollut.*, 2021, **291**, 118076.
132. Z. Zhang, D. Zhang, C. Shi, W. Liu, L. Chen, Y. Miao, J. Diwu, J. Li and S. Wang, *Environ. Sci. Nano*, 2019, **6**, 1457–1465.
133. M. Wang, B. Zheng, F. Yang, J. Du, Y. Guo, J. Dai, L. Yan and D. Xiao, *Analyst*, 2016, **141**, 2508–2514.
134. S. Dolai, S. K. Bhunia, L. Zeiri, O. Paz-Tal and R. Jelinek, *ACS Omega*, 2017, **2**, 9288–9295.
135. X. Gao, C. Du, Z. Zhuang and W. Chen, *J. Mater. Chem. C*, 2016, **4**, 6927–6945.
136. J. H. Bang and K. S. Suslick, *Adv. Mater.*, 2010, **22**, 1039–1059.
137. P. Das, S. Ganguly, P. P. Maity, M. Bose, S. Mondal, S. Dhara, A. K. Das, S. Banerjee and N. C. Das, *J. Photochem. Photobiol. B Biol.*, 2018, **180**, 56–67.
138. M. Boyle, C. Sichel, P. Fernández-Ibáñez, G. B. Arias-Quiroz, M. Iriarte-Puña, A. Mercado, E. Ubomba-Jaswa and K. G. McGuigan, *Appl. Environ. Microbiol.*, 2008, **74**, 2997–3001.
139. E. dal Lago, E. Cagnin, C. Boaretti, M. Roso, A. Lorenzetti and M. Modesti, *Polymers (Basel).*, 2020, **12**, 29.
140. B. L. Rivas, B. F. Urbano and J. Sánchez, *Front. Chem.*, 2018, **6**, 1–13.
141. V. K. Periya, I. Koike, Y. Kitamura, S. I. Iwamatsu and S. Murata, *Tetrahedron Lett.*, 2004, **45**, 8311–8313.
142. P. Rajapaksha, A. Elbourne, S. Gangadoo, R. Brown, D. Cozzolino and J. Chapman, *Analyst*, 2019, **144**, 396–411.

143. M. Camps Arbestain, F. Macías, W. Chesworth, Encyclopedia of *Soil Science*, 2016, April 7.
144. U. Chadha, S. K. Selvaraj, H. Ashokan, S. P. Hariharan, V. Mathew Paul, V. Venkatarangan and V. Paramasivam, *Adv. Mater. Sci. Eng.*, 2022, Article ID 1552334.
145. N. N. Sudareva, A. V. Penkova, T. A. Kostereva, A. E. Polotskii and G. A. Polotskaya, *Express Polym. Lett.*, 2012, **6**, 178–188.
146. N. E. Podolsky, M. A. Marcos, D. Cabaleiro, K. N. Semenov, L. Lugo, A. V. Petrov, N. A. Charykov, V. V. Sharoyko, T. D. Vlasov and I. V. Murin, *J. Mol. Liq.*, 2019, **278**, 342–355.
147. V. V. Yashin and A. C. Balazs, *J. Chem. Phys.*, 2004, **121**, 11440–11454.
148. L. Wang, C. Cheng, S. Tapas, J. Lei, M. Matsuoka, J. Zhang and F. Zhang, *J. Mater. Chem. A*, 2015, **3**, 13357–13364.
149. L. Riva, A. Fiorati and C. Punta, *Materials (Basel).*, 2021, **14**, 1–22.
150. D. R. MacHado, D. Hasson and R. Semiat, *J. Memb. Sci.*, 1999, **163**, 93–102.
151. W. Zhu, X. Jiang, F. Liu, F. You and C. Yao, *Polymers (Basel).*, 2020, **12**, 2869.
152. Z. Wang, L. Zhang, K. Zhang, Y. Lu, J. Chen, S. Wang, B. Hu and X. Wang, *Chemosphere*, 2022, **287**, 132313.

13 Organic Pollutant Sensing and Degradation over Polymer–Carbonaceous Composites

Chandra Shekhar Kushwaha,
Manoj K. Tiwari, and Saroj Kr. Shukla
University of Delhi

CONTENTS

13.1 INTRODUCTION

Exponential increment of multi-faceted organic pollutants in water, air and soil demands extensive efforts to monitor and eliminate them for sustainability of humans, animal and settlements [1,2]. Furthermore, the monitoring of organic compounds is very old practice after using the different sophisticated instruments, like high-performance liquid chromatography, gas liquid chromatography, mass spectrometers and nuclear magnetic resonance spectroscopy with limitations of high cost and trained man power. Therefore, the wide range of optical, electrical and mechanical sensors are develops to precisely monitor organic pollutants after using different polymeric and non-polymeric sensing materials [3–5]. In general, the polymer composite-based sensor exhibits synergised feature like quick responsiveness, electrical conductivity and surface functionality due presence of heterogeneity and doping effect [6]. Further, among the different kind of polymer composites, the polymer carbonaceous composites (PCC) yield many advantageous features to be use in sensing

DOI: 10.1201/9781003328094-13

organic pollutants like pharmaceutical, agrochemical and organic dyes due to functionality and responsive nature [7]. The catalytic functionality of polymer composite also degrades the organic pollutant in presence of lights, which also explored for removal of organic pollutants. In the line of above the developments, present chapter describes the synthesis and advantageous features of polymer carbonaceous composites for their sensing applications.

13.2 OVERVIEW OF POLYMER CARBONACEOUS COMPOSITES

In general, the PCC was comprised of polymer as matrix and carbon-based particles as dispersed phase. The important carbon particles are carbon black, activated carbon, carbon nanofiber, fullerene, graphite, graphene and carbon nanotubes. These carbon structures have a wide range of structures from amorphous to crystalline, zero dimension, one dimension, two dimension and three dimensions. This basic structure of representative carbon nanostructure is shown in Figure 13.1, along with structure.

The presence of carbon nanostructure in polymer matrix develops the excellent dimensional stability along with magnetism, conductivity, functionality and optical responsiveness [9] [SK Shukla, *Intelligent Nanomaterials*]. The dimensional structure of carbon nanostructure also aligned the polymeric chain for novel features along with nanoconfinements and crystallinity. These two features of PCCs are responsible for high surface-to-volume ratio, improved adsorption capacity along with the better signal communication. However, the dispersion of carbon nanostructure in polymer is an important issue for materials scientist. In this regard, the preparative methods are integrated with the use of surfactants, templates and other dispersing aids. In this regard different biopolymers, petro polymer and conducting polymer are used, making the composite for different sensing applications. The simple example of carbon nanofiber decorated chitosan was prepared by facile sonicating methods along with immobilisation of copper nanoparticle. The composite is found suitable for sensing of carbendazim after electrochemical oxidation on surface with sensing parameters,

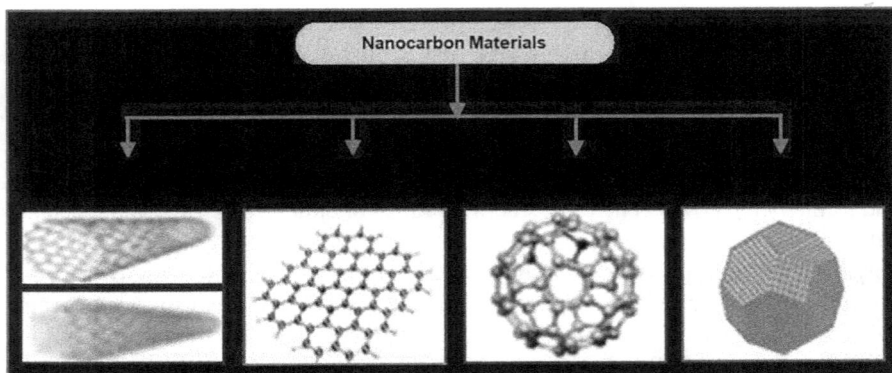

FIGURE 13.1 Structure of representative carbon nanostructure [8].

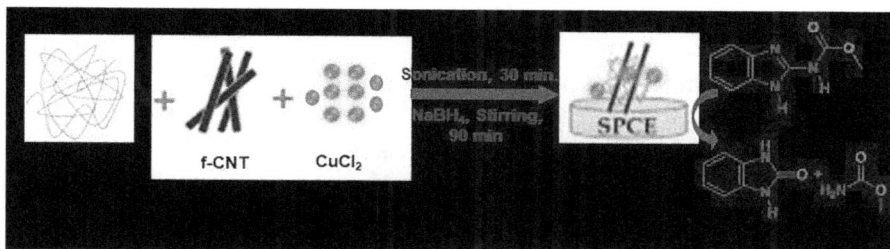

FIGURE 13.2   Synthesis of carbon fibre and chitosan composite for sensing of carbendazim [10].

i.e. sensing range 0.8–277.0 μM and limit of detection of 0.028 μM. The basic preparative methods and sensing behaviours are shown in Figure 13.2 [10].

The presence of the metal and metal oxides in PCC develops interacting sites and catalytic properties for sensing and degradation of organic pollutants. For example, the composite, comprised of polypyrrole, titanium oxide and reduced graphene oxide exhibits photocatalytic nature to degrade methyl orange under irradiation of visible light. The developed PCC-based catalyst exhibits the 93% removal efficiency in 50 minutes after using xenon lamp with wavelength of higher than 420 nm. The improvement in removal condition is due to hybrid oxidation process using • OH and •O_2^- ions [11].

13.3   ORGANIC POLLUTANTS, SENSING PRINCIPLE AND DEGRADATION

In general, all the organic molecules are related to the living organism and biological system; however, their relative activities, abundance and chemical functionality harms the physiological process of human body, functioning and progress of biochemical reactions. Also, it is well known that nature takes care of their preservation and decomposition, but the fact that industries, agriculture, and health care use them so much and dump their waste in the soil, water, and air is a serious problem. [12]. Some of the important organic pollutants are listed in Table 13.1, along with their sources and important properties.

Further, the accumulating persistence of these organic pollutants also percolates into fruits, vegetable, fishes and sea foods, which are responsible for biomagnification as well many epidemic diseases. However, the atmospheric ageing of organic pollutants after light and heat also degrades into more serious and harmful secondary pollutants. In example, the 2,3,7,8-Tetrachlorodibenzo-p-dioxin (TCDD) is by product of the burning of solid wastes, herbicides, pesticides and disinfectants. TCDD enters into food chain with long environmental life and is found in tissue, serum, milk and ovarian fluid. The impact of TCDD in life is represented in Figure 13.3 [13].

The basic principle of sensing of polymer composites is shown in Figure 13.4, which reveals that surface interaction of organic molecules either changes the resistance or generates ions to develop potential to sense the interacting organic pollutants.

TABLE 13.1
List of Representative Organic Pollutants

S. N.	Organic Pollutants and Examples	Sources	Properties
1	Herbicides, i.e. dacthal, bensulide, bentazon, bicyclopyrone and atrazine	Industrial affluent and agriculture practices	Toxic towards non-target organisms, including a variety of species inhabiting the ecosystems.
2	Pesticides: glyphosate, acephate, propoxur, metaldehyde, diazinon, dursban, DDT, malathion	Industrial affluent and agriculture practices	Causes dermatological, gastrointestinal, neurological, carcinogenic, respiratory, reproductive, and endocrine effect
3	Insecticides: organochlorine, carbamates, organosulphur and pyrethroids	Anthropogenic activity, agriculture practice and industrial discharge	Negative effect on human and animal health
4	Pharmaceuticals: antibiotic, antipyretic and antimicrobial	Discharge from pharmaceutical manufacturing facilities, waste treatment plants and human excreta	Impact on human health, animal, plants and ecosystems
5	Hydrocarbon: polycyclic hydrocarbon, polychlorinated biphenyl, phenanthrene	Automobile, oil spillage, petroleum discharge and urban stormwater, burning of fuel	Increase the cardio vascular disease risk, i.e. atherosclerosis, hypertension, thrombosis, and myocardial infarction
6	Dyes: azo dyes, anthraquinone dyes, indigo dyes, phthalocyanine dyes	Dye and textile industries	Highly toxic for human, animals and plants

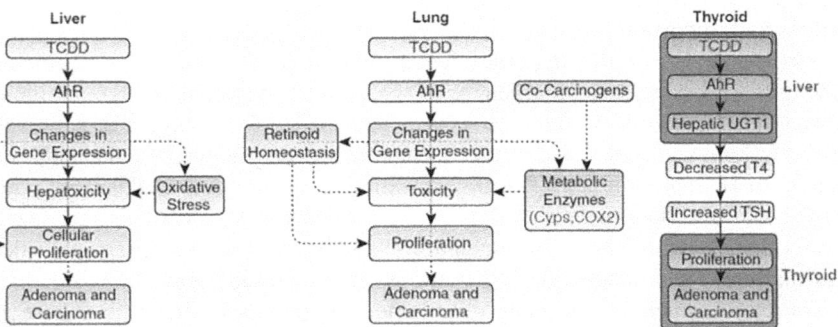

FIGURE 13.3   Site specific action TCDD in rodents [13].

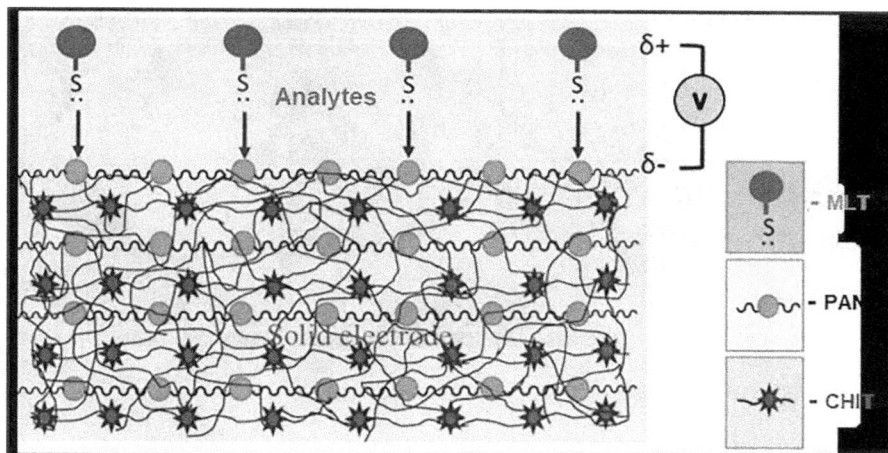

FIGURE 13.4 Potentiometric sensing mechanism of malathion over polymer composite [14].

However, in optical sensing interaction yields the luminescence after interaction with chromophores present in interacting sites. The change in intensity is indicator of relative concentration of the pollutants like pesticides, pharmaceuticals and hydrocarbon. The basic properties responsible for sensing are catalytic sites, electrical conductivity, doping effect, florescence, frequency and adsorption capacity. In this regard, the encapsulation of carbon-based structure generates the functional group as well as porosity to better adsorption of pollutants for sensing applications. The representative sensing mechanism of malathion of polymer composite is shown in Figure 13.4.

The graph indicates that the surface interaction of malathion generates ion and electron to support potentiometric sensing of residual malathion with competitive parameters like sensitivity, response and recovery time. Similarly, surface–catalytic interaction generates ions along with induced current and potential. Thus, the monitoring of current is used to quantify the absolute presence of organic pollutants along with their removals. The brief of mechanism is depicted in Figure 13.5.

13.4 SENSING OF ORGANIC POLLUTANTS

The surface interactive nature of PCC have been used for sensing different residual organic pollutants, i.e. agrochemicals, pharmaceuticals, dyes and hydrocarbon. In this context, both enzymatic and non-enzymatic principles are used after immobilising selective enzyme through suitable techniques or adding suitable chemical catalysts. Although, the enzymes are highly selective towards the selective interaction with organic molecules, but their basic limitations are of high cost and are instable under harsh physical sand chemical conditions. Thus, several studies on polymer composites including polymer and carbon nanostructure composite are devoted for

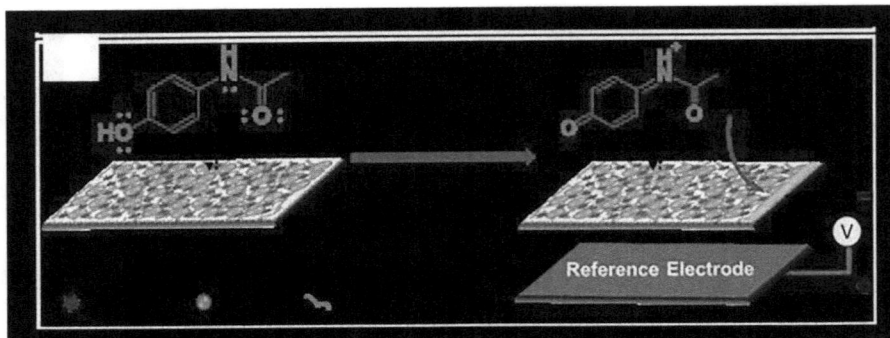

FIGURE 13.5   Potentiometric sensing of paracetamol.

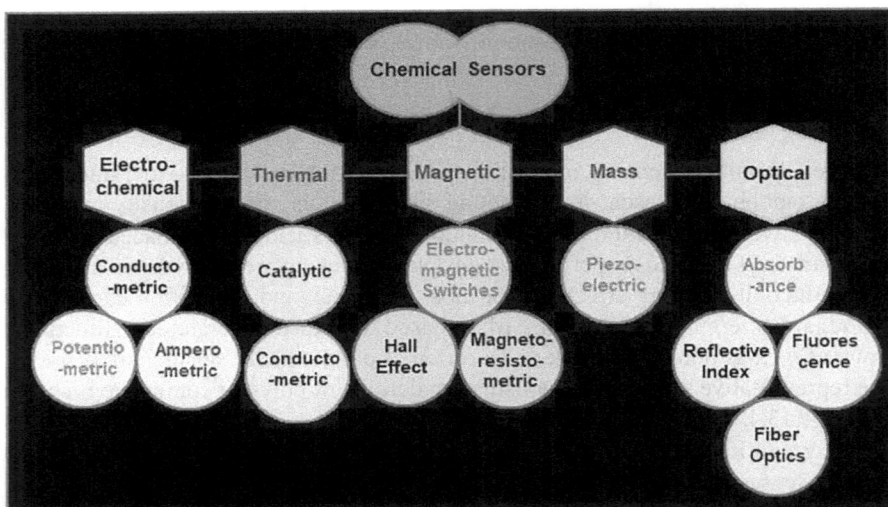

FIGURE 13.6   Different types of chemosensors and their subgroups.

non-enzymatic sensing of different organic pollutants after using different transducers, i. e. optical, electrical and mechanicals (Figure 13.6).

The same are discussed here under different subheading; the drug and chemical-based care products section comes in this category.

13.4.1   PHARMACEUTICALS

Overwhelming consumption and production of pharmaceuticals have led to serious water pollutions from active pharmaceutical ingredient (API) due to discharge from company, hospital and individuals. The exposure of API causes several deleterious

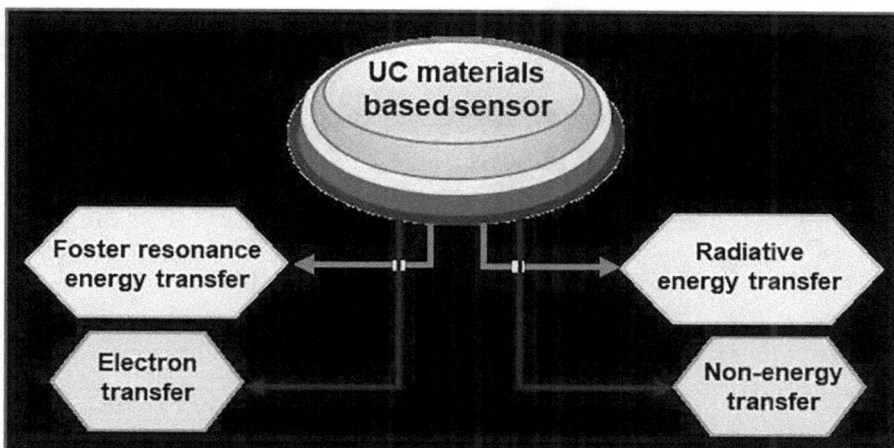

FIGURE 13.7   Sensing mechanism of pharmaceuticals [17].

negative effects on the health of ecosystem and biosystems [15]. It is reported that after very serious water purification and supply system in advance countries, like England, several pharmaceuticals are detected in tap water beyond the permissible limits. Further, biological active nature of pharmaceutical nature poses a serious challenge for environments and doctors for controlling their negative features. Thus, their significance for their controlled disposal and release in soil and water bodies is demanding serious efforts for their monitoring and removal. The regular accumulation of pharmaceutical is not a class of persistent organic compounds, since they degrade and inters into environment. The monitoring of pharmaceuticals is a well-stablished technique for related industry, but their non-portable nature demands the development of effective sensing tool and materials [16]. The carbonaceous and polymer nano-composites are used for sensing different pharmaceuticals after exploiting electrical, optical and mechanical transducers. The basic principle of sensing is increased conversion energy after active interaction between analytes and sensing materials. The simple illustration is shown in Figure 13.7 along with different transient energies.

The sensing principle are also depending on chemical structure of pharmaceuticals like antibiotic, antiseptic and antipyretic. For example, surface interaction of active group of sarafloxacin hydrochloride with functional group of reduced graphene like OH and COOH. This active surface interaction responsible for surface oxidation of sarafloxacin hydrochloride along with generation of induced currents in the range of 1.0×10^{-2} to 1.0×10^{-7} mol/L under optimum condition. The mechanism of surface interactions is shown in Figure 13.8, along with preparation of sensing electrode [18].

Similarly, the optical and mechanical sensors are also used for sensing pharmaceutical after exploring sensing electrodes. The significant polymer and carbon nanostructured composite-based pharmaceutical sensors are listed in Table 13.2.

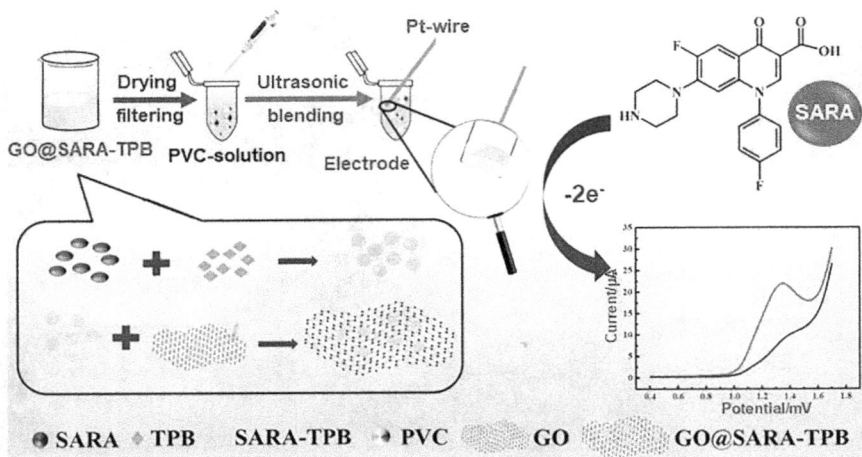

FIGURE 13.8 Sensing scheme of SARA over reduced graphene [18].

TABLE 13.2
Polymer–Carbon Nanostructured Pharmaceutical Sensors

S. N.	Composition	Analytes	Parameters	Ref.
1	Carbon nanotubes and polyvinylpyrrolidone	Paracetamol	Sensing range of linear response from 1 to 500 μM and limit of detection 0.38 μM	[19]
2	MWCNT-functionalised copolymer	Dopamine	Range 5–1000 μM and limit of detection 2.3 μM	[20]
3	Graphene supported poly-(3-aminophenylboronic acid)	Paracetamol	Sensing range of 0.15–100 μM with low detection limit of 0.028 μM	[21]
4	Polypyrrole and carbon fibre	Clozapine	Detection limit 6 μM and sensitivity of 15 μA/cm^2 μmol L.	[22]
5	Multiwalled carbon nanotubes/ polyaniline	Delafloxacin	Response ranging 1×10^{-3} to 1×10^{-8} mol/L and sensitivity of 3.5×10^{-9} mol/L	[23]
6	Polyaniline and carbon	Alfuzosin	Sensitivity of 57.16 mV/decade	[24]
7	Graphene	Amoxicillin	Good sensing parameters	[25]
8	Polyaniline/hemin/reduced graphite oxide	Dopamine	Limit of detection 0.02 μA/μM with larger storage capacity	[26]

13.4.2 AGROCHEMICALS

Exponential increase of agrochemicals, i.e. pesticides, herbicides, rodenticides and insecticides-based pollution in water, soil and edible item has attracted the attention of chemist and engineer to develop sensor for their onsite monitoring. In general, these chemicals are organic compounds of phosphorous and sulphur, which are

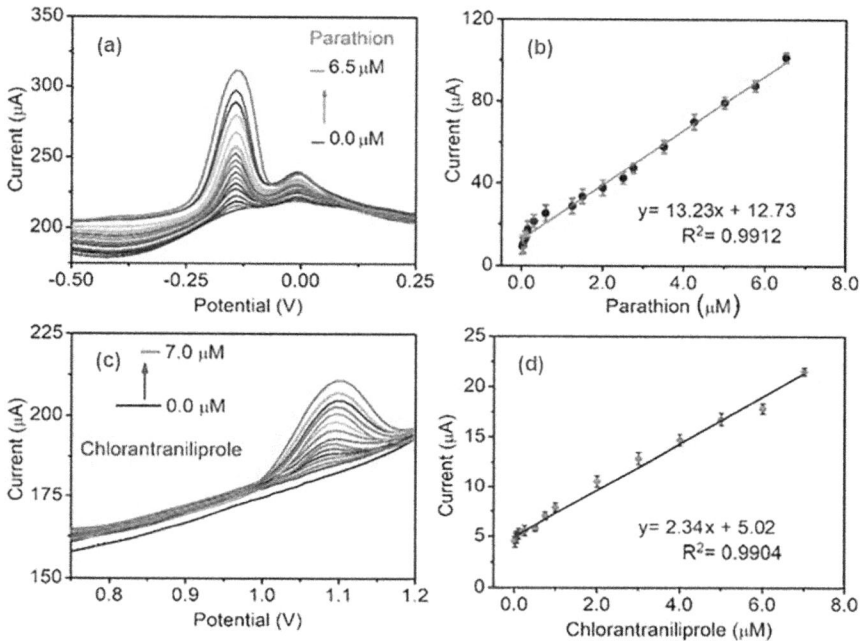

FIGURE 13.9 DPV responses on CNT-based sensor for different concentrations of (a) parathion and (c) chlorantraniliprole (c) and linear response change of (b) parathion and (d) chlorantraniliprole [29].

chemically stable and persist longer in normal conditions with toxic effects. In this context, the surface reactivity of carbonaceous nanocomposite decomposes at different electrode potentials along with generation of induced current directly proportional to the pesticide's concentrations. The interaction of agrochemical on carbon nanostructure develops the optical impulse for sensing of different agrochemicals. The several optical sensors are develops for sensing of agrochemical after using carbon nanostructure and polymer composites using different properties like quenching of florescence, change in colour and surface plasma resonance parameters [27,28].

The electrochemical sensing is another important method for residual sensing agrochemicals after monitoring induced potential and currents. For example, parathion interacted with pristine CNT and reduces due to presence NO_2 group present on parathion, and can be used for sensing application. The DPV curve of parathion and chlorantraniliprole is shown in Figure 13.9. The figure is depicting the distinct presence of two distinct peaks for both the pesticides at different voltage between 0.00 to -0.25 V and 1.0 to 1.2 V along with generation-induced current. This intensity can be used for making a calibration curve for pesticides sensing [29].

Similarly, the redox and electroactive nature of PCN was also being explored in optical sensing of different pesticides after exploring different strategies. Figure 13.10 is explaining different types of optical sensing of pesticides present in agriculture produces in onsite practices [30].

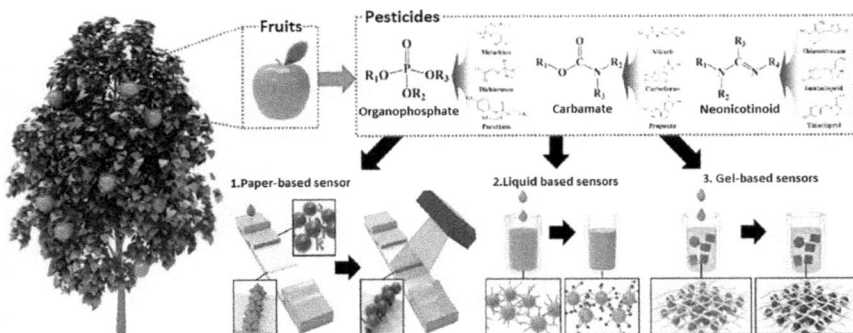

FIGURE 13.10   Outline strategy for optical sensing of onsite pesticides [30].

TABLE 13.3
Polymer–Carbon Composite-Based Agrochemical Sensors

S. N.	Composition	Analytics	Parameters	Ref.
1	Polyamide 6 (PA6)/polypyrrole (PPy)	Malathion	0.8 ng/mL	[31]
2	Polyaniline and hollow carbon sphere	Malathion	1.0 ng/mL to 10 µg/mL	[32]
3	Hybrid ferrocene-thiophene modified by CNT	Chlorantraniliprole	5.3 and 8.1 nmol/L	[29]
4	Magnetic Fe_3O_4 and polydopamine molecularly imprinted polymer magnetic nanoparticles	DDT	1×10^{-11} to 1×10^{-3} mol L	[33]
5	Graphene Oxide and graphitic carbon nitride nanohybrid electrode	Methyl parathion	8.0×10^{-8} to 1.0×10^{-4} M	[34]
6	Graphene oxide and graphitic carbon nitride nanohybrid electrode	Carbendazim	1.0×10^{-8} to 2.5×10^{-4} M	[34]
7	Hybrid nanostructures based on Poly(3,4-ethylenedioxythiophene) (PEDOT) nanofibers and gold nanoparticles	Methyl orange	10^{-8} to 10^{-4} M	[35]
8	L-lysine enzyme electrodes using poly(vinylferrocene)–multiwalled carbon nanotubes–gelatine–graphene	Lysine oxidase	9.9×10^{-7} to 7.0×10^{-4} M with detection limit of 9.2×10^{-8} M.	[36]

The other significant polymer–carbon composite-based sensors are listed in Table 13.3 along with their properties.

The structural comparison of agrochemicals is revealing the comparative structural variation between them for effective applications, which are also supporting that the relative limit of detection varies for deferent class of chemicals.

FIGURE 13.11 Average limit of detections of agrochemicals [37].

The comparative limit of detection of herbicides, fungicides and insecticides are shown in Figure 13.11.

The scientific contributions are confirming the several achievement in sensing science towards monitoring of different agrochemicals; but several problems are still needed to be solved like detoxifications along with sensing as well as multiple sensing of agrochemicals.

13.4.3 DYES AND HYDROCARBON

These are another alarming class of organic pollutants in most part of the world, since the dyes are used from small-scale industries to large scale after inculcating tanneries, food, cosmetic, textile, medicinal sectors with the approximate production of 1,000,000 tons in over the world [38]. It reported that only dyes industries discharge about 7.5 metric tons complex dyes annually. Further, the large spectrum of dyes is carcinogenic in nature as well as has severe implications on aquatic organisms that directly comes in contact. The photoactive nature of dyes also leads to its degradation into smaller molecules and are responsible for secondary pollutants as well as bio magnifications.

In this regard, surface adsorptive and responsive nature of polymer and carbon nanostructure composite has been explored for both sensing as well as degradation after using associative and dissociative interactions. All the types of sensors, i.e. optical, mechanical and electrical, are used using specific properties. For example,

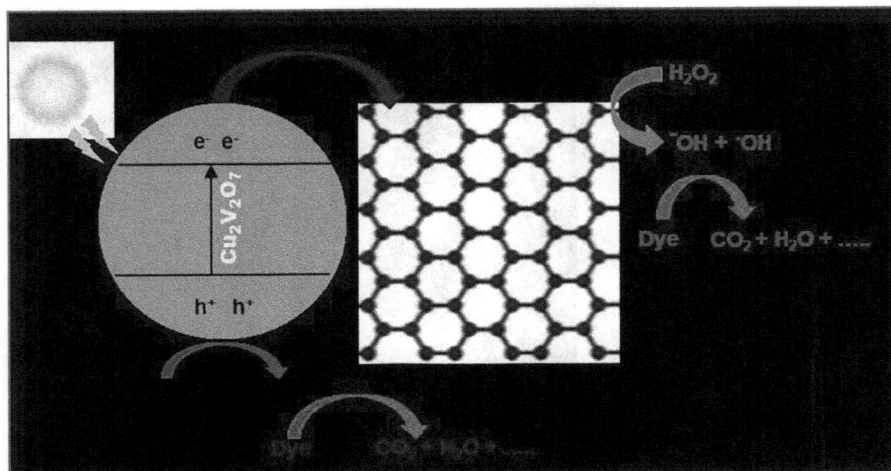

FIGURE 13.12   Degradation mechanism of dyes over graphene nanocomposite [40].

the associative interaction is suitable for conductometric sensor, while the oxo-reductive interaction helps in ampere-metric sensing as well as removal [39]. The catalytic properties are prominent properties of metals and their metal oxides. Boroujeni et al. have reported the degradation of organic, i.e. methylene blue and methyl orange on $Cu_2V_2O_7$ and graphene nanocomposite due to photocatalytic degradation. The band gap engineering in composite plays the important role during degradation of dyes on graphene metal oxide composite by 97% and 96% in less than 8 minutes [40]. The degradation mechanism is shown in Figure 13.12 along with resultant compounds.

The surface of interaction of dyes as well as metal ions on PCNs also explored for sensing application after monitoring the optical and electrical properties. For example rhodamine B is a florescent dyes with application in soft drink, biomarker and rose milk (popular Indian drink with several negative features like carcinogenic nature). He et al. have used polyethylenimine–carbon nanotubes composite for electrochemical sensing of rhodamine B present in soft drink in the concentration range of 0.01–10 μM. The electrode exhibits the improve selectivity, reproducibility and low detection limit due to synergistic interaction between constituents [41]. The schematic about preparation and sensing is shown in Figure 13.13 along with brief results and steps.

The evolution of functionality in polymer during composite formation in the presence of different carbon nanoparticles induces the properties of composite for effective interaction of residual hydrocarbon. In general, the hydrocarbon are non-functional in nature; hence their surface π–π stacking is principle for their sensing, and the effective interaction-oriented electrical and optical responses are principle for sensing of different hydrocarbon, like liquefied petroleum gas, methane, xylene and 2-nitrophenol. The functional hydrocarbon, like nitrophenol, oxidises on the

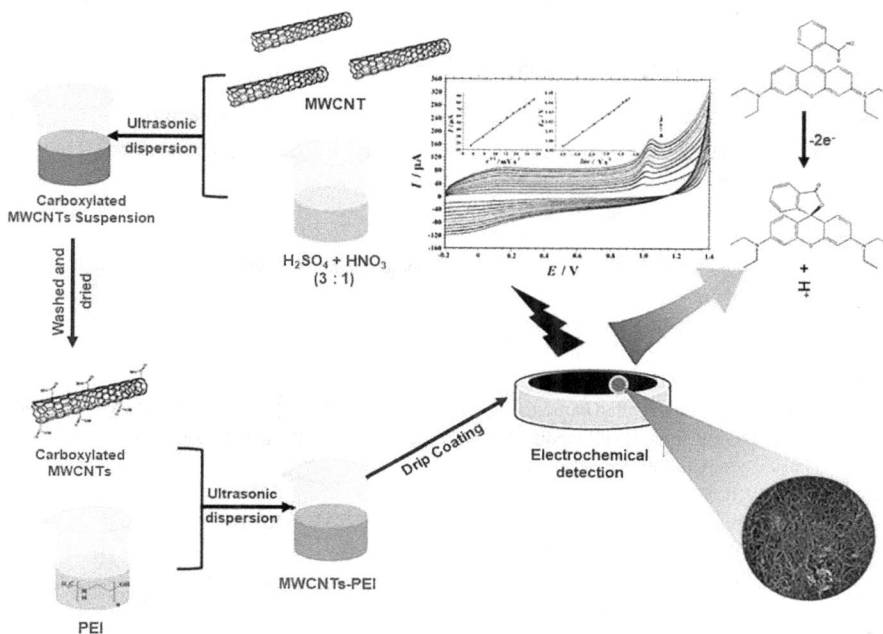

FIGURE 13.13   Electrochemical sensing of rhodamine B over carbon nanotube and polymer composites [41].

composite surface along generation of electron and current for precise sensing [42]. The brief report about the use of CPNCs in dyes and hydrocarbon sensing are listed in Table 13.4.

13.5   CONCLUSION

The significance and novel features of polymers and different carbon nanostructure composites are discussed along with basic properties, brief structural details of carbon nanostructure along with brief preparation methods. Further, the application of polymers and carbon nanostructure composites for sensing of different organic pollutants, i.e. pharmaceuticals, agrochemicals, dyes and hydrocarbon are discussed with reported sensing parameters. Further, potential of PCCs-based sensing material are also discussed with suitable scheme and sensing mechanism along with their use in removal of pollutants.

ACKNOWLEDGEMENT

The authors are thankful to Principal, Bhaskaracharya College of Applied Sciences, University of Delhi, Delhi for encouraging scientific interaction and academic activity in college.

TABLE 13.4
Polymer Composite Sensor for Dyes and Hydrocarbon

S. N.	Composition	Analytes	Parameters	Ref.
1	Polyethylenimine and multiwalled carbon nanotubes composite modified glassy carbon electrode (MWCNTs-PEI/GCE)	Rhodamine B (RhB)	Range 0.01–10 μM	[41]
2	Magnetic poly (styrene-co-divinylbenzene) (PS-DVB) and magnetic nanoparticles (MNPs) synthetised on the surface of multiwalled CNTs	Rhodamine B	Range 1.44 and 4.81 ng/mL	[43]
3	Carboxylated multiwalled carbon nanotube and ionic liquid modified pencil-graphite electrode (MWCNTs-COOH/IL/PGE)	Rhodamine B	Range 0.005–2.0 μM and 2.0–60.0 μM	[44]
4	Electrochemically polymerised glutamic acid (GA) functionalised multiwalled carbon nanotubes (MWCNTs) and graphite (GT) composite paste sensor (CPS)	Vanillin (VL)	Range 0.50–18.0 μM	[45]
5	Glassy carbon electrode with CoTTIMPPc	4-Nitrophenol	DPV method 100–1200 nM. Amperometry technique 100–800 nM	[46]
6	Polyaniline/sulfation carboxymethyl cellulose/multi-carbon nanotubes PANI/S-CMC/MWCNTs nanocomposite	2-Nitrophenol (2-NP)	0.33 M, and the limit of quantification is 1.1 μM	[42]
7	$[Ce_2(WO_4)_3 (CWO)]$. PANI@G/CWO	2-Nitrophenol	Range 1.0 nM–1.0 mM	[47]
8	Poly(acrylic acid)/multiwalled carbon nanotube (PAA/MWCNTs)	2-Nitrophenol	sensitivity of 325.24 μA m/M·cm²	[48]

REFERENCES

1. M. Jain, S.A. Khan, K. Sharma, P.R. Jadhao, K.K. Pant, Z.M. Ziora, M.A.T. Blaskovich, Current perspective of innovative strategies for bioremediation of organic pollutants from wastewater, *Bioresour. Technol.* 344 (2022) 126305. https://doi.org/10.1016/J.BIORTECH.2021.126305.

2. M.H. Hazaraimi, P.S. Goh, W.J. Lau, A.F. Ismail, Z. Wu, M.N. Subramaniam, J.W. Lim, D. Kanakaraju, The state-of-the-art development of photocatalysts for the degradation of persistent herbicides in wastewater, *Sci. Total Environ.* 843 (2022) 156975. https://doi.org/10.1016/J.SCITOTENV.2022.156975.

3. S. Muralikrishna, S. Kempahanumakkagari, R. Thippeswamy, W. Surareungchai, Functional polymer materials for environmental monitoring and safety applications, in: *Energy, Environ. Sustain*, Springer Nature, 2022: pp. 177–204. https://doi.org/10.1007/978-981-16-8755-6_9/COVER.

4. B.M. da Costa Filho, A.C. Duarte, T.A.P.R. Santos, Environmental monitoring approaches for the detection of organic contaminants in marine environments: A critical review, *Trends Environ. Anal. Chem.* 33 (2022) e00154. https://doi.org/10.1016/J.TEAC.2022.E00154.

5. M. Patel, R. Kumar, K. Kishor, T. Mlsna, C.U. Pittman, D. Mohan, Pharmaceuticals of emerging concern in aquatic systems: Chemistry, occurrence, effects, and removal methods, *Chem. Rev.* 119 (2019) 3510–3673. https://doi.org/10.1021/ACS.CHEMREV.8B00299/ASSET/IMAGES/LARGE/CR-2018-00299Y_0035.JPEG.

6. C.S. Kushwaha, P. Singh, S.K. Shukla, M.M. Chehimi, Advances in conducting polymer nanocomposite based chemical sensors: An overview, *Mater. Sci. Eng. B.* 284 (2022) 115856. https://doi.org/10.1016/J.MSEB.2022.115856.

7. E. Dhandapani, S. Thangarasu, S. Ramesh, K. Ramesh, R. Vasudevan, N. Duraisamy, Recent development and prospective of carbonaceous material, conducting polymer and their composite electrode materials for supercapacitor — A review, *J. Energy Storage.* 52 (2022) 104937. https://doi.org/10.1016/J.EST.2022.104937.

8. Z. Khorsandi, M. Borjian-Boroujeni, R. Yekani, R.S. Varma, Carbon nanomaterials with chitosan: A winning combination for drug delivery systems, *J. Drug Deliv. Sci. Technol.* 66 (2021) 102847. https://doi.org/10.1016/J.JDDST.2021.102847.

9. S.C. Tjong, Polymer composites with carbonaceous nanofillers: Properties and applications, *Polym. Compos. Carbonaceous Nanofillers Prop. Appl.* (2012). https://doi.org/10.1002/9783527648726.

10. P. Sundaresan, C.C. Fu, S.H. Liu, R.S. Juang, Facile synthesis of chitosan-carbon nanofiber composite supported copper nanoparticles for electrochemical sensing of carbendazim, *Colloids Surf. A Physicochem. Eng. Asp.* 625 (2021) 126934. https://doi.org/10.1016/J.COLSURFA.2021.126934.

11. S. Liu, X. Jiang, G.I.N. Waterhouse, Z.M. Zhang, L. min Yu, Efficient photoelectrocatalytic degradation of azo-dyes over polypyrrole/titanium oxide/reduced graphene oxide electrodes under visible light: Performance evaluation and mechanism insights, *Chemosphere.* 288 (2022) 132509. https://doi.org/10.1016/J.CHEMOSPHERE.2021.132509.

12. L. Karadurmus, A. Cetinkaya, S.I. Kaya, S.A. Ozkan, Recent trends on electrochemical carbon-based nanosensors for sensitive assay of pesticides, *Trends Environ. Anal. Chem.* 34 (2022) e00158. https://doi.org/10.1016/J.TEAC.2022.E00158.

13. D. Wikoff, L. Fitzgerald, L. Birnbaum, Persistent organic pollutants: An overview, in: *Dioxins Heal. Incl. Other Persistent Org. Pollut. Endocr. Disruptors*, Third Ed., John Wiley & Sons, Ltd, 2012: pp. 1–35. https://doi.org/10.1002/9781118184141.CH1.

14. C.S. Kushwaha, S.K. Shukla, Non-enzymatic potentiometric malathion sensing over chitosan-grafted polyaniline hybrid electrode, *J. Mater. Sci.* 54 (2019) 10846–10855. https://doi.org/10.1007/s10853-019-03625-2.

15. O.A. Jones, J.N. Lester, N. Voulvoulis, Pharmaceuticals: A threat to drinking water? *Trends Biotechnol.* 23 (2005) 163–167. https://doi.org/10.1016/J.TIBTECH.2005.02.001.

16. M.M. Mastrángelo, M.E. Valdés, B. Eissa, N.A. Ossana, D. Barceló, S. Sabater, S. Rodríguez-Mozaz, A.D.N. Giorgi, Occurrence and accumulation of pharmaceutical products in water and biota of urban lowland rivers, *Sci. Total Environ.* 828 (2022) 154303. https://doi.org/10.1016/J.SCITOTENV.2022.154303.

17. A. Jouyban, E. Rahimpour, Sensors/nanosensors based on upconversion materials for the determination of pharmaceuticals and biomolecules: An overview, *Talanta.* 220 (2020) 121383. https://doi.org/10.1016/J.TALANTA.2020.121383.

18. M. Liu, M. Jia, Y. E, D. Li, A novel ion selective electrode based on reduced graphene oxide for potentiometric determination of sarafloxacin hydrochloride, *Microchem. J.* 170 (2021) 106678. https://doi.org/10.1016/J.MICROC.2021.106678.

19. P. Pinyou, V. Blay, K. Chansaenpak, S. Lisnund, Paracetamol sensing with a pencil lead electrode modified with carbon nanotubes and polyvinylpyrrolidone, *Chemosensors*. 8 (2020) 133. https://doi.org/10.3390/CHEMOSENSORS8040133.
20. R. Liu, X. Zeng, J. Liu, J. Luo, Y. Zheng, X. Liu, A glassy carbon electrode modified with an amphiphilic, electroactive and photosensitive polymer and with multi-walled carbon nanotubes for simultaneous determination of dopamine and paracetamol, *Microchim. Acta*. 183 (2016) 1543–1551. https://doi.org/10.1007/S00604-016-1763-1/TABLES/2.
21. S. Gürsoy, F. Kuralay, Graphene supported poly(3-aminophenylboronic acid) surface via constant potential electrolysis for facile and sensitive paracetamol determination, *Colloids Surf. A Physicochem. Eng. Asp.* 633 (2022) 127846. https://doi.org/10.1016/J.COLSURFA.2021.127846.
22. S. Sriprasertsuk, S.C. Mathias, J.R. Varcoe, C. Crean, Polypyrrole-coated carbon fibre electrodes for paracetamol and clozapine drug sensing, *J. Electroanal. Chem.* 897 (2021) 115608. https://doi.org/10.1016/J.JELECHEM.2021.115608.
23. N.A. Abdallah, Y.M. Alahmadi, R. Bafail, M.A. Omar, Multi-walled carbon nanotubes/polyaniline covalently attached 18-crown-6-ether as a polymeric material for the potentiometric determination of delafloxacin, *J. Appl. Electrochem.* 52 (2022) 311–323. https://doi.org/10.1007/S10800-021-01636-Z/TABLES/6.
24. M. Wadie, H.M. Marzouk, M.R. Rezk, E.M. Abdel-Moety, M.A. Tantawy, A sensing platform of molecular imprinted polymer-based polyaniline/carbon paste electrodes for simultaneous potentiometric determination of alfuzosin and solifenacin in binary co-formulation and spiked plasma, *Anal. Chim. Acta*. 1200 (2022) 339599. https://doi.org/10.1016/J.ACA.2022.339599.
25. A. Hrioua, A. Loudiki, A. Farahi, M. Bakasse, S. Lahrich, S. Saqrane, M.A. El Mhammedi, Recent advances in electrochemical sensors for amoxicillin detection in biological and environmental samples, *Bioelectrochemistry*. 137 (2021) 107687. https://doi.org/10.1016/J.BIOELECHEM.2020.107687.
26. K.P. Aryal, H.K. Jeong, Simultaneous determination of ascorbic acid, dopamine, and uric acid with polyaniline/hemin/reduced graphite oxide composite, *Chem. Phys. Lett.* 768 (2021) 138405. https://doi.org/10.1016/J.CPLETT.2021.138405.
27. Z. Li, H. Lin, L. Wang, L. Cao, J. Sui, K. Wang, Optical sensing techniques for rapid detection of agrochemicals: Strategies, challenges, and perspectives, *Sci. Total Environ.* 838 (2022) 156515. https://doi.org/10.1016/J.SCITOTENV.2022.156515.
28. B.K. John, T. Abraham, B. Mathew, A review on characterization techniques for carbon quantum dots and their applications in agrochemical residue detection, *J. Fluoresc.* 32 (2022) 449–471. https://doi.org/10.1007/S10895-021-02852-8/TABLES/2.
29. S.O. Tümay, A. Şenocak, E. Sarı, V. Şanko, M. Durmuş, E. Demirbas, A new perspective for electrochemical determination of parathion and chlorantraniliprole pesticides via carbon nanotube-based thiophene-ferrocene appended hybrid nanosensor, *Sensors Actuators B Chem.* 345 (2021) 130344. https://doi.org/10.1016/J.SNB.2021.130344.
30. R. Umapathi, B. Park, S. Sonwal, G.M. Rani, Y. Cho, Y.S. Huh, Advances in optical-sensing strategies for the on-site detection of pesticides in agricultural foods, *Trends Food Sci. Technol.* 119 (2022) 69–89. https://doi.org/10.1016/J.TIFS.2021.11.018.
31. G. Adolfo Palacios-Rodriguez, L. Hinojosa-Reyes, J. Luis Guzmán Mar, al -, Y. Man, K. Guo, B. Jiang, F.L. Migliorini, R.C. Sanfelice, L.A. Mercante, M.H. M Facure, D.S. Correa, Electrochemical sensor based on polyamide 6/polypyrrole electrospun nanofibers coated with reduced graphene oxide for malathion pesticide detection, *Mater. Res. Express.* 7 (2019) 015601. https://doi.org/10.1088/2053-1591/AB5744.
32. L. He, B. Cui, J. Liu, Y. Song, M. Wang, D. Peng, Z. Zhang, Novel electrochemical biosensor based on core-shell nanostructured composite of hollow carbon spheres and polyaniline for sensitively detecting malathion, *Sensors Actuators, B Chem.* 258 (2018) 813–821. https://doi.org/10.1016/j.snb.2017.11.161.

33. J. Miao, A. Liu, L. Wu, M. Yu, W. Wei, S. Liu, Magnetic ferroferric oxide and poly-dopamine molecularly imprinted polymer nanocomposites based electrochemical impedance sensor for the selective separation and sensitive determination of dichlo-rodiphenyltrichloroethane (DDT), *Anal. Chim. Acta.* 1095 (2020) 82–92. https://doi.org/10.1016/J.ACA.2019.10.027.

34. D. Ilager, N.P. Shetti, Y. Foucaud, M. Badawi, T.M. Aminabhavi, Graphene/g-carbon nitride (GO/g-C3N4) nanohybrids as a sensor material for the detection of methyl para-thion and carbendazim, *Chemosphere.* 292 (2022) 133450. https://doi.org/10.1016/J.CHEMOSPHERE.2021.133450.

35. S. Ghosh, A.K. Mallik, R.N. Basu, Enhanced photocatalytic activity and photores-ponse of poly(3,4-ethylenedioxythiophene) nanofibers decorated with gold nanopar-ticle under visible light, *Sol. Energy.* 159 (2018) 548–560. https://doi.org/10.1016/J.SOLENER.2017.11.036.

36. C. Kaçar, P.E. Erden, E. Kılıç, Amperometric L-lysine enzyme electrodes based on carbon nanotube/redox polymer and graphene/carbon nanotube/redox polymer composites, Anal. *Bioanal. Chem.* 409 (2017) 2873–2883. https://doi.org/10.1007/S00216-017-0232-Y/TABLES/2.

37. H. Xiang, Q. Cai, Y. Li, Z. Zhang, L. Cao, K. Li, H. Yang, Sensors applied for the detec-tion of pesticides and heavy metals in freshwaters, *J. Sensors.* 2020 (2020) 8503491. https://doi.org/10.1155/2020/8503491.

38. K. Maheshwari, M. Agrawal, A.B. Gupta, Dye pollution in water and wastewater, in: *Nov. Mater. Dye. Wastewater Treat.*, Springer, Singapore, 2021: pp. 1–25. https://doi.org/10.1007/978-981-16-2892-4_1.

39. A. Piras, C. Ehlert, G. Gryn'ova, Sensing and sensitivity: Computational chemistry of graphene-based sensors, *Wiley Interdiscip. Rev. Comput. Mol. Sci.* 11 (2021) e1526. https://doi.org/10.1002/WCMS.1526.

40. K.P. Boroujeni, Z. Tohidiyan, Z. Hamidifar, M.M. Eskandari, Rapid and facile micro-wave-assisted synthesis of Cu2V2O7/graphene nanocomposite as a novel catalyst for degradation of organic dyes, *Indian J. Chem. Technol.* 29 (2022) 279–287. http://nopr.niscpr.res.in/handle/123456789/59934 (accessed July 25, 2022).

41. P. Deng, J. Xiao, J. Chen, J. Feng, Y. Wei, J. Zuo, J. Liu, J. Li, Q. He, Polyethylenimine-carbon nanotubes composite as an electrochemical sensing platform for sensitive and selective detection of toxic rhodamine B in soft drinks and chilli-containing products, *J. Food Compos. Anal.* 107 (2022) 104386. https://doi.org/10.1016/J.JFCA.2022.104386.

42. R.H. Althomali, K.A. Alamry, M.A. Hussein, R.M. Guedes, Highly sensitive detec-tion analytical performance of 2-nitrophenol pollution in various water samples via polyaniline/sulfation carboxymethylcellulose/multi carbon nanotubes nanocomposite-based electrochemical sensor, *J. Electrochem. Soc.* 169 (2022) 046518. https://doi.org/10.1149/1945-7111/AC3778.

43. Y. Benmassaoud, K. Murtada, R. Salghi, M. Zougagh, Á. Ríos, Surface polymers on multiwalled carbon nanotubes for selective extraction and electrochemical determina-tion of rhodamine B in food samples, *Molecules.* 26 (2021) 2670. https://doi.org/10.3390/MOLECULES26092670.

44. X. Zhu, G. Wu, C. Wang, D. Zhang, X. Yuan, A miniature and low-cost electrochemical system for sensitive determination of rhodamine B, *Measurement.* 120 (2018) 206–212. https://doi.org/10.1016/J.MEASUREMENT.2018.02.014.

45. N. Hareesha, J.G. Manjunatha, B.M. Amrutha, N. Sreeharsha, S.M. Basheeruddin Asdaq, M.K. Anwer, A fast and selective electrochemical detection of vanillin in food samples on the surface of poly(glutamic acid) functionalized multiwalled carbon nano-tubes and graphite composite paste sensor, *Colloids Surf. A Physicochem. Eng. Asp.* 626 (2021) 127042. https://doi.org/10.1016/J.COLSURFA.2021.127042.

46. V.A. Sajjan, S. Aralekallu, M. Nemakal, M. Palanna, C.P. Keshavananda Prabhu, L. Koodlur Sannegowda, Nanomolar detection of 4-nitrophenol using Schiff-base phthalocyanine, *Microchem. J.* 164 (2021) 105980. https://doi.org/10.1016/J.MICROC.2021.105980.

47. A. Khan, A.A.P. Khan, M.M. Rahman, A.M. Asiri, Inamuddin, K.A. Alamry, S.A. Hameed, Preparation and characterization of PANI@G/CWO nanocomposite for enhanced 2-nitrophenol sensing, *Appl. Surf. Sci.* 433 (2018) 696–704. https://doi.org/10.1016/J.APSUSC.2017.09.219.

48. S.H. Kim, A. Umar, R. Kumar, H. Algarni, M.S. Al-Assiri, Poly(acrylic acid)/multi-walled carbon nanotube composites: Efficient scaffold for highly sensitive 2-nitrophenol chemical sensor, *Nanosci. Nanotechnol. Lett.* 8 (2016) 200–206. https://doi.org/10.1166/NNL.2016.2110.

14 Heavy Metal Sensing and Absorption by Polymer–Carbonaceous Composites

Asit Kumar Das
Krishnath College

Amit Pramanik
A.B.N Seal College

Avishek Ghatak
Dr. A. P. J. Abdul Kalam Government College

CONTENTS

14.1 INTRODUCTION

Heavy metal ions are the foremost interests of the chemists not only due to their widespread applicability in different industries, but also they can play pivotal role in metabolic system of human body [1,2]. But so far as the issue of the environmental pollution is concerned the heavy metals like As(II), Cd(II), Hg(II), Pd(II), Cu(II), etc. are more responsible than traditional organic pollutants awing to their non-biodegradability and rampant use over the last few decades [3]. Depending upon the exposure to these toxic ions, the different health hazards are observed like extensive use of Cd(II) can cause renal and skeletal disease, whereas excess Hg(II) can create nervous system disorder. Even at very low concentration of Pd(II), myocardial infarction and gastro-intestinal disorders can be observed [4,5]. As a result, these

DOI: 10.1201/9781003328094-14

metals are actually a great threat to the human health. Furthermore, this is the era in which most of the protocols are eco-compatible [6–12]. For this reason, there is a high demand of efficient, eco-compatible and potent heavy metal ions sensors that can detect ions with precise accuracy. Earlier gold, silver electrodes, electrode made of bismuth, boron and bio molecules were successfully used for sensing [13]. But all they have the limitations like high cost, stability and most importantly selectivity. In the last few years, carbon-based nanocomposites that are cost-effective and highly stable in ambient atmosphere extensively used for rapid detection of heavy metal ions with high selectivity and efficacy. In this chapter we have briefly demonstrated widespread applications of heavy metal sensing and absorption by different polymer–carbon-based composites.

14.2 APPLICATIONS OF ORGANIC CONDUCTING POLYMERS FOR THE SENSING OF HEAVY METAL IONS

Organic conducting polymers, *viz.* polyacetylene, polythiophene, polypyrrole, polyphenylene, polyaniline, and many others represent superior chemical, physical as well as electrical properties [14,15]. Moreover, organic conducting polymers also attracts immense attention to the scientific communities due to their multifarious convenient applications in modern technologies, *viz.* batteries, transducers, membranes, supercapacitors, sensors, photodiodes, etc.[16]. The affinity of conducting polymers towards metal ions is represented in Figure 14.1.

The most attractive attributes of the organic conducting polymers are chemical sensors that are sensitive to negligible electrochemical perturbations due to their cooperative redox properties [17]. Organic conducting polymers could be synthesised efficiently [18,19] in a cost-effective [20–22], environmentally feasible [23–26] and biocompatiblemanner [27]. Few conjugated conducting polymers are represented in Table 14.1 along with their conductivities found to the maximum values [28].

Most of them are insulators that demonstrated excellent electrical resistance. But suitably synthesised or modified organic conducting polymers can have superior conducting or semiconducting properties [30]. This fact is well-known as doping,

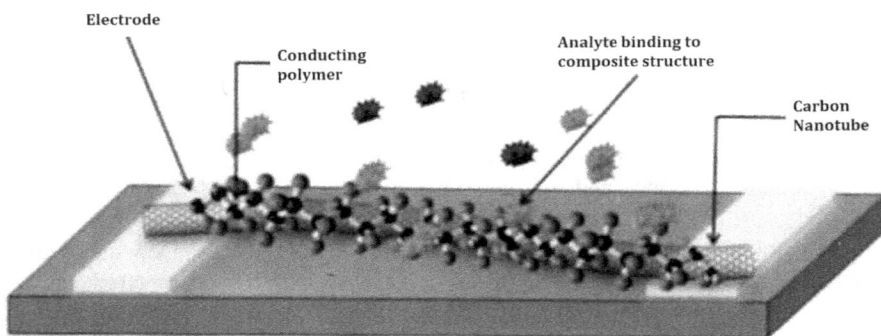

FIGURE 14.1   Representation of organic conducting polymer-based heavy metal ion sensor.

TABLE 14.1
Structures and Conductivities of Some Common Conducting Polymers [29]

Sl. No.	Polymers	Structure	Highest Conductivity (S/cm)
1.	Polyacetylene and analogues polyacetylene		$10^3 - 1.7 \times 10^5$
2.	Polypyrrole		$10^2 - 7.5 \times 10^3$
3.	Polythiophene		$10 - 10^3$
4.	Polyphenyleneand analogues Poly(paraphenylene)		$10^2 - 10^3$
5.	Poly(p-phenylenevinylene)		$3 - 5 \times 10^3$
6.	Polyaniline		$30 - 200$

which considerably enhances the electrical conductivity of some organic conducting polymers. The electrical conductivity of organic conducting polymers makes them valuable materials, which showed lower ionisation potential, smaller energy optical transitions, and higher electron affinity. Therefore, organic conducting polymers are found to be appropriate materials for the development of sensitivity and enhanced performance of various sensors. Lately, environmentally benign conducting polymers modified by carbon-based electrodes are used to replace mercury electrodes [31]. Various organic conducting polymers in bio-electrochemical systems are working as redox mediators [32], which assists the shuttling of protons as well as electrons between electrodes and electro-active reactants. Thus, organic conducting polymer-based electrodes can control the restrictions of plain materials, such as graphite-, glassy carbon-, platinum-, gold-, etc. based electrodes. Moreover, several conducting polymers are biocompatible [33,34], and, therefore, they are appropriate for the development of implantable sensors. Accordingly, organic conducting polymer-modified electrodes gained significant interest in the detection of heavy metal ions due to their enhancement of the stripping-based electrochemical responses [35]. A recent literature survey about organic conducting polymers concluded that the functional groups in the organic conducting polymers were highly responsible for heavy metal ion binding. These binding groups acted as the chelating sites for the binding of the particular metal ions. Therefore, organic conducting polymers showed significant potential for the detection and purification of metal ions [36,37]. For example, poly(1,5-diami-nonaphthalene) (PDAN) is one such conducting polymers, which is responsible to chelate the metal ions through its extra free amine group and another amine group used for the coupling with another monomer unit [38]. The applications of organic conducting polymers-based heavy metal ion sensing are represented in Table 14.2.

14.3   REMOVAL OF HEAVY METAL IONS BY POLYMER–CARBONACEOUS COMPOSITES

Carbonaceous materials such as carbon nanotubes (CNTs), graphene, and activated carbons (ACs) represent one of the most significant counterparts in the polymer-based composites, because of their high aspect ratio, good mechanical potency, and particularly the compatibility of carbon matrix through the polymeric structure for excellent dispersion of carbon counterparts into the polymers and sustenance of powerful interaction and adhesion [39]. Novel polymeric membranes were efficiently synthesised by the addition of polyethylene glycol (PEG) and ACs into polyetherim-ide (PEI)/polyphenyl sulfone (PPSU) polymers, which simultaneously improve the filtration efficiency and permeability of polymeric membranes. The authors observed that the addition of ACs considerably affected the chemical property, pore size dis-tribution, membrane morphology, and porosity, whereas PEG significantly increased the surface porosity as well as the hydrophilicity of the AC/PPSU/PEI/PEG mem-branes due to the generation of the hydrophilic pore. However, Wang and co-workers firstly synthesised carboxymethyl cellulose (CMC)/chitosan (CS) -grafted multi-walled carbon nanotubes (MWCNTs) by N2-plasma-induced grafting technique [40]. The synthesised MWCNT-g-CMC was successfully applied in the removal of U(VI) from wastewater solutions. It was found that the sorption percentage of U(VI) improved from 23% to 98%, with the raise of MWCNT-g-CMC content improved from 0.1 to 1.0 g/L, whereas only raw MWCNTs showed that the sorption propor-tion of U(VI) increased only from ~8% to ~19%. The synthesised MWCNT-g-CS were also efficiently applied to remove U(VI), Pb(II), and Cu(II) metal ions from the aqueous solutions. The results demonstrated that MWCNT-g-CS had superior sorp-tion capability than MWCNTs, which also further represents the applicability of such grafted MWCNTs-polymer composites in the preconcentration and solidification of metal ions from wastewater solutions [40].

Nyairo et al. [41] synthesised polypyrrole (PPy)-coated oxidised MWCNTs (oMWCNT/PPy), which were applied for the removal of Pb(II) and Cu(II) ions from wastewater solutions. The results showed that the sorption maximum capaci-ties for Pb(II) and Cu(II) ions were achieved at 26.32 and 24.39 mg/g, respectively. Graphene, especially graphene oxide (GO), is extensively used to make polymers for enhancing the performance of the removal of heavy metal ions due to high sur-face areas, high water solubility, and presence of abundant functional surface groups [39]. GO has also been commonly combined with chitosan (which is natural amino polysaccharide) to form hydrogel composites [42]. He et al. have prepared porous GO/chitosan (PGOC) material, and the addition of GO improved the compressive power of PGOC significantly. When 5 wt% GO was added, the sorption capacity of

TABLE 14.2

Applications of Organic Conducting Polymers for the Sensing of Heavy Metal Ions [29]

Sl. No.	Electrode/Modified Materials	Targeted Metal Ion	Detection Limit
1.	Langmuir–Blodgett films of functionalised polythiophenes	Hg^{+2}	0.1–100 ppm
2.	Poly(3-methylthiophene) modified gold electrode	Hg^{+2}	3×10^{-10} mol/L
3.	Sonogel–carbon electrode modified with poly-3- methylthiophene	Hg^{+2}	1.4×10^{-3}/mol
4.	Polythiophene–quinoline/glassy carbon-modified electrode	Hg^{+2}	0.4 ppb
5.	Polyaniline and polyaniline–methylene blue-coated screen-printed carbon electrode	Hg^{+2}	$54.27 + 3.28$ µg/L
6.	Surface ion imprinting strategy in electropolymerised microporous poly (2-mercaptobenzothiazole) films-modified glassy carbon electrode	Hg^{+2}	0.1 nM
7.	Polyaniline Sn (IV) tungstomolybdate nanocomposite cation exchange material	Pb^{+2}	1×10^{-6} M
8.	PEDOT: PSS-coated graphite carbon electrode	Pb^{+2}	0.19 nmol/L
9.	Carbon paste electrode bulkmodified with the conducting polymer poly (1,8 diaminonaphthalene): application to lead determination	Pb^{+2}	30 ng/mL
10.	Polypyrrole-modified electrode	Pb^{+2}	3×10^{-9} M
11.	A polystyrene-based membrane electrode	Cd^{+2}	3.16×10^{-6} mol/L
12.	Poly(4-vinylpyridine-co-aniline) (poly(4VP-co-Ani)	Cd^{+2}	7.94×10^{-7} M
13.	Polyaniline Langmuir–Blodgett film-modified glassy carbon electrode	Ag^{+}	4.0×10^{-10} mol/L
14.	Polypyrrole film with electrochemically induced recognition sites	Ag^{+}	6×10^{-9} M
15.	N,N-dichromone-p-phenylenediamine modified carbon paste electrode	Cu^{+2}	10^{-5} mol/L
16.	Polyviologen-modified glassy carbon electrode (PVGCE)	Cu^{+2}	0.02 ppm
17.	PANI-modified ITO electrode	Cu^{+2}	0.02 mM
18.	Polyaniline-co poly(dithiodianiline)-modified carbon paste electrode	Pb^{+2}, Cd^{+2}	0.03, 0.09 µg/L
19.	Glassy carbon electrode modified by conductive polyaniline coating	Pb^{+2}, Cd^{+2}	0.1, 0.13 µM
20.	Nafion-coated microelectrodes	Pb^{+2}, Cd^{+2}	0.3, 0.14 ppb

(Continued)

TABLE 14.2 (*Continued*)
Applications of Organic Conducting Polymers for the Sensing of Heavy Metal Ions [29]

Sl. No.	Electrode/Modified Materials	Targeted Metal Ion	Detection Limit
21.	Poly(diphenylamine-co-2-aminobenzonitrile) (P(DPA-co-2ABN))	Pb^{+2}, Cd^{+2}	0.165, 0.255 ppm
22.	Glassy carbon electrodes (GCE) modified with poly (3,4-ethylenedioxythiophene) (PEDOT)	Pb^{+2}, Cd^{+2}	1.15 and 1.47 µg/mL
23.	Peptide-functionalised conducting polymer junction array	Cu^{+2}, Ni^{+2}	63 pM (4 ppt), 0.4 nM (23 ppt)
24.	Polyacrylic acid/glassy carbon electrode (PAA/GCE	Pb^{+2}, Cd^{+2}, Co^{+2}	0.9 nM, 1.9 nM, 11.0 µM
25.	EDTA bonded 30,40 -diamino-2,20;50,200-terthiophene on GCE	Pb^{+2}, Cu^{+2}, Hg^{+2}	6.0×10^{-10}, 2.0×10^{-10}, $5.0 \times 10-10M$
26.	Polyacrylic acid/glassy carbon electrode (PAA/GCE)	Pb^{+2}, Cd^{+2}, Co^{+2}	0.9 nM, 1.9 nM, 11.0 µM
27.	Polymeric and chalcogenide glass membranes	Pb^{+2}, Cd^{+2}, Cu^{+2}, Zn^{+2}	0.4, 0.06, 0.2, 0.2 nmol/L

Pb^{2+} enhanced to ~31% (up to 99 mg/g) [43]. Li et al. synthesised magnetic cyclo-dextrin–chitosan/GO (CCGO) composites, which exhibited superior Cr(VI) removal efficiency in acidic solutions [44].

14.4   COMPOSITES BASED ON ORGANIC CONDUCTING POLYMERS FOR THE DETECTION OF HEAVY METAL IONS

The bond structure of carbon nanotubes (CNTs) is mainly based on sp^2–sp^3 hybrid-isations [45], and these bonding natures are closely related to chirality, which are formed by vacancies in CNTs structure. CNTs strongly interact with metal ions due to the presence of small defects on CNTs surface, inner layers, as well as the inner cavity [46]. The binding affinity of CNTs towards heavy metal ions strongly depends on their isoelectric point as well as the pH value of the solution. The surface of CNTs are being negatively charged when the pH value of the solution is increased above the isoelectric point; and, therefore, the CNTs are not suitable for electrostatic interaction with cationic species, which leads to desorption of metal ions. As a result, the detection of heavy metal ions (HMI) through the sorption method are being dependent on the pH value of the analyte, which is responsi-ble for the affluence of CNTs surface [47]. The hydrophobicity of CNTs and the absence of binding groups on pristine CNTs resulted in weak adsorption of metal ions. CNTs modified with functional binding groups are one of the most suitable approaches to solve these issues. The schematic diagram of the response of carbon nanotubes and conducting polymer nanocomposite are illustrated in Figure 14.2.

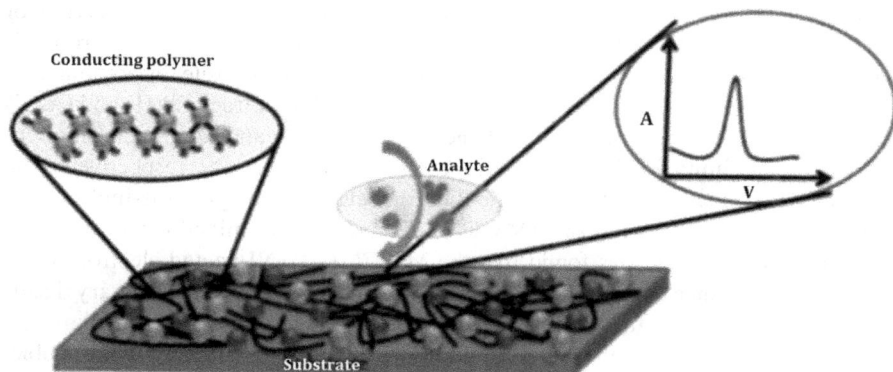

FIGURE 14.2 Heavy metal ion sensor based on carbon nanotube decorated with organic conducing polymer.

Sensor electrode modified by multiwalled carbon nanotubes (MWCNTs) decorated by dithizone for the investigation of Cd(II) ions is reported [48]. Importantly, it has been theoretically and experimentally established that organic conducting polymers are valuable materials in the field of material science because of their high pseudo-capacitance and high conductivity [49]. The main essential challenge in doped organic conducting polymers is the design of a short bandgap of the organic conjugated polymers' coatings. The extended p-conjugation in organic conducting polymers confers the required mobility to charges that are present on the polymer backbone that make them electrically conducting. Many researchers have planned to develop the ultimate properties of these two organic materials, such as organic conducting polymers and carbon nanotubes for the detection of heavy metal ions despite their several limitations and drawbacks. Organic conducting polymers are broadly used in the composite formation with carbon nanotubes due to their different beneficial properties, like high conductivity, environmental stability, flexibility, simple synthesis by chemical and electrochemical methods, as well as good procedural simplicity. Moreover, the composite of organic conducting polymers (OCPs) with carbon nanotubes can stimulate synergistic effects appropriate for advanced applications [50]. Furthermore, CNTs/OCPs can perform as a proficient filler material because of their high surface-to-volume ratio, when compared to that of other carbon materials [51]. The composites formation between CNTs and OCPs follows the p–p interactions, which often lead to collaborative properties in improving electrochemical performances. It is very important to note that CNTs and OCPs are easy to bind, and demonstrate strong interfacial coupling through donor–acceptor binding as well as *p–p* interaction. Notably, p-electrons of carbon nanotubes can efficiently interact with the quinoid ring of the organic conducting polymers [52]. Nafion is the most extensively studied and reported polymer that can smoothly be formed composite with carbon nanotubes and can be used for the detection of metal ions. Moreover, Nafion increases the solubilisation of carbon nanotubes [53], and it enhances the cation exchangeability and environmental stability. One of the most valuable applications of Nafion/carbon-nanotubes composite is the detection of metal ions through the replacement of mercury-based electrodes, which are environmentally toxic. Moreover, Nafion/carbon-nanotubes composite in combination with bismuth has been successfully used for the detection of trace Pb(II) and Cd(II) ions [54]. This composite structure enhanced the detection and reproducibility with good detection limits of 25 and 40 ng/L for Pb(II) and Cd(II) ions, respectively. The performance of glassy carbon electrode (GCE) coated with polyaniline-multiwalled carbon nanotubes nanocomposite (MWCNTs/PANI) was examined for the determination of Pb(II) ions. It was found that the MWCNTs/PANI-coated electrodes had superior performance compared to the bare GCEs [55]. Moreover, the primary disadvantage of CNTs is that these CNTs are functionalised with some functional groups, such as –COOH, –C=O, and –OH. In addition, CNTs are extremely hydrophobic. As a result, bare CNTs have no definite affinity to the metal ions due to the lack of sufficient functional groups on their surfaces, and, therefore, they cannot provide an excellent electrode material for the sensing of metal ions [56]. But organic conducting polymers have excellent environmental feasibility and chelating properties, and intrinsic affinity towards heavy metal ions [57]. Therefore, the combination of carbon nanotubes with organic conducting polymers became an excellent combination, where carbon nanotubes enable the quicker transfer of electrical signal; because of their high conductivity and organic conducting, polymers demonstrate superior affinity towards metal ions due to the aforementioned reasons. Accordingly, the combination of carbon nanotubes and organic conducting polymers became appropriate for the progress of highly selective and sensitive metal ion sensors.

14.5   CONCLUSION

Different organic as well as other carbonaceous composites for accurate and rapid sensing of toxic heavy metal ions have been accentuated in this chapter. The role of numerous solid polymeric supports like carbon nanotubes, graphene, polyacetylene, polythiophene, polypyrrole, polyphenylene, and polyaniline in terms of the stability of sensors and their corresponding activities have been demonstrated. Keeping in mind of the present environmental scenario, the applications of some eco-compatible carbon-based sensors for detection of heavy metals have also been highlighted.

REFERENCES

1. Liu, Z. J.; Chen, L.; Zhang, Z. C.; Li, Y. Y.; Dong, Y. H.; Sun, Y. B.Synthesis of multi-walled carbon nanotube-hydroxyapatite composites and its application in the sorption of Co(II) from aqueous solutions.*J. Mol. Liq.*2013, 179, 46–53.DOI:10.1016/j.molliq.2012.12.011.
2. Ren, X. M.;Yang, S. T.;Shao, D. D.;Tan, X. L.Retention of Pb(II) by alow-cost magnetic composite prepared by environmentally-friendly plasma technique.*Sep. Sci. Technol.*2013, 48, 1211–1219.DOI:10.1080/01496395.2012.726307.
3. Hu, J.;Yang, S. T.;Wang, X. K.Adsorption of Cu(II) on β-cyclodextrin modified multiwall carbon nanotube/iron oxides in the absence/presence of fulvic acid.*J. Chem. Technol. Biotechnol.*2012, 87, 673–681.DOI:10.1002/jctb.2764.

4. Hu, R.;Wang, X. K.;Dai, S. Y.;Shao, D. D.;Hayat, T.;Alsaedi, A.Application of graphitic carbon nitride for the removal of Pb(II) and aniline from aqueous solutions. *Chem. Eng. J.*2015, 260, 469–477.DOI:10.1016/j.cej.2014.09.013.

5. Arshadi, M.;Soleymanzadeh, M.;Salvacion, J. W. L.;SalimiVahid, F.Nanoscale zero-valent iron (NZVI) supported on sineguelas waste for Pb(II) removal from aqueous solution: Kinetics, thermodynamic and mechanism. *J. Colloid Interface Sci.*2014, 426, 241–251.DOI:10.1016/j.jcis.2014.04.014.

6. Ghatak, A.;Bhar, S.Chemoselective reduction of nitroaromatics using recyclable alumina-supported nickel nanoparticles in aqueous medium—Exploration to one pot synthesis of benzimidazoles. *Synth. Commun.*2022, 52, 368–379.DOI:10.1080/00397911.2 021.2024853.

7. Sinha, D.;Biswas, S.;Das, M.;Ghatak, A.An eco-friendly, one pot synthesis of tri-substituted imidazoles in aqueous medium catalyzed by RGO supported Au nano-catalyst and computational studies. *J. Mol. Struct.*2021, 1242, 130823.DOI:10.1016/j.molstruc.2021.130823.

8. Pramanik, A.;Ghatak, A.;Khan, S.;Bhar, S.Hydroarylation of alkynes and alkenes through alumina-sulfuric acid catalyzed regioselective C-C bond formation. *Tetrahedron Lett.*2019, 60, 1091.DOI:10.1016/j.tetlet.2019.03.006.

9. Ghatak, A.;Khan, S.;Bhar, S.Catalysis by β–cyclodextrin hydrate – synthesis of 2, 2-disubstituted-2H-chromenes in aqueous medium. *Adv. Synth. Catal.*2016, 358, 435–443.DOI:10.1002/adsc.201500358.

10. Ghatak, A.;Khan, S.;Roy, R.;Bhar, S.Chemoselective and ligand-free synthesis of diaryl ethers in aqueous medium using recyclable alumina-supported nickel nanoparticles. *Tetrahedron Lett.*2014, 55, 7082–7088. DOI:10.1016/j.tetlet.2014.10.144.

11. Manivannan, A.;Ramakrishnan, L.;Seehra, M. S.;Granite, E.;Butler, J. E.;Tryk, D. A.;Fujishima, A.Mercurydetection at boron doped diamond electrodes using a rotating disk technique. *J. Electroanal. Chem.*2005, 577, 287–293. DOI:10.1016/j.jelechem.2004.12.006.

12. Dragoe, D.;Sparatu, N.;Kawasaki, R.;Manivannan, A.;Sparatu, T.;Tryk, D. A.;Fujishima, A.Detection of trace levels of Pb^{2+} in tap water at boron-doped diamond electrodes with anodic stripping voltammetry. *Electrochim. Acta.*2006, 51, 2437–2441.DOI:10.1016/j.electacta.2005.07.022.

13. Kukla, A. L.;Kanjuk, N. I.;Starodub, N. F.;Shirshov, Y. M.Multienzyme electrochemical sensor array for determination of heavy metal ions. *Sens. Actuators B.*1999, 57, 213–218. DOI: 10.1016/S0925–4005(99)00153–7.

14. Stenger-Smith, J. D.Intrinsically electrically conducting polymers. Synthesis, characterization, and their applications. *Prog. Polym. Sci.*1998, 23, 57–79.DOI:10.1016/ S0079–6700(97)00024–5.

15. ALOthman, Z.; Alam, M.; Naushad, M.; Bushra, R. Electrical conductivity and thermal stability studies on polyaniline Sn(IV) tungstomolybdate nanocomposite cation-exchange material: Applicatinas Pb(II) ion-selective membrane electrode. *Int. J. Electrochem. Sci.*2015, 10, 2663–2684.DOI:10.1016/S1452–3981(23)04876–9.

16. Ratautaite, V.;Plausinaitis, D.;Baleviciute, I.;Mikoliunaite, L.;Ramanaviciene, A.;Ramanavicius, A.Characterization of caffeine imprinted polypyrrole by a quartz crystal microbalance and electrochemical impedance spectroscopy. *Sensors Actuators BChem.* 2015, 212, 63–71.DOI:10.1016/j.snb.2015.01.109.

17. Tarabek, A.;Rapta, P.;Jahne, E.;Ferse, D.;Adler, H.;Maumy, M.;Dunsch, L.Spectro electrochemical and potentiometric studies of functionalised electroactive polymers. *Electrochim. Acta.*2005, 50, 1643–1651.DOI: 10.1016/j.electacta.2004.10.020.

18. Kausaite-Minkstimiene,A.;Mazeiko,V.;Ramanaviciene,A.;Ramanavicius,A.Evaluation of chemical synthesis of polypyrrole particles. *Colloids Surf. A: Physicochem. Eng. Aspects.*2015, 483, 224–231.DOI: 10.1016/j.colsurfa.2015.05.008.

19. Leonavicius, K.;Ramanaviciene, A.;Ramanavicius, A.Polymerization model for hydrogen peroxide initiated synthesis of polypyrrole nanoparticles. *Langmuir.*2011, 17, 10970–10976.DOI:10.1021/la201962a.

20. Ramanavicius, A.;Rekertaite, A. I.; Valiu _ nas, R.; Valiuniene, A. Single-step procedure for the modification of graphite electrode by composite layer based on polypyrrole, prussian blue and glucose oxidase. *Sensors Actuators B Chem.*2017, 240, 220–223.DOI: 10.1016/j.snb.2016.08.142.

21. Das, A. K.;Sepay, N.;Nandy, S.;Ghatak, A.;Bhar, S.Catalytic efficiency of β-cyclodextrin hydrate-chemoselective reaction of indoles with aldehydes in aqueous medium. *Tetrahedron Lett.*2020, 61, 152231–152237.DOI:10.1016/j.tetlet.2020.152231.

22. Das, A. K.;Nandy, S.;Bhar, S.Chemoselective and ligand-free aerobic oxidation of benzylic alcohols to carbonyl compounds using alumina supported mesoporous nickel nanoparticle as an efficient recyclable heterogeneous catalyst. *Appl Organomet Chem.*2021, 35, e6282.DOI:10.1002/aoc.6282.

23. Ratautaite, V.;Ramanaviciene, A.;Oztekin, Y.;Voronovic, J.;Balevicius, Z.;Mikoliunaite, L.;Ramanavicius, A.Electrochemical stability and repulsion of polypyrrole film. *Colloids Surf. A: Physicochem. Eng. Aspects.*2013, 418, 16–21.DOI: 10.1016/j. colsurfa.2012.10.052.

24. Das, A. K.;Nandy, S.;Bhar, S.Cu(OAc)$_2$catalysed aerobic oxidation of aldehydes to nitriles under ligand-free condition. *RSC Adv.*2022, 12, 4605–4614.DOI:10.1039/ DIRA07701E.

25. Nandy, S.;Ghatak, A.;Das, A. K.;Bhar, S.Chemoselective and metal-free synthesis of aryl esters from the corresponding benzylic alcohols in aqueous medium using TBHP/TBAI as an efficient catalytic system. *Synlett.*2018, 29, 2208–2212. DOI:10.1055/s–0037–1610247.

26. Nandy, S.;Das, A. K.;Bhar, S.Chemoselective formation of C-N bond in wet acetonitrile using Amberlyst®-15(H) as a recyclable catalyst. *Syn Commun.*2020, 50, 3326–3336. DOI:10.1080/00397911.2020.1801745.

27. Ramanavicius, A.;Andriukonis, E.;Stirke, A.;Mikoliunaite, L.;Balevicius, Z.;Ramanaviciene, A.Synthesis of polypyrrole within the cell wall of yeast by redox-cycling of [Fe(CN)6]3-/[Fe(CN)6]4-. *Enzyme Microb. Technol.*2016, 83, 40–47.DOI:10.1016/j. enzmictec.2015.11.009.

28. Kanatzidis, M. G.Conducting polymers.*Chem. Eng. News.*1990, 68, 36–54.DOI: 10.1021/cen-v068n049.p036.

29. Deshmukh, M. A.;Shirsat, M. D.;Ramanaviciene, A.;Ramanavicius, A.Composites based on conducting polymers and carbon nanomaterials for heavy metal ion sensing (review). *Crit.Rev.Anal. Chem.*2018, 48, 293–304. DOI:10.1080/10408347 .2017.1422966.

30. Kumari, K.;Ali, V.;Kumar, A.;Kumar, S.;Zulfequar, M. D. C.Conductivity and spectroscopic studies of polyaniline doped with binary dopant ZrOCl2/AgI.*Bull. Mater. Sci.*2011, 34, 1237–1243.DOI: 10.1007/s12034-011-0238–6.

31. Somerset, V.;Silwana, B.;Horst, C.;Iwuoha, E.Construction and evaluation of a carbon paste electrode modified with polyanilineco-poly(dithiodianiline) for enhanced stripping voltammetric determination of metal ions.*Sens. Electroanal.*2014, 8, 143–154. DOI: 10.204/7930.

32. Oztekin, Y.;Ramanaviciene, A.;Yazicigil, Z.;Solak, A. O.;Ramanavicius, A.Direct electron transfer from glucose oxidase immobilized on polyphenanthroline modified-glassy carbon electrode.*Biosens. Bioelectron.*2011, 26, 2541–2546.DOI:10.1016/j. bios.2010.11.001.

33. Chen, S.;Wen, T.;Gopalan, A.Electrosynthesis and characterization of a conducting copolymer having S–S links.*Synth. Met.*2003, 132, 133–143.DOI: 10.1016/ S0379–6779(02)00211–4.

34. Ramanaviciene, A.;Kausaite, A.;Tautkus, S.;Ramanavicius, A.Biocompatibility of polypyrrole particles: An in vivo study in mice. *J. Pharm. Pharmacol.*2007, 59, 311–315.DOI: 10.1211/jpp.59.2.0017.

35. Vaitkuviene, A.;Ratautaite, V.;Mikoliunaite, L.;Kaseta, V.;Ramanauskaite, G.;Biziuleviciene, G.;Ramanaviciene, A.;Ramanavicius, A.Some biocompatibility aspects of conducting polymer polypyrrole evaluated with bone marrow-derived stem cells.*Coll. Surf. A.*2014, 442, 152–156.DOI: 10.1016/j. colsurfa.2013.06.030.

36. Oztekin, Y.;Yazicigil, Z.;Ramanaviciene, A.;Ramanavicius, A.Polyphenol-modified glassy carbon electrodes for copper detection.*Sens. Actuators, B Chem.*2011, 152, 37–48.DOI:10.1016/j. snb.2010.09.057.

37. Laskar, M. A.;Siddiqui, S.;Islam, A.Reflection of the physiochemical characteristics of 1-(2-pyridylazo)-2-naphthol on the pre-concentration of trace heavy metals. *Crit. Rev. Anal. Chem.*2016, 46, 413–423.DOI: 10.1080/10408347.2016.1140019.

38. Palys, B. J.; Skompska, M.; Jackowska, K.SERS of 1,8-diaminonaphthalene on gold, silver and copper electrodes polymerisation and complexes formed with the electrode material. *J. Electroanal. Chem.* 1997, 428, 19–24. DOI: 10.1016/S0022–0728(97)00029–6.

39. Hupe, J.; Wolf, G. D.; Jonas, F.DMS-E – A recognised principle with a novel basis Through-hole contacting of printed circuit boards using conductive polymers. *Galvanotechnik.* 1995, 86, 3404.

40. Chowdhury, S.; Balasubramanian, R.Recent advances in the use of graphene-family nanoadsorbents for removal of toxic pollutants from wastewater. *Adv. Colloid. Interf. Sci.*2014, 204, 35–56. DOI:10.1016/j.cis.2013.12.005.

41. Shao, D.; Hu, J.; Wang, X.Plasma induced grafting multiwalled carbon nanotube with chitosan and its application for removal of UO, Cu^{2+}, and Pb^{2+} from aqueous solutions. *Plasma Process. Polym.*2010, 7, 977–985. DOI:10.1002/ppap.201000062.

42. Nyairo, W. N.; Eker, Y. R.; Kowenje, C.; Akin, I.; Bingol, H.; Tor, A.; Ongeri, D. M.Efficient adsorption of lead (II) and copper (II) from aqueous phase using oxidized multiwalled carbon nanotubes/polypyrrole composite. *Sep. Sci. Technol.*2018, 1498–1510. DOI:10.1080/01496395.2018.1424203.

43. He, Y. Q.; Zhang, N. N.; Wang, X. D.Adsorption of graphene oxide/chitosan porous materials for metal ions. *Chinese Chem. Lett.*2011, 22, 859–862. DOI:10.1016/j. cclet.2010.12.049.

44. Li, L.; Fan, L.; Sun, M.; Qiu, H.; Li, X.; Duan, H.; Luo, C.Adsorbent for chromium removal based on graphene oxide functionalized with magnetic cyclodextrin-chitosan. *Colloid. Surface B.* 2013, 107, 76–83. DOI:10.1016/j.colsurfb.2013.01.074.

45. Blase, X.; Benedict, L.; Shirley, E.; Louie, S.Hybridization effects and metallicity in small radius carbon nanotubes. *Phys. Rev. Lett.*1994, 72, 1878–1881. DOI: 10.1103/PhysRevLett.72.1878.

46. Li, Y.; Wang, S.; Luan, Z.; Ding, J.; Xu, C.; Wu, D.Adsorption of cadmium(II) from aqueous solution by surface oxidized carbon nanotubes. *Carbon.* 2003, 41, 1057–1062. DOI: 10.1016/S0008–6223 (02)00440–2.

47. McPhail, M.; Sells, J.; He, Z.; Chusuei, C.Charging nanowalls: Adjusting the carbon nanotube isoelectric point via surface functionalization. *J. Phys. Chem. C.*2009, 113, 14102–14109. DOI: 10.1021/jp901439g.

48. Karimi, M.; Aboufazeli, F.; Zhad, H.; Sadeghi, O.; Najafi, E.Determination of cadmium (II) ions in environmental samples: A potentiometric sensor. *Curr World Environ.* 2012, 7, 201–206. DOI: 10.12944/CWE.7.2.02.

49. Peng, C.; Zhang, S.; Jewell, D.; Chen, G.Carbon nanotube and conducting polymer composites for supercapacitors. *Prog.Nat. Sci.*2008, 18, 777–788. DOI: 10.1016/j. pnsc.2008.03.002.

50. Bal, S.; Samal, S. S.Carbon nanotube reinforced polymer composites – A state of the art. *Bull. Mater. Sci.*2007, 30, 379–386. DOI: 10.1007/s12034-007-0061–2.

51. Sharma, A.; Sharma, Y.P-toluene sulfonic acid doped polyaniline carbon nanotube composites: Synthesis via different routes and modified properties. *J. Electrochem. Sci. Eng.* 2013, 3, 47–56. DOI: 10.5599/96.
52. Huang, J. E.; Li, X. H.; Xu, J. C.; Li, H. L.Well-dispersed singlewalled carbon nanotube/polyaniline composite films. *Carbon.* 2003, 41, 2731–2736. DOI: 10.1016/S0008–6223(03)00359–2.
53. Wang, J.; Musameh, M.; Lin, Y.Solubilization of carbon nanotubes by nafion toward the preparation of amperometric biosensors. *J. Am. Chem. Soc.* 2003, 125, 2408–2409. DOI: 10.1021/ja028951v.
54. Xu, H.; Zeng, L.; Xing, S.; Xian, Y.; Shi, G.; Jin, L.Ultrasensitive voltammetric detection of trace lead(II) and cadmium(II) using MWCNTs-nafion/bismuth composite electrodes. *Electroanalysis.*2008, 20, 2655–2662. DOI: 10.1002/elan.200804367.
55. Wang, Z.; Liu, E.; Gu, D.; Wang, Y.Glassy carbon electrode coated with polyaniline-functionalized carbon nanotubes for detection of trace lead in acetate solution. *Thin Solid Films.*2011, 519, 5280–5284. DOI: 10.1016/j.tsf.2011.01.175.
56. Lau, K. K. S.; Bico, J.; Teo, K. B. K.; Chhowalla, M.; Amaratunga, G. A. J.; Milne, W. I.; McKinley, G. H.; Gleason, K. K.Superhydrophobic carbon nanotube forests. *Nano Lett.*2003, 3, 1701–1705. DOI: 10.1021/nl034704t.
57. Lange, U.; Roznyatovskaya, N. V.; Mirsky, V. M.Conducting polymers in chemical sensors and arrays. *Anal Chim Acta.*2008, 614, 1–26. DOI: 10.1016/j.aca.2008.02.068.

15 Catalytic Degradation-Based Water Pollution Remediation

Tushar Kanti Das
Silesian University of Technology

CONTENTS

15.1 INTRODUCTION

Water is an integral part of any living being support systems, and no one can think about living without water in the universe. So, it is very much essential to clean the water and supply high quality of water to the society. Due to urbanisation, accelerated growth of various industries and the paucity of people's knowledge about water pollution, people of many countries are falling to get fresh quality of water and suffering from various diseases (Schwarzenbach et al. 2010). Most of the earth's surface (almost 70%) is enclosed with water, but less than 1% freshwater is available for human beings to live our lives. This water has been found in lake, river, reservoirs, and streams, and it has been found that it is very difficult to extract the freshwater at a reasonable cost from these water resources (Mishra and Dubey 2015). The ability of the atmosphere is not so much strong that it can accommodate large quantity of freshwater for a longer time, and very large amount of this freshwater is recycled at a very short time to the atmosphere (Boberg 2005). Among this quantity of freshwater, only 2.8% is available for human consumption, and rest of the amount is present in the ocean, and this amount cannot be used without further purification due to its saltiness (Viman et al. 2010). It is already reported by World Health Organization

(WHO) that almost 84 crore people worldwide are not getting enough drinking water (Wutich et al. 2020). Regular use of polluted or untreated water might cause diseases not only for human beings but also for other animals. So, it is very essential to remediate or recirculate the wastewater properly, and it has become now one of the hot topics of research (Karimi-Maleh and Arotiba 2020). On the other side for the last few decades, the remediation of waste water has become one of auspicious trend to obtain water sustainability. The recycling of wastewater not only reduces the pollution of environment but also compensates the shortage of clean water resources (Van der Bruggen 2013).

Pollution of water takes place when discharge of one or more hazardous materials into the freshwater bodies will change the quality of water, and provide a negative impact on the health of living beings and to the environment. The source of polluted materials can be off two types; one is that pollutants come from single sources such as from industries and discharge to clean water, and another one is the discharge of pollutants from multiple sources (Crini and Lichtfouse 2019). The reasons behind the pollution of water are multiple: industrial waste, mining activity, agricultural waste, domestic waste, radioactive waste, energy use, urban development, etc. Whatever the activity may be, when it produces toxic waste by-products that get mixed with water are identified as pollutants. So, a sustained effort should be done to protect clean water resources from contamination of this type of pollutants (Egbueri et al. 2022).

15.2 POLLUTANTS PRESENT IN WATER

The types of pollutants present in the wastewater can be classified based on sources, mode of their occurrences and type of activity for which the pollutants are mixed with water. According to the source, the pollutants can be divided into two categories: one is point sources and the other is non-point sources. Point source pollution takes place from a specific source, and it is very easy to recognise and monitors the contaminants. The discharge of pollutants either from the municipal wastewater plant or industrial plant into the freshwater bodies is an example of point source pollution (Viman et al. 2010). On the other hand, non-point source pollution has multiple discharge points. So, the pollution mixed with water cannot easily trace and quantify. The pollutants coming from agriculture sources are examples of non-point source, as the contaminants present in the agriculture waste is mixed up of pollutants from different sources (Moss 2008). Depending upon their mode of occurrence the water pollutants can be classified into three categories: namely physical, chemical and biological pollutants (Akinbo and Tawari-Fufeyin 2014). The example of physical pollutants is waste heat from the industries, solid particles, various types of dyes and pigments used as colourants, etc. On the other hand, chemical pollutants are divided into two types: one is organic (like plastic, pesticides etc.) and the other is inorganic (such as N, P, Cl, F, As, Pb, etc.). The biological pollutants include various microorganism, worms, and algae. Any kind of human activity can easily disturb the environmental balance, which leads to pollution in water. But, for more clarity, we can divide three sources as a primary category for the water pollution, namely, industrial waste, agricultural waste, and domestic waste.

15.2.1 INDUSTRIAL WASTE

Waste from the industries is the major source of water pollutants. In most of the countries the industries disposed pollutants to the untreated water, which got mixed with freshwater bodies and polluted water supply. Manufacturing industries such as chemical plant, steel plant, refining plant, etc. polluted water much more than others, owing to high production of toxic heavy metals and various organic chemicals. Other industries such textile, leathers, pharmaceuticals, paper, pulp, rubber industries, etc. have also an impact on the water pollution, but not so much high like that of manufacturing industries (Gambhir et al. 2012). For the treatment of industrial wastewater pollutants, it is necessary to develop a specific type of treatment process, as normal sewage treatment cannot effectively remove this type of toxic pollutants. Pollutants containing heavy metals are the major contributors of the industrial waste, and these heavy metals are non-degradable and highly toxic, which is a major concern for treatment process. So, it is better to recover these heavy materials from the wastewater and use them for other purpose of applications. This not only reduces the pollution but also increases the consumption of resources (Jumbe and Nandini 2009). The contaminants originating from industries can be introduced into the freshwater bodies through three different ways, namely, gaseous emission, liquid waste, and solid wastes. But the most serious problem will happen when the waste materials were disposed in water through uncontrolled ways. Industrial waste can be classified into four categories such as:

a. Biodegradable waste with high biological oxygen demand (BOD), which is produced from mainly food industries,
b. Waste produced with high BOD and notable toxicity, which is generated from petroleum refining, paper, and pulp industries,
c. Waste with high toxicity and low BOD produced from chemical industries,
d. Thermal waste originated from power plants and steel plants.

From the viewpoint of water pollution, (b) and (c) are considered to cause the most serious problems to living beings (Pandey 2006).

15.2.2 AGRICULTURAL WASTE

The agriculture waste is formed during cultivation of crops and nurture of various animals. The agriculture waste is the primary resource of sediment pollution in water. Owing to this type of pollution, the transport ability of freshwater bodies is reduced, and it also decreases penetration of sunlight in the water, which inhabits growth of flora inside water. This is the food of living bodies present in water, and prohibition to formation of this flora directly affects the food chain system. This type of sediment pollution is sometimes attached to living bodies in the aqueous system and has become cause of death. The sediment pollutants include pesticides, nutrients and various dangerous chemicals used during cultivations (Gambhir et al. 2012). It has been reported that globally 4.6 million tons of this type of pollutants are outspread in the environment that are gradually discharged into freshwater bodies (Zhang et al. 2011).

15.2.3 Domestic Waste

Domestic wastes are generated from the household. The domestic waste mixes up with freshwater either through leakage of tank or unlawful dumping of waste materials near river or lake. The most important concern is that the detergents used for cleaning of household products are manufactured from petrochemical products and can affect the human life. Hence, the discharge of this chemical-contaminated water without treatment into the river, lake or sea can cause serious problems for the living bodies. This type of discharge is not only the threat to human life but also affects the physicochemical and biological quality of the water bodies (Kuhn et al. 2022).

15.3 CATALYTIC TREATMENT OF WASTEWATER CONTAINING POLLUTANTS

Catalysis is a process that accelerates the rate of chemical reaction without affecting itself. The primary function of catalyst is to reduce the activation energy of any reaction through formation of complex by the active site of the catalyst and reactants to form a surface-bound intermediate states. The main advantage of catalytic treatment of pollutants presents in wastewater is that the time required for the treatment is less, as well as it can be recycled number of times. And sometimes the material converted into other products can be useful for some other purpose (Das et al. 2018; Heck et al. 2019). The advancement of polymer science and carbonaceous materials allow us to fabricate advanced composites materials based on polymer- and carbon-based materials for an extended range of application beginning from energy storage to environmental protection. Polymer composites are a combination of polymer and other materials in which basically polymer serves as matrix, and other materials are dispersed throughout the polymer matrix. These materials exhibit distinguishable characteristic properties that are hard to recognise in either of the single component, and the unique characteristic properties of the polymer composites are build up through synergetic effects of both polymer and other materials (such carbon-based materials) (Gupta et al. 2021).

Recently, waste water contains various toxic and carcinogenic pollutants that cannot be removed by conventional techniques. So, advanced technologies are essential for the treatment of water before it is discharged into main stream water bodies. Among those various advanced technologies, advanced oxidation process (AOPs) has been found to be more effective than any other process. In this process, reactive oxygen species are created in the presence of sunlight or ultra violet light by the catalytic material, which is responsible for degradation of toxic pollutants such as phenols, pesticides, pharmaceuticals, dyes, and petrochemicals products (Ateia et al. 2020). Various AOPs such as photocatalysis, Fenton-like processes and ozonisation are engaged in the treatment of pollutants present in water, but among them the photocatalytic treatment of pollutants are most popular than other AOPs (Ghime and Ghosh 2020). Though the process is less efficient in the removal of total organic carbon (TOC) in seawater, but the advantages of photocatalysis process such as low cost, reusable, eco-friendly and complete degradation of pollutants makes them one of the potential process for treatment of water pollutants (Kumar et al. 2014). Owing to these advantageous points of photocatalytic process, researchers are more concentrated on the modification of these photocatalytic materials to obtain better performance in the purification of water.

15.4 POLYMER–CARBONACEOUS FILLER-BASED COMPOSITES FOR CATALYTIC DEGRADATION-BASED WATER POLLUTION REMEDIATION

At present, wide ranges of materials are available to treat the pollutants present in water bodies. Among them, polymer–carbonaceous filler-based composites have been found to be much more prominent and gained popularity than others owing to facile synthesis, high stability, superior catalytic activities, unique electrochemical properties, high strength-to-weight ratio, and easy-to-fabricate device for catalytic applications (Ismail and Goh 2018; Das et al. 2019). Here, we discuss few interesting works on polymer–carbonaceous filler-based composites for catalytic degradation of water pollutants and various aspects of different fabricated polymer–carbonaceous filler-based composites towards degradation of different water contaminants. Alpatova et al. fabricated Fe_2O_3 nanoparticles and multiwall carbon nanotubes (MWCNTs) impregnated polyvinylidene fluoride (PVDF) composite membrane for Fenton-like catalytic degradation of cyclohexanoic acid and humic acids in the presence of hydrogen peroxide. They performed the catalytic degradation in batch mode by polymer composites' sample as well as flow mode through the fabricated membranes, and studied the catalytic degradation of pollutants through the membranes. The membrane used in flow mode for catalytic degradation was fabricated by specific concentration of Fe_2O_3 nanoparticles and multiwall carbon nanotubes, and this concentration was judged by the water permeability test (Alpatova et al. 2015). Recently, graphene oxide (GO)-based polymer composite has been found to be one of the promising materials for treatment of water pollutants. This type of composites are prepared by dispersing exfoliated GO in the polymer matrix resulting in development of attractive properties compared to that of pure polymers (Kuilla et al. 2010). On the other side, Polyaniline (PANI), a conducting polymer has also been found an effective photocatalytic material, as it contains benzenoid and quinonoid ring structure, which holds highly mobilised charge carriers upon being excited with visible light. Based upon this, Shin et al. prepared graphene–PANI nanocomposites by *in-situ* polymerisation of aniline and used them as photocatalytic material for degradation toxic Rose Bengal (RB) dye. The nanocomposites have high adsorption capacity of RB owing to π–π^* interaction between RB molecules and graphene, and upon irradiation with light the photogenerated charged carriers are formed in PANI, and presence of graphene in the nanocomposites help to interrupt the combination of charge carriers (electrons and hole). The generated electrons get transferred from PANI to graphene surface, where the electrons are reacted with dissolved oxygen to form active radicals on its surface, which is primarily responsible for the degradation of toxic RB dyes into CO_2 and H_2O (Ameen et al. 2012). Moon et al. prepared pH-sensitive poly(vinyl alcohol)/poly(acrylic acid)/TiO$_2$/graphene oxide (GO) nano-composite hydrogels through radical and condensation polymerisation for photocatalytic degradation of toxic methylene blue (MB) and coomassie brilliant blue R-250 (CBB). They revealed that the photocatalytic degradation behaviour of the polymeric composites towards these pollutants was significantly improved by the introduction of carbonaceous GO due to more facile electron transfer between TiO$_2$ nanoparticles and GO upon irritated with UV light and good adsorption property GO towards these water pollutants (Moon et al. 2013). Alshabanat et al. prepared eco-friendly polymer nanocomposites films using chitosan and polyvinyl alcohol (PVA) as matrix and carbon black as a reinforcement, and the films were engaged in photocatalytic degradation of toxic pollutants Congo red (CR). They reported that pH of the solution, pollutants concentration, intensity of light and contact time with the polymer composites films affect the degradation rate of pollutants, and the presence of carbon black particles make the polymeric films from insulating to conductive nature, which is responsible for functioning the polymer nanocomposites film as a photocatalyst (Alshabanat and AL-Anazy 2018). Carbon nitride (g-C$_3$N$_4$)-based photocatalyst has been found to be gaining attention owing to its two-dimensional (2D) structure, adjustable electronic properties, and excellent chemical stability. But still its application as alone is very much limited due to ease of recombination of photogenerated electron and hole, and low surface area resulting less adsorption of pollutants. For this reason researchers are trying to combine with various polymers that remove the barrier for performing as an active photocatalyst towards degradation of various toxic pollutants (Lee and Chang 2019). Zhang et al. fabricated hierarchical nanocomposites of polyaniline (PANI) nanorods developed on the surface of carbon nitride (g-C$_3$N$_4$) sheets via dilute polymerisation, and its photocatalytic activity was tested against two toxic pollutants methylene blue (MB) and methyl orange (MO) as model pollutants. They reported that the composites exhibit better catalytic activity towards degradation of MO than MB due to opposite nature of charge of MO, and fabricated composites enhanced more adsorption of pollutants on composites surfaces. They believed that the improved activity of composites owing to synergistic effect of PANI and g-C$_3$N$_4$, and demonstrated three reasons behind its superior photocatalytic activity: (i) at the interface of g-C$_3$N$_4$ and PANI a heteroconjugation is formed that helps in active separation of photogenerated electron–hole pairs and reduces the probability of their combinations, (ii) a wide range of adsorption band of PANI in the visible region makes the composite more efficiently used in the solar spectrum, and (iii) good adsorption properties of composites towards MO and MB pollutants. The schematic mechanism of photocatalytic degradation of dye molecules through formation of active radical species such as holes, $\cdot$O2$-$, $\cdot$OH is represented in the Figure 15.1 (Zhang et al. 2014).

Lei et al. prepared a binary polymer composite based on poly(diphenylbutadiyne) (PDPB) and g-C$_3$N$_4$, and examined its photocatalytic activity towards degradation of toxic phenol and Rhodamine B dyes. The increase catalytic activity of the composite is mainly due to high visible light adsorption property of PDPB as well as its effectiveness for functioning as charge separator in the composite (Lei et al. 2017). Wang et al. fabricated palladium nanoparticles (PdNPs)-decorated nitrogen-doped carbon nanotubes (CNTs) through variation of polyvinylpyrrolidone (PVP)

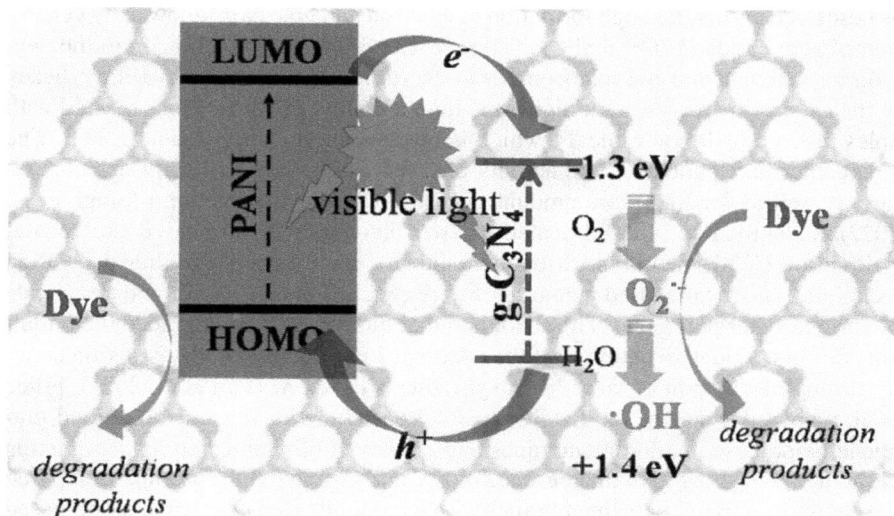

FIGURE 15.1 Schematic diagram illustrating the mechanism of photodegradation over CN–PANI under visible light. [Adapted from Zhang et al. (2014) with permission from Elsevier.]

amount, and they found that amount of PVP play a significant role in the fabrication of nanostructured catalyst. The fabricated catalyst exhibits good catalytic activity towards conversion of iodobenzene (a chemical that is also found in wastewater), and the activity of catalyst increases with increase in temperature as well as content of PdNPs in the fabricated nanostructured materials (Wang et al. 2017). Yu et al. fabricated polydopamine/graphitic carbon nitride (PDA/g-C$_3$N$_4$) composites by modified dopamine polymerisation and the used for catalytic degradation of toxic methylene blue (MB), Rhodamine B (RhB) and phenol in presence visible light. In this composite, polydopamine (PDA) functions as a light adsorption material, and the ample of quinone functional groups help in transfer of photo-induced electrons and accept the electron from the semiconductors g-C$_3$N$_4$ materials that reduce the probability of electron–hole recombination resulting in increase in photocatalytic activity. They also studied the effect of PDA ratio on composites towards photocatalytic degradation, and they revealed that the degradation of toxic materials increases with increment of PDA in the composites owing to the above mentioned properties (Yu et al. 2017). Singh et al. fabricated trinary-nanocomposites based on zinc oxide nanoparticles (ZnO NPs), activated carbon (AC) and polypyrrole (PPy) through *in-situ* synthesis of AC from biowaste rice husk followed by polymerisation of pyrrole monomer in acid medium, and later activated carbon (AC) and PPy-based composites combined with ZnO NPs via synthesis in basic medium. The prepared nanocomposite was engaged in photocatalytic degradation of methylene blue (MB), and they studied the effect of temperature, pH of the medium, catalyst dosages and recyclability of the fabricated nanocomposite towards degradation of MB. They proposed similar mechanism of photocatalytic degradation of pollutant, as described earlier, through formation of electron and hole pair followed by generation of active radicals O^{2-} and OH. They revealed that at low pH medium the degradation efficiency of the nanocomposite decreases, while at high pH the efficacy of the catalyst increases, as at higher pH high amount of hydroxyl ions reacted with holes to from hydroxyl radicals, which is responsible for degradation of MB. The nanocomposites photocatalytic activity also increases with increment of temperature as well as loading more amount of catalyst in reaction medium (Singh et al. 2022). Recently, Pan et al. fabricated polypyrrole-coated carbon-based electrocatalytic (PPy@CCM) membrane through monitoring electro polymerisation deposition technique, and the prepared membranes were engaged in electrocatalytic treatments of various water pollutants. They explored that the PPy@CCM degraded the pollutants through oxidative action, and the coating of PPy increases the oxidation activity through formation of more hydroxyl radicals on CCM (Pan et al. 2022). Filice et al. prepared graphene oxide (GO) and sulfonic functional groups-modified graphene oxide-based nafion membranes and employed in photocatalytic degradation of methyl orange (MO) in the presence of UV and visible light. They also compared the catalytic activity of the membranes with commonly used titanium dioxide-based nanfion membranes, and reported that the activity of sulfonic functional groups-modified graphene oxide-based nafion membranes towards degradation of MO is much higher than titanium dioxide (TiO_2)-based nanfion membranes (Filice et al. 2015). Here we described few research works and their applications in treatments of water contaminants, but still researchers have continued their study to develop highly effective, catalytic, carbon-based polymer composites by either modification of already developed materials or fabrication of new polymer–carbonaceous filler-based composites for treatment of water pollutants.

15.5   SUMMARY

Based on the above discussion on catalytic degradation of water pollutants by the carbon-based polymer composites, one can conclude that various water pollutants can be efficiently degraded by catalytic process. The catalytic treatments are very much important when water is contaminated with highly toxic materials that are difficult to remove by conventional techniques. The application of carbon-based polymer composites towards treatment of water pollutants are still in the lab scale, and researchers are paid more attention to make such devices so that it can be useful in the commercial scale. Besides, the production of carbon-based materials still face the problem such as high production cost and mass production resulting in the research based on carbon-based polymer composites for catalytic degradation of water pollutants remains in the lab scale. On the other side, the toxicity of carbon-based polymer composites has to be considered during fabrication of water treatment catalytic devices, as there is always a chance to release of such toxic materials in the environment. Though lots of challenges still exist based on carbon-based polymer composites, but based on the advantageous point of carbon-based polymer composite over others, it can be an excellent candidate in catalytic degradation of water pollutants in future, and the researchers should be paid more attention on this field for future growth of carbon-based polymer composites.

ABBREVIATIONS

2D	Two-dimensional
AC	Activated carbon
AOPs	Advanced oxidation process
BOD	Biological oxygen demand
CBB	Coomassie brilliant blue R-250
CNTs	Carbon nanotubes
CR	Congo red
g-C$_3$N$_4$	Carbon nitride
GO	Graphene oxide
MB	Methylene blue
MO	Methyl orange
MWCNTs	Multiwall carbon nanotubes
PANI	Polyaniline
PDA	Polydopamine
PDA/g-C$_3$N$_4$	Polydopamine/graphitic carbon nitride
PdNPs	Palladium nanoparticles
PDPB	Poly(diphenylbutadiyne)
Ppy	Polypyrrole
PPy@CCM	Polypyrrole-coated carbon-based electrocatalytic
PVA	Polyvinyl alcohol
PVDF	Polyvinylidene fluoride
PVP	Polyvinylpyrrolidone
RB	Rose bengal
RhB	Rhodamine B
TiO$_2$	Titanium dioxide
TOC	Total organic carbon
WHO	World Health Organization
ZnO NPs	Zinc oxide nanoparticles

REFERENCES

Akinbo, M. I. and P. Tawari-Fufeyin (2014). "Physical, chemical and biological parameters of water from medical waste dumpsites in South-Western Niger Delta, Nigeria." *Asian Journal of Water, Environment and Pollution* 11(4): 83–88.

Alpatova, A., M. Meshref, K. N. McPhedran and M. G. El-Din (2015). "Composite polyvinylidene fluoride (PVDF) membrane impregnated with Fe2O3 nanoparticles and multiwalled carbon nanotubes for catalytic degradation of organic contaminants." *Journal of Membrane Science* 490: 227–235.

Alshabanat, M. N. and M. M. AL-Anazy (2018). "An experimental study of photocatalytic degradation of congo red using polymer nanocomposite films." *Journal of Chemistry* 2018: Article ID 9651850.

Ameen, S., H.-K. Seo, M. S. Akhtar and H. S. Shin (2012). "Novel graphene/polyaniline nanocomposites and its photocatalytic activity toward the degradation of rose Bengal dye." *Chemical Engineering Journal* 210: 220–228.

Ateia, M., M. G. Alalm, D. Awfa, M. S. Johnson and C. Yoshimura (2020). "Modeling the degradation and disinfection of water pollutants by photocatalysts and composites: A critical review." *Science of the Total Environment* 698: 134197.

Boberg, J. (2005). *Liquid Assets: How Demographic Changes and Water Management Policies Affect Freshwater Resources*, RAND Corporation.

Crini, G. and E. Lichtfouse (2019). "Advantages and disadvantages of techniques used for wastewater treatment." *Environmental Chemistry Letters* 17(1): 145–155.

Das, T. K., S. Ganguly, P. Bhawal, S. Mondal and N. C. Das (2018). "A facile green synthesis of silver nanoparticle-decorated hydroxyapatite for efficient catalytic activity towards 4-nitrophenol reduction." *Research on Chemical Intermediates* 44(2): 1189–1208.

Das, T. K., P. Ghosh and N. C. Das (2019). "Preparation, development, outcomes, and application versatility of carbon fiber-based polymer composites: A review." *Advanced Composites and Hybrid Materials* 2(2): 214–233.

Egbueri, J. C., D. A. Ayejoto and J. C. Agbasi (2022). "Pollution assessment and estimation of the percentages of toxic elements to be removed to make polluted drinking water safe: A case from Nigeria." *Toxin Reviews*: 1–15. doi: 10.1080/15569543.2021.2025401.

Filice, S., D. D'Angelo, S. Libertino, I. Nicotera, V. Kosma, V. Privitera and S. Scalese (2015). "Graphene oxide and titania hybrid Nafion membranes for efficient removal of methyl orange dye from water." *Carbon* 82: 489–499.

Gambhir, R. S., V. Kapoor, A. Nirola, R. Sohi and V. Bansal (2012). "Water pollution: Impact of pollutants and new promising techniques in purification process." *Journal of Human Ecology* 37(2): 103–109.

Ghime, D. and P. Ghosh (2020). *Advanced Oxidation Processes: A Powerful Treatment Option for the Removal of Recalcitrant Organic Compounds. Advanced Oxidation Processes-Applications, Trends, and Prospects*, IntechOpen.

Gupta, N., R. P. Behere, R. K. Layek and B. K. Kuila (2021). "Polymer nanocomposite membranes and their application for flow catalysis and photocatalytic degradation of organic pollutants." *Materials Today Chemistry* 22: 100600.

Heck, K. N., S. Garcia-Segura, P. Westerhoff and M. S. Wong (2019). "Catalytic converters for water treatment." *Accounts of Chemical Research* 52(4): 906–915.

Ismail, A. F. and P. S. Goh (2018). *Carbon-Based Polymer Nanocomposites for Environmental and Energy Applications*, Elsevier.

Jumbe, A. S. and N. Nandini (2009). "Heavy metals analysis and sediment quality values in urban lakes." *American Journal of Environmental Sciences* 5(6): 678.

Karimi-Maleh, H. and O. A. Arotiba (2020). "Simultaneous determination of cholesterol, ascorbic acid and uric acid as three essential biological compounds at a carbon paste electrode modified with copper oxide decorated reduced graphene oxide nanocomposite and ionic liquid." *Journal of Colloid and Interface Science* 560: 208–212.

Kuhn, R., I. M. Bryant, R. Jensch and J. Böllmann (2022). "Applications of environmental nanotechnologies in remediation, wastewater treatment, drinking water treatment, and agriculture." *Applied Nano* 3(1): 54–90.

Kuilla, T., S. Bhadra, D. Yao, N. H. Kim, S. Bose and J. H. Lee (2010). "Recent advances in graphene based polymer composites." *Progress in Polymer Science* 35(11): 1350–1375.

Kumar, S., W. Ahlawat, G. Bhanjana, S. Heydarifard, M. M. Nazhad and N. Dilbaghi (2014). "Nanotechnology-based water treatment strategies." *Journal of Nanoscience and Nanotechnology* 14(2): 1838–1858.

Lee, S. L. and C.-J. Chang (2019). "Recent developments about conductive polymer based composite photocatalysts." *Polymers* 11(2): 206.

Lei, J., F. Liu, L. Wang, Y. Liu and J. Zhang (2017). "A binary polymer composite of graphitic carbon nitride and poly (diphenylbutadiyne) with enhanced visible light photocatalytic activity." *RSC Advances* 7(44): 27377–27383.

Mishra, R. and S. Dubey (2015). "Fresh water availability and it's global challenge." *International Journal of Engineering Science Invention Research & Development* 2(VI).

Moon, Y.-E., G. Jung, J. Yun and H.-I. Kim (2013). "Poly (vinyl alcohol)/poly (acrylic acid)/ TiO2/graphene oxide nanocomposite hydrogels for pH-sensitive photocatalytic degradation of organic pollutants." *Materials Science and Engineering: B* 178(17): 1097–1103.

Moss, B. (2008). "Water pollution by agriculture." *Philosophical Transactions of the Royal Society B: Biological Sciences* 363(1491): 659–666.

Pan, Z., S. Xu, H. Xin, Y. Yuan, R. Xu, P. Wang, X. Yan, X. Fan, C. Song and T. Wang (2022). "High performance polypyrrole coated carbon-based electrocatalytic membrane for organic contaminants removal from aqueous solution." *Journal of Colloid and Interface Science*. 626: 283–295.

Pandey, S. (2006). "Water pollution and health." *Kathmandu University Medical Journal (KUMJ)* 4(1): 128–134.

Schwarzenbach, R. P., T. Egli, T. B. Hofstetter, U. Von Gunten and B. Wehrli (2010). "Global water pollution and human health." *Annual Review of Environment and Resources* 35: 109–136.

Singh, A. R., P. S. Dhumal, M. A. Bhakare, K. D. Lokhande, M. P. Bondarde and S. Some (2022). "In-situ synthesis of metal oxide and polymer decorated activated carbon-based photocatalyst for organic pollutants degradation." *Separation and Purification Technology* 286: 120380.

Van der Bruggen, B. (2013). "Integrated membrane separation processes for recycling of valuable wastewater streams: Nanofiltration, membrane distillation, and membrane crystallizers revisited." *Industrial & Engineering Chemistry Research* 52(31): 10335–10341.

Viman, O. V., I. Oroian and A. Fleşeriu (2010). "Types of water pollution: Point source and nonpoint source." *Aquaculture, Aquarium, Conservation & Legislation* 3(5): 393–397.

Wang, L.-l., L.-p. Zhu, N.-c. Bing and L.-j. Wang (2017). "Facile green synthesis of Pd/N-doped carbon nanotubes catalysts and their application in Heck reaction and oxidation of benzyl alcohol." *Journal of Physics and Chemistry of Solids* 107: 125–130.

Wutich, A., A. Y. Rosinger, J. Stoler, W. Jepson and A. Brewis (2020). "Measuring human water needs." *American Journal of Human Biology* 32(1): e23350.

Yu, Z., F. Li, Q. Yang, H. Shi, Q. Chen and M. Xu (2017). "Nature-mimic method to fabricate polydopamine/graphitic carbon nitride for enhancing photocatalytic degradation performance." *ACS Sustainable Chemistry & Engineering* 5(9): 7840–7850.

Zhang, S., C. B. Qiu, Y. Zhou, Z. P. Jin and H. Yang (2011). "Bioaccumulation and degradation of pesticide fluroxypyr are associated with toxic tolerance in green alga Chlamydomonas reinhardtii." *Ecotoxicology* 20(2): 337–347.

Zhang, S., L. Zhao, M. Zeng, J. Li, J. Xu and X. Wang (2014). "Hierarchical nanocomposites of polyaniline nanorods arrays on graphitic carbon nitride sheets with synergistic effect for photocatalysis." *Catalysis Today* 224: 114–121.

16 Challenges and Future Prospects

Aparajita Pal
Centre of Rubber Technology

CONTENTS

16.1 INTRODUCTION

Water is the most abundant natural resource on earth and yet only 1% of this huge resource is available for the safe consumption of human beings. But, unfortunately, this available 1% is under continuous threat of contamination by organic and inorganic substances. According to the current UN report, in the 21st century, fulfilling the demand of 7.5 billion population on earth for clean and affordable water resources has been a major concern. The removal of these toxic substances from water with a 99% purification guarantee is a challenging task. Various carbonaceous materials, their derivatives, and composites were extensively studied for application in water treatment and purification performance. Carbon atom has a very unique electronic structure. It can form covalent bonds with different metal and non-metal elements. Carbon can also be present into different molecular forms. Carbon in nanoscale possesses extraordinary properties compared to other absorbent materials that existed. Carbonaceous nanomaterials are classified based on their shape and geometrical structures. Activated carbon, carbon nanotubes (CNTs), graphene, graphene oxide, graphene nitride, carbon dots, MXenes are some of the well-known

DOI: 10.1201/9781003328094-16

nanomaterials in the field of water filtration research. The polymer composites, surface functionalisation against each of the nanomaterials possess significant benefit in terms of increase in dispersibility and absorption performance. Additionally, numerous methods and techniques are there, which included photocatalytic degradation, adsorption, filtration, membrane separation, reverse osmosis, forward osmosis, nano and microfiltration, biosorption, coagulation and electrochemical operations. However, there are several challenges associated with different processing routes. For example, nanomaterials like CNT when used at higher than their safe limit of exposure can pose a serious threat to the human immune system. Other than that, CNT faces issues like poor dispersion in an aqueous medium, leading to loss of adsorption efficiency. Furthermore, filtration is mainly involved with the accumulation of organic and inorganic material on the membrane surface, resulting in the formation of a fouling layer. This again hampers the purification ability of the composite membrane. Graphene—a sp^2 hybridised 2D nanostructure poses interplanar interaction that limits its ability to disperse homogeneously. Whereas the application of CNF is limited due to the non-viability of commercialisation of the electrospinning process to match the bulk-scale production. Natural polymer–graphene oxide composite promotes comparatively better life cycle assessment and biodegradability, but lags in the establishment of an optimised purification route. Hence, this current chapter introduces a wide array of challenges associated with various carbonaceous material–polymer composite systems and also depicts the future research area that needs urgent attention.

16.2   CHALLENGES AND FUTURE PROSPECTS IN USING CNT/POLYMER COMPOSITES

16.2.1   AGGLOMERATION

The polymer/CNT composites are known for their thermal resistance, dimensional stability, good electrical property and high strength. The nano-structure-reinforced polymer composites show high mechanical strength, which is suitable for different orthopaedic implants and other engineering applications. The electrical conductivity of the composites makes them even more suitable for pollutants adsorption applications. However, the main challenge in using polymer/CNT nanocomposite is the high cost of CNT. Additionally, CNTs show poor dispersibility in the polymer matrix, leading to clustering of the nanotubes (Arora and Attri, 2020). This weakens the interfacial interactions between the nanotubes and the polymer. The agglomeration of CNTs generates many defect sites on the polymer matrix. The processing of purified pristine CNT is also very time-consuming, adding to the negativity of the situation. Hence, the main challenges of using CNT/polymer composites are the alignment of the nanotubes in the polymer matrix and the homogeneous dispersion of fillers in the polymeric matrix. There are two ways to bring homogeneity in dispersion. They are mechanical and chemical processes. The mechanical process includes ultrasonication and mixing at a high shear rate. However, these processes can damage the nanostructures, reducing the length of the nanorods, nanotubes, etc. Therefore, chemical modification to mitigate the chance of agglomeration is much preferred.

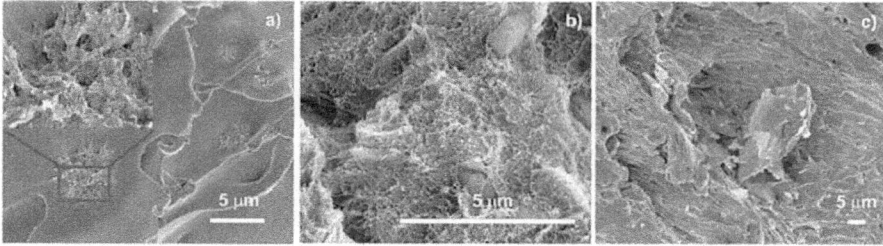

FIGURE 16.1 SEM image depicting the agglomeration behaviour of (a) CNT in polyamide 12 and (b) oxidised CNT in polyamide 12 at 5 wt% loading of nanotubes. (Source: Chatterjee et al., 2011.)

The main issues with SWCNT/polymer nanocomposite materials are the inability to form homogeneous dispersion in the polymer solution, high interfacial bonding due to van der Waals attraction between the nanotubes and the non-reactive surface of the nanotubes.

In Figure 16.1a, certain discrete agglomeration of CNT is observed, whereas a large amount of extended agglomeration of oxidised CNTs is reported in polyamide 12 polymer matrix in Figure 16.1b. The agglomerations act as a stress concentration point for the CNT/polyamide 12 composites. The non-homogeneous distribution and random orientation of the carbon nanotubes in the polymer matrix result in an inefficient transfer of load causing mechanical failure.

Another study by Garzia Trulli et al. (2017) reported that pristine CNTs face issues like poor dispersion in an aqueous medium resulting in inferior adsorption ability and chemical reactivity to be used for bulk-scale production of CNT/polymer matrix. The agglomeration lowers the sorption affinity of nanotubes towards the contaminants. The poor dispersion is mainly caused by the π–π interactions between the two atomic planes and the presence of van der Waals force among the CNTs. The tendency of agglomeration increases with decreasing number of graphitic sp^2 hybridised layers of CNTs. The SWCNTs are more prone to convert to close-fitting bundles than the MWCNTs. However, different functionalisation of the nanotubes has been adopted to combat the issue. The functional groups grafted on the CNT surfaces enhance the repulsive force between the nanotubes in the dispersion, surpassing the π–π attraction and van der Waals force, encouraging better dispersion ability.

16.2.2 Pollution and Health Risk

Amid its immense benefits and applications, it has posed a great threat if it gets into the human immune system. Leaking from the wastewater treatment plant causes the CNTs to be getting lost in the soil, air and water. CNTs can increase the level of toxicity for the aquatic and terrestrial flora and fauna. CNT is embedded in the polymer matrix to increase the mechanical strength and adsorption

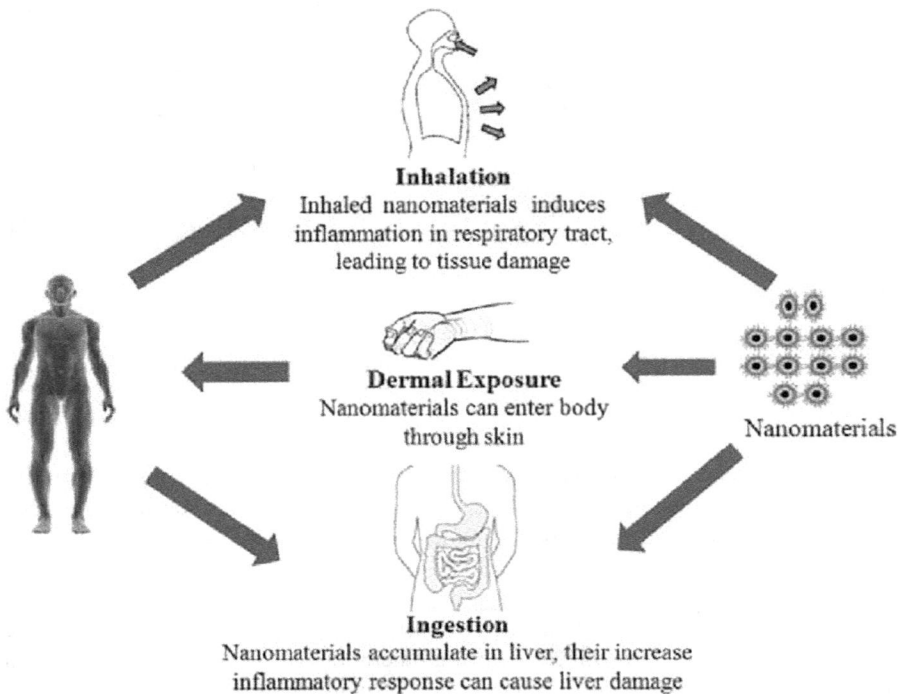

FIGURE 16.2   The potential interaction between the human immune system and the CNTs. (Source: Ali et al., 2016.)

ability of the composite material. However, the polymer may degrade due to photoreaction, hydrolysis, oxidation, pyrolysis, etc., releasing the CNTs into the environment (Sousa et. al., 2020). Prolonged exposure to CNTs can cause asthma, bronchitis and other serious lung diseases. The CNTs can affect the central nervous system as well by penetrating the semipermeable membrane between the brain and spinal cord or the blood and the brain. The nanostructures can be directly transferred to the brain. Other kinds of transport include trigeminal (nose to the brain) or axonal transport. National Institute for Occupational Safety and Health has suggested that the maximum exposure limit of CNTs is 1 $\mu g/m^3$ of air to combat the potential risk factor of the nanomaterial.

16.2.3   COST

The commercialisation of carbon nanotubes has been hampered by the high price and low production volume. CNTs can be produced at a rate of 595 kg/h through the decomposition of hydrocarbon in the process of catalytic chemical vapour deposition. The price generally lies between 25 USD and 38 USD/kg, while using plug flow or fluidised bed reactor (Akinpelu et al., 2019).

16.2.4 ANTI-FOULING CHALLENGE OF THE MEMBRANE

Filtration is associated with the accumulation of organic and inorganic matters and particles on the membrane surface. This accumulated layer is called the fouling layer. This layer acts as an additional filtration layer, restricting the desired throughput of the clean water through the membrane.

16.2.5 FUTURE PROSPECTS

Removing polycyclic aromatic hydrocarbons from wastewater is another crucial form of water purification (Boretti et. al.,2019). There are both physical and chemical approaches to polycyclic aromatic compounds removal; but they generate unwanted toxic materials in the process. The physical approaches are based on liquid–liquid solvent extraction, filtration and adsorption mechanism, which are safe, affordable and simple to exercise. However, the process of liquid–liquid solvent extraction involves the usage of a high volume of toxic, volatile and flammable solvents. Therefore, the further process needs to be studied for pollutant removal, which provides a synergistic effect. There is a major concern observed with the reverse osmosis, nanofiltration and ultrafiltration processes. The fouling of the filtration membrane is becoming a severe problem with all the three process routes (Akinpelu et al., 2019). Additionally, all of them are pressure-driven processes that require more energy and equipment. The forward osmosis process is reported to have several advantages over the existing process in terms of energy usage and anti-fouling nature. But, the commercial viability of the forward osmosis process is questionable since its slow nature. Hence, further extensive research is required to derive an eco-friendly energy-efficient process that can match the industry's needs.

16.3 CHALLENGES AND FUTURE PROSPECTS IN USING GRAPHENE/POLYMER COMPOSITES

Graphene is a sp² hybridised, two-dimensional nanostructure with a single-walled honeycomb lattice. For graphene/polymer nanocomposites, the interplanar interactions between the nanofillers make it more difficult for homogeneous dispersion. Pure graphene is hardly compatible with polymer matrix due to its layer-by-layer staking property. They form aggregation due to van der Waals interactions. However, graphene oxide (GO) comprises hydroxyl and epoxy groups on the planar sheet, and carbonyl and carboxyl groups at the edge, increasing the chances of more interactions with the polymer matrix. The surface of the graphene oxide is much easy to modify with the functional groups like amines or esters to facilitate stability in the dispersion mediums. The stability is enhanced due to increased interfacial interaction of the carbonaceous fillers with the host matrix. They are covalent and non-covalent functionalisation routes to undertake. The non-covalent process includes physical adsorption or interaction with the surface of the graphene oxide. Different techniques like ultrafiltration, flocculation and chemical oxidation are used to treat the wastewater from the industries. However, these methods are costly. Graphene-based composites mainly use photolytic processes to purify the water. Photocatalysis is a green technology that uses solar power and a simple lab set-up as a process for effluent treatment and degradation.

Recently, research related to functionalised graphene–polymer membranes has gained severe momentum in the application of water treatment and desalination. The modified membranes show better performance than the pristine graphene-based membranes. The functionalisation alters the physicochemical properties of the membrane by changing the hydrophilicity, porosity, charge density and surface roughness. Future research has been focused on the development of a novel polymeric composite having superior separation mechanism, anti-fouling property and mechanical stability (Wang et al., 2018). A study related to the size of the functionalised graphene is recommended to understand the influence of the size of the functionalised graphene on the water purification application.

16.4   CHALLENGES IN USING CNF/POLYMER COMPOSITES

For the composite applications, the mechanical properties increase with increasing CNF content in the polymer matrix. For example, the value of resistance to fracture is enhanced by 66%–78%, with the addition of CNF content in the CNF–epoxy polymer composite (Feng et al., 2014). However, the tensile property of the composite is significantly compromised in the due process with increasing wt% of CNF in the matrix due to the creation of voids. These voids act as defects in stress propagation due to the bundling of CNFs. The strain concentration leads to premature failure even in low-strain circumstances.

For the preparation of the composite, the melt mixing and solution process are the two most common techniques used today (Bhawal et al., 2019). However, due to the generation of high shear in the melt mixing process, it is very difficult to retain the high aspect ratio and shape of the CNF material.

Furthermore, the specific surface area of the CNF ranges from 10 to 30 m²/g, which is at least two orders of magnitude lower than the CNT and graphene-derived membrane (Liu et al., 2015). This affects the adsorption ability of the CNF-composite membrane. Now, macro and mesoporous structures can increase the specific surface area of the CNF. But, it will adversely affect the flexibility of the electrospun membrane.

In addition, another major issue is the mass production of CNF *vis* electrospinning technique to cater for the high commercial demand in the market. In 90% of cases, polyacrylonitrile (PAN) is used as the precursor material to prepare the spinning dope (Nie et al., 2020). The high cost of PAN again retards the process of commercialising CNF-composite membrane for wastewater treatment application. Besides, there is a very limited method available to recover the toxic unused solvent like DMF, which is the solvent material for the dope material.

The low cost and easy dispersibility make them a suitable alternative to the conventional CNT and carbon black. However, studies related to the quantitative analysis to understand the CNF distribution and fractal analysis for CNF connection in the composite are still at the initial stage, which needs further consideration. Additionally, future research should be performed in the domain related to the influence of structural characteristics of CNFs on the mechanical properties of the composite.

16.5 CHALLENGES AND FUTURE PROSPECTS IN USING NATURAL POLYMER-GRAPHENE-BASED COMPOSITE

Natural polymer-based, for example cellulose-, chitosan-, alginate-, protein- and lignin-based composites show poor mechanical properties compared to their synthetic counterparts. Cellulose–GO aerogel was fabricated for selective oil absorption. The compressive properties of the cellulose composite aerogel were evaluated under 60% strain. The recovery rate was reported as greater than 90%. The bulk density of the composite membrane was 5.9 mg/cm^3, and the surface area was 47.3 m^2/g (Mi et al., 2018). Due to its ultralight weight and high surface area, the material provided with exceptional absorption capability for oil substances.

However, on several occasions, the presence of hydroxyl and other polar groups in the structure aid in increasing the moisture content of the natural polymers. It can create more difficulty in miscibility with the hydrophobic fillers. This can lead to poor interfacial interaction of the polymer composite, leading to loss of mechanical property of the filtration membrane.

There are also other challenges like the establishment of an optimised route for production, if natural polymer composite is used.

Graphene has a large surface area of 2630 m^2/g, which makes it very deserving as absorbent material (Chen et al., 2013). But in the case of pristine graphene, there is only van der Waals force present to bind the absorbates. It is mainly due to the sp^2 hybridised C-atom in the graphene planar sheets. Hence, graphene cannot be considered a potential absorbate for the metal ions from the wastewater. However, the adsorption capacity of graphene can be significantly improved by grafting active functional groups on its surface. The modified graphene is called graphene oxide (GO), which has plenty of oxygen atoms on the basal plane and also on the edges of the sheets due to the presence of epoxy, carboxyl and hydroxyl groups. The metal ions get absorbed by a coordination bond, and the positively charged organic pollutants get scavenged by electrostatic force. But being a negatively charged absorbent, graphene exhibits comparatively lower affinity towards anionic dyes from the textile processing industries. Besides, strong interplanar interaction for graphene and its derivatives causes stacking in a layer-by-layer mode. Generally, when these 2D sheets are converted into powder, there is a significant loss of specific surfaces. The dispersion of single-walled graphene oxide helps to use the highest extent of the available specific surface area. But it is hard to collect these GO sheets from water.

To address the above issue, graphene oxide–natural polymer composite aerogels need to be investigated. One such study has been conducted by Croitoru et al. (2020), where GO–chitosan composite hydrogels were studied to check the broad spectrum of its application in water purification. These hydrogels can be prepared by the process of 3D self-assembly of graphene sheets in the presence of different cross-linking agents. The GO–chitosan composite hydrogels exhibit low aggregation and generate more interconnected pores to facilitate the adsorption mechanism. The GO–chitosan composite can even demonstrate a high capacity of adsorption towards both the anionic and cationic dyes. Future research is recommended to study the suitability of different compositions of hydrogel for various absorbate materials

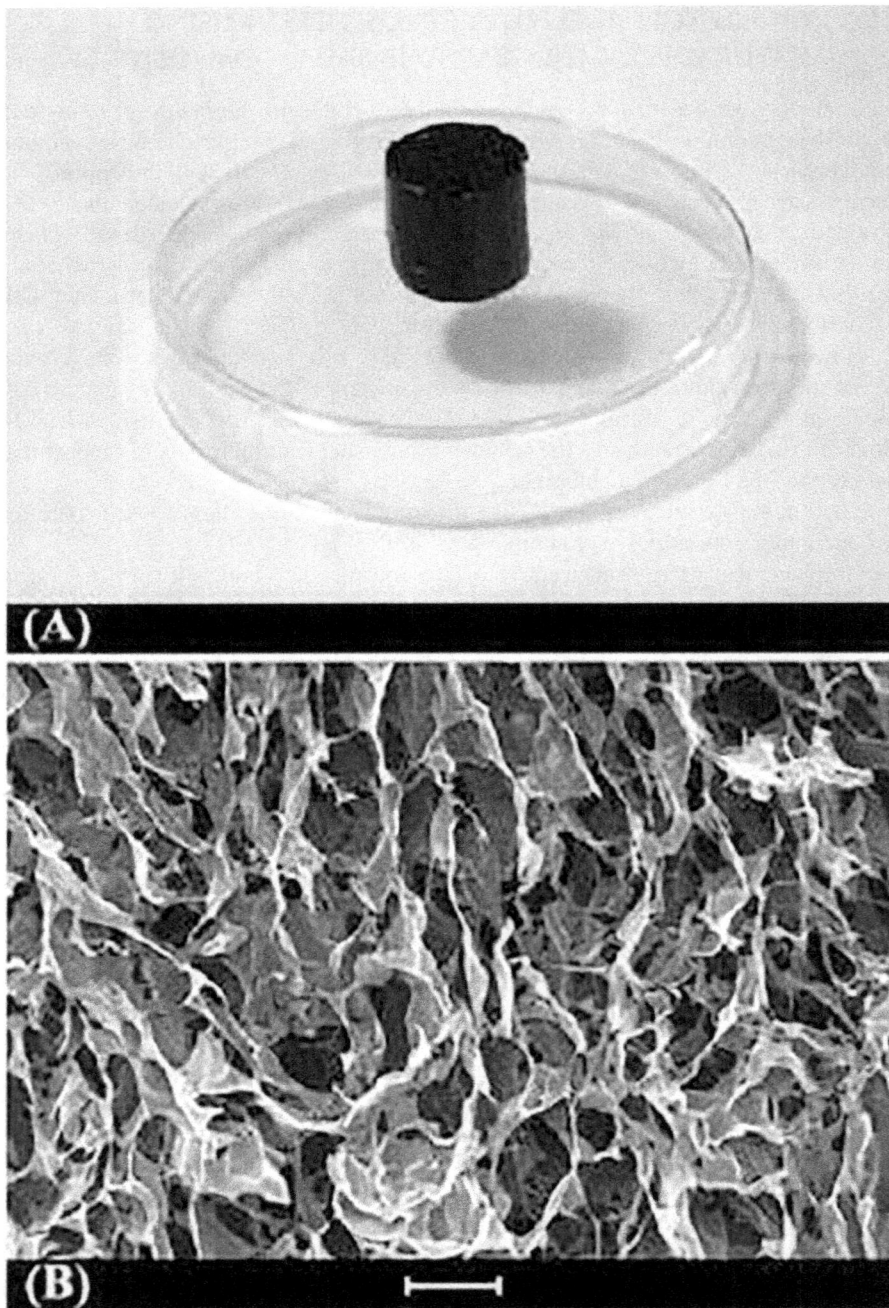

FIGURE 16.3   (a) Photograph of GO–chitosan composite hydrogel on a Petri dish (b) SEM image of GO–chitosan at a scale bar of 20 μm. (Source: Chen et al., 2013.)

16.6   CHALLENGES AND FUTURE PROSPECTS IN USING POLYMER MXene COMPOSITE

The route of wet chemistry synthesis results in the formation of layered flakes of MXene, in which the functional groups like hydroxyl and fluoride groups appear on the edges. These functional groups tend to react with the external matter, damaging the stability of the product. This has become a major threat in scaling up the production of MXene for its commercialisation (Mashtalir et al., 2014). The pristine MXene flakes can undergo degradation by the formation of oxides, while being exposed to water or at high temperatures (Figure 16.4).

The general formula of MXene is $M_{x+1}X_nT_x$ ($n = 1$–3), where M refers to the early transition metals, X refers to carbon or nitrogen, and T_x describes the surface termination groups (–OH, –O, –F, etc.) (Anasori and Gogotsi, 2019).

The main adsorption characteristic of MXene is due to its layered structure. The gradual nucleation of TiO_2 at the edges of the flakes over time causes the disappearance of the layered characteristic. There is also a visual change when oxidation of MXene causes TiO_2 precipitation. The solution colour changes from transparent to cloudy.

There are different approaches under investigation to address the issue of MXene agglomeration in an aqueous medium. Development of several composites is underway, which can introduce a synergistic effect in terms of reduction of agglomeration, providing improved stability for scaling up production in the bulk application. Designed MXene structure showcases the hydrophilic nature of the surface, metallic conductivity and rich surface chemistry. However, surface functionalisation can influence the stability of the MXene products. Different silylation reagents are reported to be used to combat the continuous degradation in the stability of the MXenes by oxidation. Besides, dispersibility, thermal stability and optical properties can also be improved with MXene functionalisation (Wang et al., 2020). The aggregation of MXene is a critical issue, causing degradation in the adsorption performance. Hence, this needs serious attention in future observations.

During functionalisation with the polysulfone groups, the aryl groups are attached to the Ti_3C_2 MXene surfaces in form of Ti–O–C or MXene–aryl linkages (Khatami

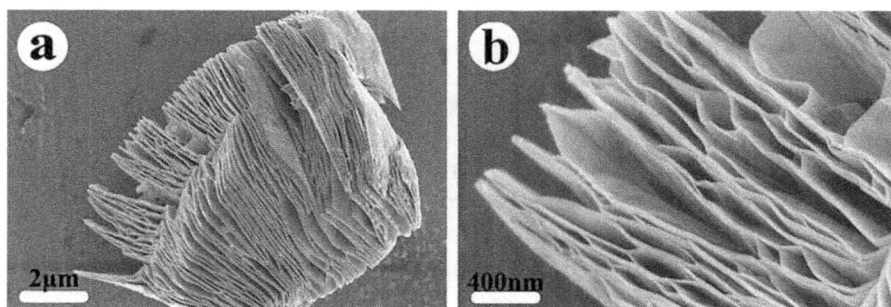

FIGURE 16.4   SEM images showing the layered structure of MXene–Ti_3C_2 (a) at low magnification (b) at high magnification. (Source: Wang et al., 2014.)

and Iravani, 2021). This makes the multilayer structure of MXene get delaminated, which increases further dispersion ability and improves the specific surface area. Additionally, MXene can be used to functionalise other carbonaceous materials like graphene and graphene derivatives.

Recently, an experiment shows improvement in the anti-fouling property of the membrane when the MXene composite is modified using silver nanoparticles (AgNPs). The AgNP-MXene composite membrane is prepared using the process of self-reduction of silver nitrate on the MXene surface. In a study, the cross-section of the PVDF substrate: pristine MXene and AgNP-MXene composite material is observed. 21% Ag@MXene exhibits layered stacking of MXene nanosheets with AgNPs embedded in between the layers. This creates slit interspacing, resulting in additional pores (Pandey et al., 2018). The water flux is improved significantly ($\sim$420 L/m^{2}h bar). Nevertheless, further investigation on the antibacterial mechanism of the MXene and its derivatives needs to be conducted. There is also no data available on the interaction of thiol with $Ti_3C_2T_x$ MXene, which requires to be considered for future research. The biosafety and biocompatibility of the MXenes require further evaluation.

Besides, the traditional synthesis process uses hazardous and toxic chemicals that can arise environmental concerns (https://www.cdc.gov/niosh). Therefore, a simple, cost-effective, efficient and green route for bulk production and synthesis of MXene is very important for future development.

Furthermore, the CVD growth of MXene should be further studied so that combination of MXene with other 2D structures can be promoted and different van der Waals heterostructures can be synthesised.

It is vital to develop various MXene heterogeneous structures that can be proved as next-generation materials for water purification applications with proper recyclability, improved stability and optimisation. The antimicrobial property due to its unique physicochemical properties and ultrathin lamellar structure can be utilised in the application of wastewater treatment.

16.7 CONCLUSION

To summarise, the chapter provides the challenges in existing technology and prospective future research to address the critical issues. CNT-based composites have been considered to pose huge potential in the separation and purification of water applications for complex effluent treatment processes. Whereas, there is a very limited study available for CNF–polymer composites, additionally, the commercial application of CNF-based composites is impossible due to process constraints. Graphene–polymer hydrogels exhibit a low aggregation tendency in comparison to graphene oxide nanosheets. Hence, further research is recommended to be performed in order to study the different compositions of graphene–polymer hydrogel for the adsorption of various absorbate materials. MXene poses a unique array of properties. Its layered structure is very suitable for the adsorption mechanism. Future research and development on different MXene-based polymer composites are very important to introduce a synergistic effect in terms of reduction in the aggregation of MXene sheets, providing improved stability in an aqueous medium and its efficiency in water purification mechanism.

REFERENCES

Akinpelu, A.A., Ali, M.E., Johan, M.R., Saidur, R., Qurban, M.A. and Saleh, T.A., 2019. Polycyclic aromatic hydrocarbons extraction and removal from wastewater by carbon nanotubes: A review of the current technologies, challenges and prospects. *Process Safety and Environmental Protection*, *122*, pp. 68–82.

Ali, A., Suhail, M., Mathew, S., Shah, M.A., Harakeh, S.M., Ahmad, S., Kazmi, Z., Rahman Alhamdan, M.A., Chaudhary, A., Damanhouri, G.A. and Qadri, I., 2016. Nanomaterial induced immune responses and cytotoxicity. *Journal of Nanoscience and Nanotechnology*, *16*(1), pp. 40–57.

Anasori, B. and Gogotsi, Û.G., 2019. *2D Metal Carbides and Nitrides (MXenes)* (Vol. 416). Berlin: Springer.

Arora, B. and Attri, P., 2020. Carbon nanotubes (CNTs): A potential nanomaterial for water purification. *Journal of Composites Science*, *4*(3), p. 135.

Bhawal, P., Ganguly, S., Das, T.K., Mondal, S., Nayak, L. and Das, N.C., 2019. A comparative study of physico-mechanical and electrical properties of polymer-carbon nanofiber in wet and melt mixing methods. *Materials Science and Engineering: B*, *245*, pp. 95–106.

Boretti, A. and Rosa, L., 2019. Reassessing the projections of the world water development report. *NPJ Clean Water*, *2*(1), pp. 1–6.

Chatterjee, S., Nüesch, F.A. and Chu, B.T., 2011. Comparing carbon nanotubes and graphene nanoplatelets as reinforcements in polyamide 12 composites. *Nanotechnology*, *22*(27), p. 275714.

Chen, Y., Chen, L., Bai, H. and Li, L., 2013. Graphene oxide–chitosan composite hydrogels as broad-spectrum adsorbents for water purification. *Journal of Materials Chemistry A*, *1*(6), pp. 1992–2001.

Croitoru, A.M., Ficai, A., Ficai, D., Trusca, R., Dolete, G., Andronescu, E. and Turculet, S.C., 2020. Chitosan/graphene oxide nanocomposite membranes as adsorbents with applications in water purification. *Materials*, *13*(7), p.1687.

Feng, L., Xie, N. and Zhong, J., 2014. Carbon nanofibers and their composites: A review of synthesizing, properties and applications. *Materials*, *7*(5), pp. 3919–3945.

Khatami, M. and Iravani, S., 2021. MXenes and MXene-based materials for the removal of water pollutants: Challenges and opportunities. *Comments on Inorganic Chemistry*, *41*(4), pp. 213–248.

Liu, Y., Zhou, J., Chen, L., Zhang, P., Fu, W., Zhao, H., Ma, Y., Pan, X., Zhang, Z., Han, W. and Xie, E., 2015. Highly flexible freestanding porous carbon nanofibers for electrodes materials of high-performance all-carbon supercapacitors. *ACS Applied Materials & Interfaces*, *7*(42), pp. 23515–23520.

Mashtalir, O., Cook, K.M., Mochalin, V.N., Crowe, M., Barsoum, M.W. and Gogotsi, Y., 2014. Dye adsorption and decomposition on two-dimensional titanium carbide in aqueous media. *Journal of Materials Chemistry A*, *2*(35), pp. 14334–14338.

Mi, H.Y., Jing, X., Politowicz, A.L., Chen, E., Huang, H.X. and Turng, L.S., 2018. Highly compressible ultra-light anisotropic cellulose/graphene aerogel fabricated by bidirectional freeze drying for selective oil absorption. *Carbon*, *132*, pp. 199–209.

National Institute for Occupational Safety and Health, N., 2022. [online] Available at: https://www.cdc.gov/niosh/updates/upd-04-24-13.html. [Accessed 21 June 2022].

Nie, G., Zhao, X., Luan, Y., Jiang, J., Kou, Z. and Wang, J., 2020. Key issues facing electrospun carbon nanofibers in energy applications: On-going approaches and challenges. *Nanoscale*, *12*(25), pp. 13225–13248.

Pandey, R.P., Rasool, K., Madhavan, V.E., Aïssa, B., Gogotsi, Y. and Mahmoud, K.A., 2018. Ultrahigh-flux and fouling-resistant membranes based on layered silver/MXene (Ti 3 C 2 T x) nanosheets. *Journal of Materials Chemistry A*, *6*(8), pp. 3522–3533.

Sousa, S.P., Peixoto, T., Santos, R.M., Lopes, A., Paiva, M.D.C. and Marques, A.T., 2020. Health and safety concerns related to CNT and graphene products, and related composites. *Journal of Composites Science*, *4*(3), p. 106.

Trulli, M.G., Sardella, E., Palumbo, F., Palazzo, G., Giannossa, L.C., Mangone, A., Comparelli, R., Musso, S. and Favia, P., 2017. Towards highly stable aqueous dispersions of multi-walled carbon nanotubes: The effect of oxygen plasma functionalization. *Journal of Colloid and Interface Science*, *491*, pp. 255–264.

Wang, S., Liu, Y., Lü, Q.F. and Zhuang, H., 2020. Facile preparation of biosurfactant-functionalized Ti2CTX MXene nanosheets with an enhanced adsorption performance for Pb (II) ions. *Journal of Molecular Liquids*, *297*, p. 111810.

Wang, F., Yang, C., Duan, C., Xiao, D., Tang, Y. and Zhu, J., 2014. An organ-like titanium carbide material (MXene) with multilayer structure encapsulating hemoglobin for a mediator-free biosensor. *Journal of the Electrochemical Society*, *162*(1), p. B16.

Wang, X., Zhao, Y., Tian, E., Li, J. and Ren, Y., 2018. Graphene oxide-+based polymeric membranes for water treatment. *Advanced Materials Interfaces*, *5*(15), p. 1701427.

17 Polymer Composites
Processing, Safety, and Disposal

Ashish Kaushal
National Institute of Technology

Rahul Sharma
CSIR-National Physical Laboratory Delhi

Vishal Singh
National Institute of Technology

CONTENTS

DOI: 10.1201/9781003328094-17

17.1   INTRODUCTION

Composite materials have carved a niche for themselves in a multitude of different materials due to their numerous advantages, such as corrosion resistance, low weight, high fatigue strength, etc. A composite is a mixture of at least two materials that differ in physicochemical properties where the different components retain their distinct identities; they do not dissolve or mix [1–5]. Metal matrix composites have attracted the attention of scientists around the world due to their wide application in transportation, cutting tools, consumer electronics, defence, aerospace, marine, and other fields [6,7]. The drawbacks of metal-based composites such as corrosion, heavyweight, and high cost make polymer composites (PCs) the material of better choice. Certain advantages of PCs such as low specific weight, ease of shaping, high stability against corrosion, and high fatigue resistance make them preferred over other composite materials [8,9]. Despite the many advantages of PCs, their major disadvantage is that they are very difficult to dispose of [10]. Today, the post-disposal of PCs has become a major concern, because plastics contain toxic ingredients, including phthalates, and polyfluorinated chemicals can pose serious environmental hazards and public health [10,11]. Polymeric waste can potentially carry these chemicals into clean environments and, when ingested by organisms, can cause the transfer of chemicals into their systems [12]. The presence of hazardous metallic substances such as chromium (VI), lead, and mercury additives in plastics are also very hazardous to human health and have adverse effects on fertility and sexual functions [12–14]. Apart from these, some other human health problems like burning eyes, difficulty breathing, liver disease, cancer, skin diseases, lung problems, headache, dizziness, and gastrointestinal problems also result from using toxic plastics [15,16]. A report by Doughty et al. [16] showed that prolonged exposure to even small amounts of styrene can be neurotoxic, causing adverse effects such as cytogenetic, carcinogenic, and haematologic. Therefore, the recycling of polymer-based composites is considered a favourable solution to the problem of their disposal. Several studies have focused on mechanical recycling [17,18] and chemical recycling [17,19], and other methods [20,21] of recycling plastic waste.

Walter Kaminsky [22] found in his study of plastic recycling that not all plastic waste could be mechanically recycled; he also explained that pyrolysis is an effective way to recycle plastics and save the environment. Rai et al. [23] concluded in their study of waste plastics that the compressive strength of plastic mix concrete could be increased by the addition of a superplasticiser. Payne et al. [24] developed a new technique using zinc-based catalysts and methanol (ZCM) for the recycling of plastic waste at room temperature. They found that ZCM was able to convert the waste into its chemical constituents. Zhao et al. [25] expanded a framework to include several mathematical models and methodologies in one of their research, ranging from the methodology, recycling techniques, and energy sources to ecological effects.

17.2   POLYMER COMPOSITES: PROPERTIES AND APPLICATIONS

A polymer matrix composite is a composite material formed by the addition of filler material to the matrix of the polymer material (Figure 17.1). Generally, polymers

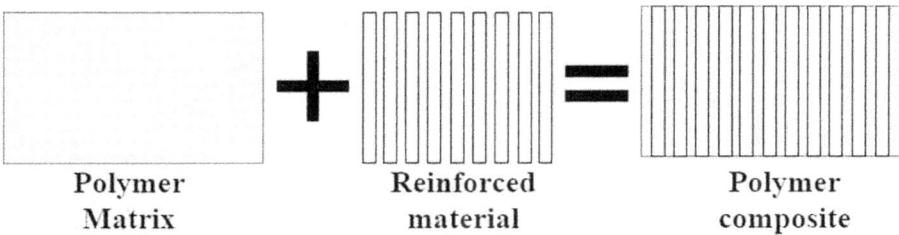

FIGURE 17.1   Schematic illustration of polymer composite.

are divided into two categories called thermosets and thermoplastics, according to the type of matrix [26]. Thermoset polymers are polymers that harden irreversibly when exposed to heat. Thermosetting polymers cannot be melted down, so it is given their final shape. Epoxy, phenolic, polyurethane, and polyamide are such types of thermosetting polymers. Thermoplastics, other than thermosets, are generally characterised by their properties such as flexibility, high strength, and resistance to shrinkage depending on the type of resin [26,27]. Thermoplastics become soft and malleable when heat is applied, and become hard again once cooled. Polystyrene (PS), polypropylene (PP), and polyethylene (PE) are the most commonly used thermoplastic polymers [28]. PCs have several remarkable properties such as corrosion resistance, impact resistance, fatigue resistance, abrasion resistance, high creep resistance, etc. that make them a better choice than other materials [29,30].

Some of the above-mentioned properties provide insight into the wide range of applications for PCs shown in Figure 17.2. The automobile industry continues to use polymer composites for other applications such as fuel tanks, rear-view mirrors, and engine parts [31]. PCs have also been used in rubber industries for their lightweight and cost savings [32]. PCs can provide an even better strength-to-weight ratio of up to 20% than metals. The significant weight, high strength, and high resistance of fibre-based PCs have increased their use in the aviation industry [33]. Due to their versatility, PCs are more widely used in the medical industries for the development and production of a wide range of products, such as surgical instruments, X-ray tables, orthopaedic products, wheelchairs, etc. [34,35]. PCs are used to perform various functions in various electrical and electronic devices due to their remarkable versatility, which allows the development and production of a variety of products at an affordable price to meet various application requirements [36]. PCs are also used in a wide range of sporting goods and their associated industries. Fibre-infused PCs can be used as a building material for sports equipment like tennis racks, snowboards, baseball bats, and bicycle frames. PCs are effectively used to cushion motorcycle handlebars, muscle rollers, exercise mats, floating duck bats, and a variety of equipment. Moisture in sports equipment is prone to bacterial and fungal growth. PCs can absorb excess moisture and eliminate the risk of bacterial and fungal growth [37,38].

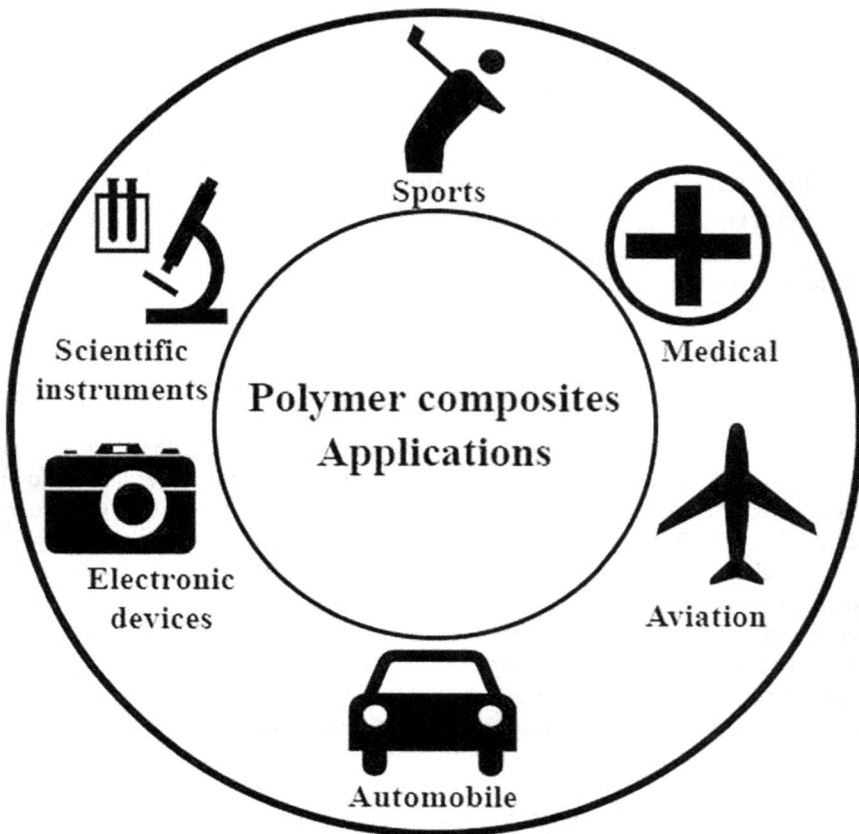

FIGURE 17.2 Schematic illustration of polymer composites applications.

17.3 PROCESSING TECHNIQUES OF POLYMER COMPOSITES

The processing of PCs depends upon the type of plastic material being used. For fibre-reinforced thermosets, various processing techniques are used that are considered under the following three headings [39]: (i) manual process; (ii) semi-automatic process; and (iii) automated process. These processes are placed under two main headings, the first being the open mould system, in which the material comes into contact with the mould on only one surface during the moulding operation. The second process is the closed mould technique in which composites are shaped between two matching dies.

17.3.1 OPEN MOULDING

Open moulding is the most flexible and less expensive process, because part sizes and design options are virtually unlimited, and the resin does not have to be isolated

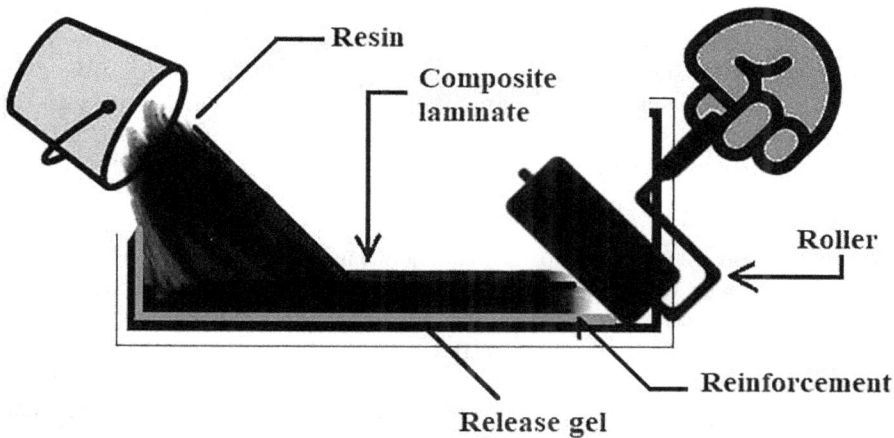

FIGURE 17.3 Schematic illustration of open mould process (hand lay-up).

from the environment. Typically, this process is used for large parts that cannot be manufactured in more automated processes [39,40].

17.3.1.1 Hand Lay-Up

Hand lay-up composite is one of the simplest moulding processing techniques, offering simple processing and a wide range of part sizes. The advantage of this technique is that the design is easily modified, and the investment in equipment is also minimal. This process is mainly divided into four stages, which include mould preparation, gel coating, lay-up, and curing [39–41]. Figure 17.3 shows the schematic illustration of the hand lay-up technique. In the first step of the method, the mould surface is treated with an anti-adhesive agent to prevent the resin from sticking to the surface, and then a thin plastic sheet is applied over and under the mould plate to obtain a smooth surface.

The woven reinforcement layers are then cut to the required size and placed on the surface of the mould. After that, the resin is mixed with the other ingredients and applied to the surface of the reinforcement, and the brush is used to spread it evenly. Another pad is then placed over the polymer layer, and pressed with a roller to remove air bubbles and excess polymer. The pressure is then released by closing the mould. After curing at room temperature, the mould is opened and the composite is removed from the mould surface. The low production efficiency of this method makes it unsuitable for bulk production [42,43]. The difference in the level of production personnel and environmental conditions creates instability in the product quality, which is one of the disadvantages of this method.

17.3.1.2 Spray up

Spray up technique uses special equipment, specifically a chopper gun to cut the reinforcement material into short fibres before adding it to the resin mixture on the surface of the mould. This technique is more automated than hand lay-up and is generally used for large quantities of production. In this technique, the gel coat is applied to the mould and allowed to cure through a spray gun. The roller is driven over the sprayed material to remove the air trapped in the lay-up. After the fibre and resin have been sprayed to the required thickness, the product is processed at elevated temperatures. After curing, the mould is opened, and further processing is done by removing the developed composite part [43].

17.3.1.3 Filament Winding

This technique is primarily used for manufacturing hollow, circular parts, such as pipes and tanks, in which fibres are cast into a cylindrical mould that continues to spin. Filament winding uses continuous reinforcement to maximise the strength of the fibre. In this technique, fibres are passed over a heated roller until the roller becomes sticky and then wound on a rotating mandrel. After winding, the assembly is either subjected to room temperature or cured in an oven. After curing, the mandrel is removed [44,45].

17.3.2 Closed Moulding

Closed moulding is used to produce large quantities of precision parts that require specialised and automated equipment. In closed compression moulding, dry reinforcement is placed in a basic mould, the mould is closed, and resin is injected into the closed cavity with a pressure pump or vacuum. When the laminate hardens, the mould is opened, and the part is removed. Close moulding benefits include lower material and disposal costs; a more consistent, repeatable process; and shorter cycle times, resulting in increased productivity and reduced labour costs [39,46].

17.3.2.1 Pultrusion

Pultrusion is a continuous process used in the plastics industry. In this process, the fibres are extracted from the spool that is passed through a resin bath with the help of a tension roller using a heated dye, and the strands are stretched to form a coherent section shape. The mould completes the impregnation of the fibres, controls the resin material, and cures the material to its final shape (Figure 17.4). Pultrusion allows the manufacture of long, continuously shaped objects such as hollow bars, tubing, channels, pipes or rods, etc. [47,48].

17.3.2.2 Resin Transfer Moulding

Resin transfer moulding is a low-temperature process for transferring a liquid thermosetting resin into a closed mould. In this process, a preform is first formed and packed into a mould cavity containing a part of the desired shape; then the mould cavity is closed and clamped. The less viscous resin is poured into the hot mould until the mould is full. The injection stage must ensure complete impregnation of the workpiece. After the injection phase, the curing cycle begins, and the resin polymerises into a rigid plastic [49,50]. This process allows fabricators to produce complex parts with smooth surfaces and also allows for an unlimited number of combinations and orientations, including 3D reinforcement. It is used for a variety of applications in the military, aerospace, and construction industries.

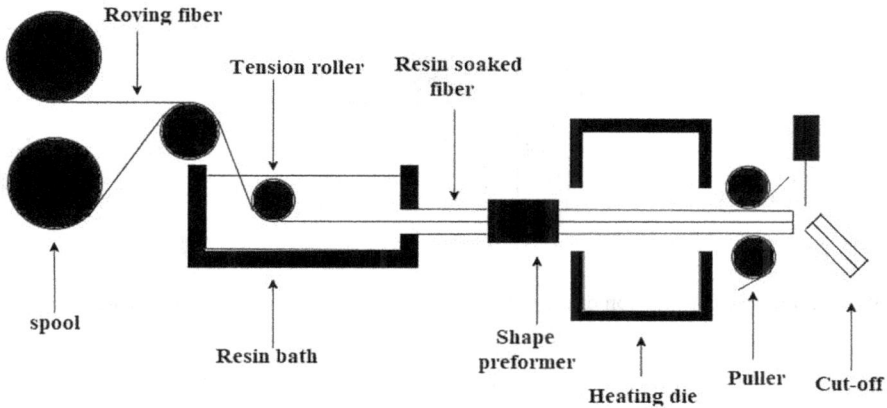

FIGURE 17.4 Schematic illustration of closed mould process (pultrusion).

17.3.2.3 Compression Moulding

Compression moulding involves sandwiching the composite material between two matching moulds, using heat and intense pressure until the part is cured. This process is used to manufacture complex polymer components that are very difficult to manufacture by other processes. This process requires manual processing of the mould and product, resulting in longer cycle times [51,52]. The disadvantages of this process include poor product stability and low compatibility with some parts.

The processing methods for thermoplastic PCs generally require a single operation. The processing stages of thermoplastic PCs contain heating, forming, and cooling in continuous cycles.

17.3.3 SOLUTION CASTING

Solution casting is a widely used method for preparing polymer nanocomposites. This method consists of three steps, namely, dispersion of the filler in a suitable solvent, mixing with the polymer, and recovery of the composite by precipitation.

In this method, the polymer and prepolymer are mixed evenly and then removed by drying water or solvent to form a solid layer on the carrier. The resulting cast layer can be stripped from the carrier substrate to form a film [53]. Polymer solution castings are used to deliver high-quality films with their superior optical, mechanical, and physical film properties.

17.3.4 IN-SITU POLYMERISATION

The in-situ polymerisation process consists of a series of polymerisation steps that lead to the formation of a composite. This process requires the use of monomers, initiators, and high temperatures. In-situ polymerisation typically deals with insoluble polymers that cannot be processed by solvent casting and melt processing methods

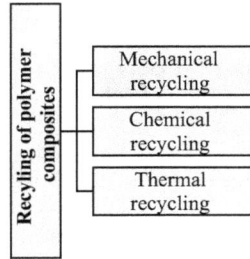

FIGURE 17.5   Classification of recycling techniques of polymer composite waste.

[54]. The main advantage of this process is the material saving and the bond between the fillers and the polymer matrix, resulting in improved mechanical properties due to strong interfacial bonds. This method is usually suitable for polymers that cannot be safely prepared in solution, because the solvents used to dissolve them are highly toxic. However, some disadvantages of this method include the limited availability of usable materials and the short time to perform the polymerisation process.

17.4   SAFETY AND DISPOSAL

With the increasing production and demand for polymer composites, the amount of waste associated with them is also increasing, most of which is non-degradable. The disposal of this waste is a serious concern, because the inability to dispose of the composite material is polluting the environment. Therefore, at present, recycling is considered a suitable method of disposal of polymer composites. The landfill is known to be one of the most traditional and widely used methods in waste management. However, disposal of polymer waste from landfills can lead to soil and groundwater pollution [55]. In addition, transporting waste to landfills is often difficult.

For the recycling of used polymer composite materials, various recycling methods have been developed and implemented, depending on their nature and recyclability. The processing of recycling plastic waste is divided into three parts (Figure 17.5): (i) mechanical processing, (ii) chemical processing, and (iii) thermal processing [56–58].

17.4.1   MECHANICAL RECYCLING

Mechanical recycling is a multi-step process to recover plastic waste through mechanical processes such as collection, sorting, shredding, washing, drying, and re-granulation (Figure 17.6). Mechanical recycling does not change the chemical composition of the material, which allows the polymer material to be reused/recycled over and over again.

This process involves a wide range of material types in different shapes, colours, and sizes. The X-ray technology is used for the proper identification of materials, which is essential to achieving maximum processing accuracy. First of all, the sorting process is done with the help of special equipment or plants to separate the different

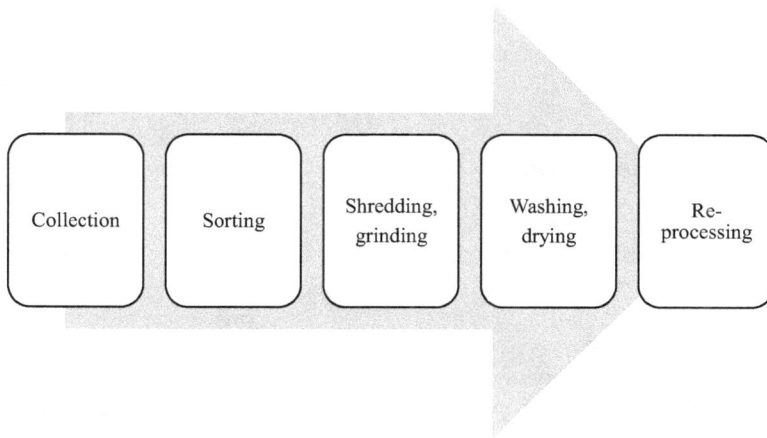

FIGURE 17.6   Schematic illustration of the mechanical recycling process.

materials of plastic waste [53–59]. However, there are problems with sorting due to separation errors in the plastic waste, making it very difficult to achieve efficiency, and due to this it often leads to contamination of recycled plastics with other plastics. The cleaning is done after sorting, and the plastic is divided into small pieces with the help of a grinder.

The material is recovered by re-melting and re-granulation after the grinding process. It has been observed that during processing, high temperatures and shear forces can degrade the thermal and mechanical properties of the polymers, which can affect the length and distribution of polymer chains, and also the crystallinity. In mechanical recycling, plastic waste is sorted by material type, washed, melted, and then dried. This recycling type is used only for the PCs that can be re-melted and recycled into products using methods such as injection moulding, compression moulding, and extrusion. The quality of the sorting processes also depends on the efficiency of the collection systems. However, the resulting quality often only allows use in low-value applications.

17.4.2   Chemical Recycling

Chemical recycling refers to the chemical or thermochemical processing of plastic waste into raw materials for the chemical industry [53,54]. Chemical recycling is the process of converting plastic polymers into individual monomers. This process recycles the chemical building blocks that make up plastic.

Figure 17.7 illustrates the steps in the chemical recycling process. Once plastics return to these basic building blocks, the material can then be used to produce new chemicals, and plastic chemical recycling is best for hard, layered, or heavily contaminated plastics. The main advantages of the chemical recycling process are that it is more resistant to contamination and produces a polymer similar to the original without decay cycles. Generally, two main recycling routes are used for chemical

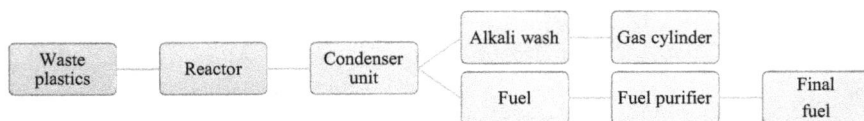

FIGURE 17.7   Schematic illustration of steps in the chemical recycling process.

recycling, *viz.*, dissolution and depolymerisation. These processes are elaborated on in subsequent sections.

17.4.2.1   Dissolution

Decomposition is a well-known way to recycle plastic waste. This process takes place in several stages, which involves first the sorting of the plastic waste and preparing it for further processing. In the second stage, the sorted plastic waste is dissolved to extract the polymer and create new recycled plastics. The additives are separated from the polymer before the polymer is removed from the solution. In the final step, new additives are added to the polymers to produce new recycled plastics that can be selectively crystallised after being dissolved in a solvent. During dissolution, the structure of the polymer does not change, making it possible to reuse them in the same form [60].

17.4.2.2   Depolymerisation

Depolymerisation is a chemical recycling technique that converts a polymer chain into an oligomer with its monomer units as well as the associated formation of gaseous products. In this process, the sorted plastic waste is broken down into monomers so that they can be fed back into plastic production. This process begins with the sorting of plastic waste, in which heat is used to break down the polymer into monomers. In the next step, the monomers are separated to remove any impurities and then being fed as recycled material. This process is only applicable to condensed polymers that make this process flawed [61].

17.4.3   Thermal Recycling

The thermal recycling process is a type of recycling in which high-temperature heat is used to break down the scrap. In this process, the temperature depends on the type of resin used in the scrap composites. Non-recyclable waste that cannot be recycled by chemical or mechanical recycling techniques is disposed with the help of thermal technology [62].

Thermal processing is mainly divided into three parts: thermal decomposition, thermal decomposition in the liquid layer, and microwave thermal decomposition. In all the three methods, the polymer matrix is destabilised by low-molecular-weight polymers such as carbon dioxide, hydrogen, methane, and gas and oil fractions. Thermal treatment reduces the volume and mass of waste, neutralises harmful components, and also helps reduce pollutant emissions. This method is used to achieve a low-cost solution for solid waste disposal.

17.5 CONCLUSION: RECENT ADVANCES AND PROSPECTS

With outstanding mechanical–thermal–electrical properties, PCs have become the first-choice materials in various applications. Particularly, the engineered PCs with hierarchical architectures have gained market maturity in the recent times. It is expected that the market of PCs will reach $9bn by 2027, which indicates the worth of PCs. Furthermore, there are several sectors where PCs are entering or just maturing to enter the commercialisation phase. Noticeably, the commercial viability of PCs chiefly depends upon their ability to access diverse properties in real-time applications, which, in turn, are governed by the chemical and structural composition of PCs. Consequently, it is critical to control the distribution of fillers in the parent polymer matrix to exploit the optimum performance of properties of PCs, because this distribution of two phases decides the overall performance. In this context, the choice of processing technology is quite important. Although, processing techniques like melt processing, solution casting, in-situ polymerisation, etc. are well established and in use for several decades for preparing PCs, they still offer a key research field for fabricating hierarchical and ordered arrays of fillers in the polymer matrix. Furthermore, the focus should be paid to processing time, processing cost, emissions of volatile compounds, and health and environmental impacts of emissions. Multifunctionality is the next key iteration for PCs, so that single material can perform additional roles beyond their primary applications. Precisely, smart and multifunctional PCs will be in high demand soon. The embedded functionality can include improved electrical–thermal–mechanical property-based sensors, actuators, energy storage, energy harvesting, data transmission, self-healing mechanisms, etc.

Failure of PCs in any form during application is always a major concern. The fire resistance, release of volatile organic compounds (VOCs), and mechanical properties have gained ample attention, owing to their critical contributions to maintaining the structural integrity of PCs. Moreover, these properties facilitate contribution to the reliability and safety of PCs in real-time applications. Specifically, in most polymers, organic moieties are the backbone that makes PCs vulnerable to flames and high-temperature condition properties. Additionally, the combustion of different types of organic compounds generates extremely harmful gaseous species. Therefore, the development of fire-resistant PCs is imperative as far as safety concerns of PCs are considered. However, most flame retardants deteriorate the mechanical strength of PCs, and hence novel hierarchical approach for the designing of PCs need to be adopted, which ensures a synergistic enhancement in mechanical and fire-safety properties. In this context, conventional surface modifications alone cannot serve the purpose of achieving desired outcomes, as interfacial compatibility between the phases rely on several factors. Therefore, strategies that can improve interfacial interactions and surface wettability characteristics need to be identified for addressing the safety issues of PCs.

Waste disposal, recycling, and upcycling have become the central theme in decision-making strategies for achieving sustainable development goals. For PCs, the major factors that affect the disposal processes are non-bio-degradable nature, risk of secondary contamination, non-polymer impurities, etc. The substandard quality of products, production yield, and high recycling cost restrict the recycling opportunities for PCs. To overcome this, research on reclamation and recycling of PCs-based wastes needs to be accelerated. However, many novel methodologies have been developed and adopted in the recent times for recycling PCs that include size reduction, separation of metals or other impurities, and increasing compatibility to reduce contaminates due to incompatible polymers. Though the future of PCs looks promising, still there are significant challenges in their realisation, including the development of facile, inexpensive, rapid processing techniques, predictability of safety and failure factors during use, and designing of environmentally benign disposal and recycling methods.

REFERENCES

1. Roylance, David. *Introduction to Composite Materials*. Department of material science and engineering, Massachusetts Institute of Technology, Cambridge, 2000.
2. Chawla, Krishan K. *Composite Materials: Science and Engineering*. Springer Science & Business Media, 2012.
3. Campbell, F. C. "Introduction to composite materials." *Structural Composite Materials* 1(2010): 1–29.
4. Tsai, Stephen W., and H. Thomas Hahn. *Introduction to Composite Materials*. Routledge, 2018.
5. Clyne, Trevor William, and Derek Hull. *An introduction to Composite Materials*. Cambridge University Press, 2019.
6. Kainer, Karl Ulrich. "Basics of metal matrix composites." *Metal Matrix Composites: Custom-made Materials for Automotive and Aerospace Engineering* (2006): 1–54.
7. Chawla, Krishan K. "Metal matrix composites." In *Composite Materials*, pp. 197–248. Springer, New York, NY, 2012.
8. Thomas, Sabu, Kuruvilla Joseph, Sant Kumar Malhotra, Koichi Goda, and Meyyarappallil Sadasivan Sreekala, eds. *Polymer Composites, Macro-and Micro Composites*, Vol. 1. John Wiley & Sons, 2012.
9. Åström, B. Tomas. *Manufacturing of Polymer Composites*. Routledge, 2018.
10. Oladele, Isiaka Oluwole, Samson Oluwagbenga Adelani, Okikiola Ganiu Agbabiaka, and Miracle Hope Adegun. "Applications and disposal of polymers and polymer composites: A." *European Journal of Advances in Engineering and Technology* 9, no. 3 (2022): 65–89.
11. Alabi, Okunola A., Kehinde I. Ologbonjaye, Oluwaseun Awosolu, and Olufiropo E. Alalade. "Public and environmental health effects of plastic waste disposal: A review." *Journal of Toxicology and Risk Assessment* 5, no. 021 (2019): 1–13.
12. Manzoor, Javid, Manoj Sharma, Irfan Rashid Sofi, and Ashaq Ahmad Dar. "Plastic waste environmental and human health impacts." In *Handbook of Research on Environmental and Human Health Impacts of Plastic Pollution*, pp. 29–37. IGI global, 2020.
13. Rahman, Zeeshanur, and Ved Pal Singh. "The relative impact of toxic heavy metals (THMs)(arsenic (As), cadmium (Cd), chromium (Cr)(VI), mercury (Hg), and lead (Pb)) on the total environment: An overview." *Environmental Monitoring and Assessment* 191, no. 7 (2019): 1–21.
14. Campanale, Claudia, Carmine Massarelli, Ilaria Savino, Vito Locaputo, and Vito Felice Uricchio. "A detailed review study on potential effects of microplastics and additives of concern on human health." *International Journal of Environmental Research and Public Health* 17, no. 4 (2020): 1212.
15. Mazumder, Debendranath Guha, and U. B. Dasgupta. "Chronic arsenic toxicity: Studies in West Bengal, India." *The Kaohsiung Journal of Medical Sciences* 27, no. 9 (2011): 360–370.

16. Proshad, Ram, Tapos Kormoker, Md Saiful Islam, Mohammad Asadul Haque, Md Mahfuzur Rahman, and Md Mahabubur Rahman Mithu. "Toxic effects of plastic on human health and environment: A consequences of health risk assessment in Bangladesh." *International Journal of Health* 6, no. 1 (2018): 1–5.
17. Ragaert, Kim, Laurens Delva, and Kevin Van Geem. "Mechanical and chemical recycling of solid plastic waste." *Waste Management* 69(2017): 24–58.
18. Schyns, Zoé OG, and Michael P. Shaver. "Mechanical recycling of packaging plastics: A review." *Macromolecular Rapid Communications* 42, no. 3 (2021): 2000415.
19. Davidson, Matthew G., Rebecca A. Furlong, and Marcelle C. McManus. "Developments in the life cycle assessment of chemical recycling of plastic waste–A review." *Journal of Cleaner Production* 293(2021): 126163.
20. Buekens, A. "Introduction to feedstock recycling of plastics." *Feedstock Recycling and Pyrolysis of Waste Plastics* 6, no. 7 (2006): 1–41.
21. Lee, Alicia, and Mei Shan Liew. "Tertiary recycling of plastics waste: An analysis of feedstock, chemical and biological degradation methods." *Journal of Material Cycles and Waste Management* 23, no. 1 (2021): 32–43.
22. Kaminsky, Walter. "Chemical recycling of plastics by fluidized bed pyrolysis." *Fuel Communications* 8 (2021): 100023.
23. Rai, B.; Rushad, S.T.; Kr, B.; Duggal, S.K. Study of waste plastic mix concrete with a plasticizer. *International Scholarly Research Notices* 2012 (2012): 469272.
24. Payne, Jack M., Muhammad Kamran, Matthew G. Davidson, and Matthew D. Jones. "Versatile chemical recycling strategies: Value-added chemicals from polyester and polycarbonate waste." *ChemSusChem* 15, no. 8 (2022): e202200255.
25. Zhao, Xiang, and Fengqi You. "Consequential life cycle assessment and optimization of high-density polyethylene plastic waste chemical recycling." *ACS Sustainable Chemistry & Engineering* 9, no. 36 (2021): 12167–12184.
26. Bîrcă, Alexandra, Oana Gherasim, Valentina Grumezescu, and Alexandru Mihai Grumezescu. "Introduction in thermoplastic and thermosetting polymers." In *Materials for Biomedical Engineering*, pp. 1–28. Elsevier, 2019.
27. Wang, Ru-Min, Shui-Rong Zheng, and Ya-Ping George Zheng. *Polymer Matrix Composites and Technology*. Elsevier, 2011.
28. Pascault, Jean-Pierre, and Roberto JJ Williams. "Thermosetting polymers." *Handbook of Polymer Synthesis, Characterization, and Processing*, pp. 519–533, 2013.
29. Koniuszewska, Anna G., and Jacek W. Kaczmar. "Application of polymer-based composite materials in transportation." *Progress in Rubber Plastics and Recycling Technology* 32, no. 1 (2016): 1–24.
30. Guo, John Zhanhu, Kenan Song, and Chuntai Liu, eds. *Polymer-Based Multifunctional Nanocomposites and Their Applications*. Elsevier, 2018.
31. Arumugaprabu, V., R. Deepak Joel Johnson, M. Uthayakumar, and P. Sivaranjana, eds. *Polymer-Based Composites: Design, Manufacturing, and Applications*. CRC Press, 2021.
32. Bledzki, Andrzej K., Omar Faruk, and Adam Jaszkiewicz. "Cars from renewable materials." *Kompozyty* 10, no. 3 (2010): 282–288.
33. Kesarwani, Shivi. "Polymer composites in the aviation sector." *International Journal of Engine Research* 6, no. 6 (2017): 518–525.
34. Ramakrishna, S., J. Mayer, E. Wintermantel, and Kam W. Leong. "Biomedical applications of polymer-composite materials: A review." *Composites Science and Technology* 61, no. 9 (2001): 1189–1224.
35. Rajak, Dipen Kumar, Durgesh D. Pagar, Pradeep L. Menezes, and Emanoil Linul. "Fiber-reinforced polymer composites: Manufacturing, properties, and applications." *Polymers* 11, no. 10 (2019): 1667.

36. Bhadra, Sambhu, Mostafizur Rahaman, and P. Noorunnisa Khanam. "Electrical and electronic application of polymer–carbon composites." In *Carbon-Containing Polymer Composites*, pp. 397–455. Springer, Singapore, 2019.
37. Abhemanyu, P. C., E. Prassanth, T. Navin Kumar, R. Vidhyasagar, K. Prakash Marimuthu, and R. Pramod. "Characterization of natural fiber reinforced polymer composites." In *AIP Conference Proceedings*, Vol. 2080, no. 1, p. 020005. AIP Publishing LLC, 2019.
38. Abbasi, Sadaf, M. H. Peerzada, Sabzoi Nizamuddin, and Nabisab Mujawar Mubarak. "Functionalized nanomaterials for the aerospace, vehicle, and sports industries." In *Handbook of Functionalized Nanomaterials for Industrial Applications*, pp. 795–825. Elsevier, 2020.
39. Hollaway, Leonard. *Polymer Composites for Civil and Structural Engineering*. Springer Science & Business Media, 2012.
40. Bader, M. G. "Open mold laminations—Contact molding." In *Processing and Fabrication Technology*, pp. 103–115. Routledge, 2017.
41. Elkington, Michael, D. Bloom, C. Ward, A. Chatzimichali, and K. Potter. "Hand layup: Understanding the manual process." *Advanced Manufacturing: Polymer & Composites Science* 1, no. 3 (2015): 138–151.
42. Andressen, F. R. *Open Molding: Hand Lay-up and Spray-up*, pp. 450–456. ASM International, Materials Park, OH, 2001.
43. Xiao, Bing, Yuqiu Yang, Xiaofeng Wu, Mengyuan Liao, Ryuiti Nishida, and Hiroyuki Hamada. "Hybrid laminated composites molded by spray lay-up process." *Fibers and Polymers* 16, no. 8 (2015): 1759–1765.
44. Peters, Stanley T., W. Donald Humphrey, and Ronald F. Foral. "Filament winding-composite structure fabrication", 1991.
45. Peters, Stanley T., ed. *Composite Filament Winding*. ASM International, 2011.
46. Woll, Suzanne LB, and Douglas J. Cooper. "Pattern-based closed-loop quality control for the injection molding process." *Polymer Engineering & Science* 37, no. 5 (1997): 801–812.
47. Peng, Xi, Mizi Fan, John Hartley, and Majeed Al-Zubaidy. "Properties of natural fiber composites made by pultrusion process." *Journal of Composite Materials* 46, no. 2 (2012): 237–246.
48. Silva, Francisco JG, F. Ferreira, M. C. S. Ribeiro, Ana CM Castro, M. R. A. Castro, M. L. Dinis, and A. Fiúza. "Optimising the energy consumption on pultrusion process." *Composites Part B: Engineering* 57 (2014): 13–20.
49. Potter, Kevin. *Resin Transfer Moulding*. Springer Science & Business Media, 2012.
50. Pascault, Jean-Pierre, and Roberto JJ Williams. "Thermosetting polymers." *Handbook of Polymer Synthesis, Characterization, and Processing*, pp. 519–533, 2013.
51. Tatara, Robert A. "Compression molding." In *Applied Plastics Engineering Handbook*, pp. 291–320. William Andrew Publishing, 2017.
52. Park, C. H., and W. I. Lee. "Compression molding in polymer matrix composites." In *Manufacturing Techniques for Polymer Matrix Composites (PMCs)*, pp. 47–94. Woodhead Publishing, 2012.
53. Inoue, Takashi, Toshiaki Ougizawa, Osamu Yasuda, and Keizo Miyasaka. "Development of modulated structure during solution casting of polymer blends." *Macromolecules* 18, no. 1 (1985): 57–63.
54. Park, Cheol, Zoubeida Ounaies, Kent A. Watson, Roy E. Crooks, Joseph Smith Jr, Sharon E. Lowther, John W. Connell, Emilie J. Siochi, Joycelyn S. Harrison, and Terry L. St Clair. "Dispersion of single wall carbon nanotubes by in situ polymerization under sonication." *Chemical Physics Letters* 364, no. 3–4 (2002): 303–308.
55. Şener, Başak, M. Lütfi Süzen, and Vedat Doyuran. "Landfill site selection by using geographic information systems." *Environmental Geology* 49, no. 3 (2006): 376–388.

56. Sahajwalla, Veena, and Vaibhav Gaikwad. "The present and future of e-waste plastics recycling." *Current Opinion in Green and Sustainable Chemistry* 13 (2018): 102–107.
57. Rahimi, AliReza, and Jeannette M. García. "Chemical recycling of waste plastics for new materials production." *Nature Reviews Chemistry* 1, no. 6 (2017): 1–11.
58. Wang, Chong-qing, Hui Wang, Jian-gang Fu, and You-nian Liu. "Flotation separation of waste plastics for recycling—A review." *Waste Management* 41 (2015): 28–38.
59. Vollmer, Ina, Michael JF Jenks, Mark CP Roelands, Robin J. White, Toon van Harmelen, Paul de Wild, Gerard P. van Der Laan, Florian Meirer, Jos TF Keurentjes, and Bert M. Weckhuysen. "Beyond mechanical recycling: Giving new life to plastic waste." *Angewandte Chemie International Edition* 59, no. 36 (2020): 15402–15423.
60. Uddin, Riaz, Nadia Saffoon, and Kumar Bishwajit Sutradhar. "Dissolution and dissolution apparatus: A review." *International Journal of Current Pharmaceutical Research* 1, no. 4 (2011): 201–207.
61. Chen, Huan, Kun Wan, Yayun Zhang, and Yanqin Wang. "Waste to wealth: Chemical recycling and chemical upcycling of waste plastics for a great future." *ChemSusChem* 14, no. 19 (2021): 4123–4136.
62. Aguado, J., D. P. Serrano, and J. M. Escola. "Fuels from waste plastics by thermal and catalytic processes: A review." *Industrial & Engineering Chemistry Research* 47, no. 21 (2008): 7982–7992.

Index

Note: **Bold** page numbers refer to tables and *italic* page numbers refer to figures.

For Product Safety Concerns and Information please contact our EU
representative GPSR@taylorandfrancis.com
Taylor & Francis Verlag GmbH, Kaufingerstraße 24, 80331 München, Germany